NEUROENDOCRI
OF GASTROINTE
ULCERATION

HANS SELYE SYMPOSIA ON NEUROENDOCRINOLOGY AND STRESS

Series Editors:

Sandor Szabo, *Harvard Medical School, Boston, Massachusetts*
Yvette Taché, *UCLA, Los Angeles, California*
Beatriz Tuchweber-Farbstein, *University of Montreal, Montreal, Quebec, Canada*

NEUROENDOCRINOLOGY OF GASTROINTESTINAL ULCERATION

Edited by

Sandor Szabo

Harvard Medical School
Boston, Massachusetts
and University of California, Irvine
VA Medical Center
Long Beach, California

and

Yvette Taché

Center for Ulcer Research and Education
University of California, Los Angeles
Los Angeles, California

Associate Editor

Gary B. Glavin

University of Manitoba
Winnipeg, Manitoba, Canada

SPRINGER SCIENCE+BUSINESS MEDIA, LLC

Library of Congress Cataloging-in-Publication Data

Neuroendocrinology of gastrointestinal ulceration / edited by Sandor
  Szabo and Yvette Taché ; associate editor, Gary B. Glavin.
      p.   cm. -- (Hans selye symposia on neuroendocrinology and
  stress ; v. 2)
    Includes bibliographical references and index.
    ISBN 978-1-4613-5759-9   ISBN 978-1-4615-1867-9 (eBook)
    DOI 10.1007/978-1-4615-1867-9
    1. Peptic ulcer--Pathophysiology. 2. Neuroendocrinology.
  I. Szabo, Sandor, 1944-   . II. Taché, Yvette. III. Glavin, Gary
  B.   IV. Series.
    [DNLM: 1. Peptic Ulcer--physiopathology. 2. Gastric Mucosa-
  -secretion. 3. Stress--physiopathology. 4. Disease Models, Animal.
  WI 350 N495 1995]
  RC821.N44 1995
  616.3'4307--dc20
  DNLM/DLC
  for Library of Congress   95-14350
                                                             CIP

Proceedings of a meeting on Neuroendocrinology of Gastrointestinal Ulceration,
held September 13–15, 1989, in Esterel, Quebec, Canada

ISBN 978-1-4613-5759-9

© 1995 Springer Science+Business Media New York
Originally published by Plenum Press, New York in 1995
Softcover reprint of the hardcover 1st edition 1995

10 9 8 7 6 5 4 3 2 1

IN MEMORIAM: MARC CANTIN, PRESIDENT HANS SELYE FOUNDATION

1933-1990

Dr. Cantin died on June 17, 1990 in Montréal at the age of 57. A native of Québec City, he completed his studies up to Doctor of Medicine (1958) from the University of Laval. Following this training, he went to the University of Montréal and received a Ph.D. in Experimental Medicine and Surgery under the leadership of Dr. Hans Selye. Subsequently, he specialized in anatomical pathology at the University of Chicago in the Department of Anatomical and Experimental Pathology. Since 1984, Dr. Cantin was Director of the Multidisciplinary Research Group on Hypertension at the Institute of Clinical Research of Montréal and Professor of Pathology at the University of Montréal. He directed a very productive group and authored 379 papers. He will be remembered for his important scientific contributions which addressed the role of the endocrine system in the pathogenesis of cardiovascular diseases.

Dr. Cantin was a recipient of several awards, including the Marcel Piché prize and the Léo Parizeau prize. In 1988 he received, together with Dr. P. Needelman, the Research Achievement Award from the American Heart Association for their work on the atrial natriuretic factor. Dr. Cantin was a member of the editorial boards of several scientific journals, president of the Medical Advisory Board of the Cancer Research Society and had been the first president of the Hans Selye Foundation, a post he held since 1982. He spent his life totally devoted to his work and scientific responsibilities. Each meeting with him revealed his commitment, enthusiasm and great internal resources for reaching goals. We will miss him as a friend and colleague. This book will be an appropriate tribute to his memory since he played an important role in the organization of this symposium in Esterel, Québec, Canada.

Yvette Taché, Ph.D.
Sandor Szabo, M.D., Ph.D.
Vice Presidents, Hans Selye Foundation

PREFACE

Selye pioneered the concept of biologic stress and contributed to the exploration of this field for over 40 years. He established the Hans Selye Foundation in 1980 to promote research and education on stress. In the spirit of the Foundation, Dr. Marc Cantin, late president, and Drs. Sandor Szabo and Yvette Taché, vice presidents, have established the Hans Selye Symposia on Neuroendocrinology stress conference and book series. So far three symposia have been held in Montreal where Selye started to work in the 1930's. These symposia bring together laboratory and clinical investigators engaged in the study of the neuroendocrinology, pathophysiology of stress, and related fields.

The first Hans Selye Symposium, held in 1986, was devoted to "Neuropeptides and Stress" and the proceedings were published by Springer Verlag. In 1992, in commemoration of the tenth anniversary of Selye's passing in October 1982, another Hans Selye Symposium was organized and was entitled "Corticotropin-Releasing Factor and Cytokines: Role in the Stress Response". The proceedings recently appeared in 1993 as volume 697 of the New York Academy of Sciences.

A substantial part of the present volume originates from the Hans Selye Symposium and devoted to the "Neuroendocrinology of Gastrointestinal Ulceration". A section on the novel therapeutic methods of approach to gastrointestinal ulceration encompasses the influence of calcium modulators, dopaminergic agents, dietary factors and growth factors in the development of experimental ulcers. Several chapters deal with the complex problems of the underlying neural mechanisms involved in the pathogenesis and prevention of gastrointestinal ulceration. In particular, emphasis has been placed on the role of the vagus and the sympatho-adrenergic regulation of defensive mechanisms such as bicarbonate secretion and sensory nerves. Advances in the understanding of the chemical coding and neuroanatomical substrata of the brain which influence gastric function and ulcer formation are summarized in the last portion of the book. More specifically covered are the role of medullary thyrotropin-releasing hormone in the vagal dependent formation of gastric lesions and of brain corticotropin-releasing factor in stress-related alterations of gastrointestinal function. Mounting evidence of the key influence of the limbic and mesolimbic system in stress ulcer formation is also presented.

The editors are exceedingly grateful to all the contributors who have enabled us to produce a comprehensive survey of knowledge in the field of the neuroendocrinology of gastrointestinal ulceration. With respect to the preparation of this volume, we would like to express our thanks to Mrs. Carroll Powney for her meticulous typing and proofreading of the manuscript.

Sandor Szabo
Yvette Taché
Gary Glavin

ACKNOWLEDGMENTS

We would like to express our deep appreciation to the patrons and sponsors of the "Hans Selye Symposia on Neuroendocrinology and Stress: Neuroendocrinology of Gastrointestinal Ulceration" for their support of the excellent international conference:

Patrons of the Symposium

Hans Selye Foundation
Medical Research Council of Canada
Angio-Medical Corporation
Kali-Chemie-Pharma
Marion Laboratories

Sponsors of the Symposium

Ayerst Laboratories
Fond de la recherche en santé du Quebec
Kotobuki Seiyaku Co., Ltd.
Nordic Laboratories, Inc.
Rorer Group, Inc.
G.D. Searle and Co.
UCLA Brain Research Institute
Université de Montreal
The Upjohn Co.

Personally, we would like to thank the late Dr. Marc Cantin, former President of the Hans Selye Foundation, Mrs. Louise Drevet-Selye, Dr. Pierre Bois, former President of the Medical Research Council of Canada, Dr. Arnold Scheibel, Director of the Brain Research Institute, UCLA, Los Angeles, as well as Joyce Fried, Assistant Dean for Educational Administration, UCLA School of Medicine, who dedicated her excellent organizational skills to coordinate all aspects of the conference. We are also grateful to Mrs. Carroll Powney for excellent manuscript preparation for this volume.

Finally, we thank the conference participants and authors of the chapters in this volume, who prepared and revised manuscripts and whose contributions represent an up-to-date survey of the neuroendocrinology of gastrointestinal ulceration.

Sandor Szabo, Boston
Yvette Taché, Los Angeles
Gary Glavin, Winnipeg

CONTENTS

NEUROENDOCRINE FACTORS IN ULCER PATHOGENESIS

BRAIN-GUT INTERACTIONS IN ULCER PATHOGENESIS

NOVEL THERAPEUTIC APPROACHES TO GASTROINTESTINAL ULCERATION

ROLE OF SENSORY NEURONS IN THE CONTROL OF GASTRIC MUCOSAL BLOOD FLOW AND PROTECTION

P. Holzer, I.Th. Lippe, M. Jocič, Ch. Wachter,
R. Amann, M.A. Pabst, A. Heinemann, B.M. Peskar,
B.A. Peskar, H.E. Raybould, E.H. Livingston,
H.C. Wong, J.H. Walsh, Y. Taché, and P.H. Guth

Nowhere else in the digestive system is the mucosa more endangered than in the gastroduodenal region, the threats being primarily of a chemical nature and of both endogenous and exogenous origin. Most tissues would rapidly disintegrate if exposed to the concentrations of hydrochloric acid that bathe but do not harm the gastric surface epithelium. This is because a multitude of mechanisms constituting the gastric mucosal barrier prevents hydrogen ions from entering the tissues at quantities that would produce cell injury[1]. Yet even a weakening of the barrier will usually not lead to any major lesion formation because protective mechanisms will become operational, limit the damage and promote repair of the injured tissue.

While a number of protective mechanisms has been identified in the gastric mucosa, it is still not clear how these mechanisms are activated and coordinated in the face of pending injury to the mucosa. A well-studied response to barrier disruption and acid backdiffusion into the gastric mucosa is a marked rise of mucosal blood flow[2-4]. This increase in blood flow is thought to be an important mechanism of protection which, by facilitating the disposal of acid, prevents the build-up of an injurious concentration of hydrogen ions in the tissue. The anatomical and functional organization of the gastric mucosal microcirculation requires submucosal arterioles to be dilated in order to increase blood flow to the mucosa[5]. Thus the message of acid backdiffusion needs to be transmitted over the distance of more than 500 μm which in the rat stomach separates the surface of the mucosa from submucosal arterioles. Considering this distance and the rapidity with which mucosal blood flow is increased following barrier disruption[3], it would seem conceivable that neurons mediate the blood flow response to acid backdiffusion. Since atropine, however, does not inhibit the vasodilatation induced by acid backdiffusion[3], this idea did not attract much attention for some time.

Meanwhile, though, a particular class of extrinsic afferent neurons has been recognized to form part of an important neural emergency system in the mucosa of the stomach and other regions of the digestive tract. This discovery was made possible by the availability of a neuropharmacological tool, capsaicin, with which the activity of primary afferent neurons with unmyelinated or thinly myelinated nerve fibres can selectively be manipulated[6,7]. Capsaicin is an excitotoxin and, depending on the dose and route of application, may cause a long-lasting defunctionalization of, and depletion of transmitter substances from, sensory neurons. Because of their selectivity, both the stimulant and neurotoxic actions of capsaicin can be utilized to examine the functional implications of thin afferent nerve fibres.

The capsaicin-sensitive afferent neurons supplying the gastrointestinal tract arise from two different sources. One group are spinal sensory neurons which originate from cell bodies in the dorsal root ganglia and reach the stomach via the splanchnic and mesenteric nerves, a route whereby they pass through the coeliac ganglion[8]. The other group are vagal sensory nerve fibres having their cell bodies in the nodose ganglion[8]. The spinal afferent neurons form a particularly dense plexus of fibres around gastrointestinal arteries and arterioles and are characterized by a rich content of peptides such as calcitonin gene-related peptide (CGRP) and substance P[8,9]. In addition, they have been found to be especially sensitive to chemical noxious stimuli including hydrogen ions[10-13]. There is now compelling evidence that perivascular afferent nerve fibres containing CGRP play an important role in regulating gastric blood flow and resistance to injury, and this review gives a brief account of the experimental findings that have led up to the discovery and elucidation of this neural emergency system. The reader is also referred to other articles in which certain aspects of the gastroprotective role of peptidergic afferent neurons have been reviewed[14-16].

ABLATION OF SENSORY NEURONS EXACERBATES GASTRIC INJURY

Chemical ablation of sensory neurons, by pretreating rats with a neurotoxic dose of capsaicin, provided the first evidence that capsaicin-sensitive afferent neurons control certain mechanisms of gastric mucosal protection[17-22]. Defunctionalization of sensory neurons does not cause damage by itself but leads to exacerbation of mucosal lesion formation in response to a variety of injurious factors including acid[17,22,23], aspirin[21,24], indomethacin[18,19,25], ethanol[19,24,26,27], taurocholate[28], endothelin-1[29] and platelet-activating factor[20]. Figure 1 (left panel) illustrates that gross injury caused by acid backdiffusion following disruption of the gastric mucosal barrier with ethanol is markedly aggravated by defunctionalization of capsaicin-sensitive afferent neurons[23]. Histological examination of the mucosa has revealed that the depth of injury is increased similarly[23]. Injury induced by cold and restraint stress, however, remains unchanged after elimination of sensory neurons[30]. Unpublished data obtained by Holzer, Kolve and Taché (Figure 1, right panel) confirm this observation and suggest that capsaicin-sensitive nerves either are not involved in the protection from stress-induced ulceration or their protective role is overridden by certain stress-induced processes.

STIMULATION OF SENSORY NEURONS PREVENTS GASTRIC INJURY

Szolcsányi and Barthó[17] were the first to propose a protective function of capsaicin-sensitive sensory neurons in the gastric mucosa when they observed that intragastric administration of small quantities of capsaicin, to stimulate sensory nerve fibres, protected from injury induced by acid accumulation in the stomach. In subsequent studies it was found that acute intragastric administration of capsaicin prevented experimental injury caused by a number of injurious factors including aspirin[31,32], ethanol[33-40], indomethacin[41] and taurocholate[42]. Figure 2 (left panel) shows that gross injury of the rat gastric mucosa evoked by ethanol is reduced by intragastric capsaicin in a dose-related manner, an effect that is associated with prevention of deep mucosal erosions[31,34,35].

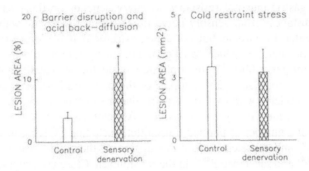

Figure 1. Effect of sensory denervation on gross damage in the rat gastric mucosa following barrier disruption and acid backdiffusion[23] or exposure to cold (8 °C) and restraint stress for 3 h. Sensory denervation was achieved by pretreating rats with a total dose of 125 mg/kg capsaicin subcutaneously 10 days before the experiments. Control rats received the respective vehicle. Injury induced by barrier disruption and acid backdiffusion was quantified after perfusion of the stomach of urethane-anesthetized rats with 15 % ethanol in 0.15 N HCl for 1 h. Gross injury is expressed either as a percentage of the total area of the corpus mucosa or in mm². Means ± SEM, n = 8-10; * P < 0.05 versus control (U test).

PATHWAYS AND MEDIATORS OF SENSORY NERVE-MEDIATED GASTRIC MUCOSAL PROTECTION

The protective effect of intragastric capsaicin involves afferent nerve fibres because it is prevented by defunctionalization of extrinsic sensory neurons[33,35]. Additional evidence for an implication of neurons has come from the observation that capsaicin's ability to reduce ethanol injury is suppressed by tetrodotoxin, a

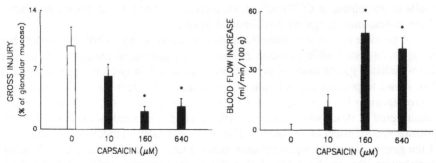

Figure 2. Concentration-dependent effect of intragastric capsaicin to reduce gross mucosal injury caused by 25 % ethanol (left panel) and to increase gastric mucosal blood flow (right panel) . Means ± SEM, n = 8; * P < 0.05 versus control (U test) (data taken from Holzer et al.)[35].

3

blocker of nerve conduction[35]. However, the pathways underlying sensory nerve-mediated gastric mucosal protection in the stomach have not yet been fully delineated. Chronic vagotomy and selective ablation of afferent nerve fibres in the vagus nerve have been reported to prevent capsaicin from protecting the gastric mucosa from ethanol injury[43]. This observation is not necessarily in contrast with the finding that acute vagotomy, acute extirpation of the coeliac ganglion and acute ligation of the blood vessels to the adrenal glands are devoid of any influence on the capsaicin's gastroprotective effect[33]. These data suggest that the ability of capsaicin to strengthen gastric mucosal resistance to injury depends on the local release of transmitter substances from sensory nerve fibres within the gastric wall, a process which is prevented by chronic, but not acute, extrinsic denervation of the stomach. Whether enteric neurons, which can be stimulated by transmitters released from extrinsic afferent nerve fibres[44], are implicated awaits further investigation.

Whilst a neurophysiological mapping of the pathways is lacking, considerable information is available with regard to the mediator substances that are responsible for afferent nerve-mediated gastroprotection. As referred to above, spinal afferent neurons contain a number of peptide transmitters including CGRP, substance P and neurokinin A[13], and capsaicin-evoked stimulation of their fibres causes release of CGRP[12,41,45,46,47] and substance P[48,49] in the rat stomach. Close arterial administration of CGRP to the rat stomach prevents gastric damage induced by ethanol, acidified aspirin or endothelin-1[29,40,50,51]. Intravenously[52,53], subcutaneously[54] and intracisternally[55] administered CGRP have also been reported to protect against gastric injury. Substance P, in contrast, seems to be devoid of a significant protective action[56] and has in fact been found to exaggerate mucosal damage by an action which involves degranulation of mast cells and may be mediated by tachykinin NK-1 and/or NK-3 receptors[57]. Neurokinin A, though, and related analogues acting preferentially on tachykinin NK-2 receptors do enhance the resistance of the gastric mucosa to injury[58,59].

There is conclusive pharmacological evidence that both CGRP and nitric oxide (NO) are essential mediators of afferent nerve-mediated gastric mucosal protection. The protective effects of both capsaicin and CGRP are blocked by the CGRP receptor antagonist $CGRP_{8-37}$[40] and by immunoneutralization of CGRP with polyclonal[40] and monoclonal[60] antibodies to CGRP. The gastroprotective action of capsaicin is furthermore prevented when the formation of NO is blocked by inhibitors of the NO synthase[36,39,40]. Since the ability of CGRP to strengthen gastric mucosal defence against injurious factors is likewise suppressed by blockers of the NO synthase[40,53] it would appear that stimulation of sensory nerve fibres by capsaicin results in the release of CGRP which, via formation of NO, augments the resistance of the gastric mucosa against experimental injury.

A possible role of substance P and neurokinin A acting on NK-2 receptors has not yet been tested while eicosanoids have been ruled out as mediators of sensory nerve-mediated gastric mucosal protection. Thus, capsaicin given intragastrically at a dose shown to prevent ethanol-induced injury does not affect the ex vivo formation of prostaglandin E_2, 6-oxo-prostaglandin $F_{1\alpha}$ and leukotriene C_4, and indomethacin administered at doses which inhibit gastric prostaglandin synthesis fails to alter the gastroprotective effect of capsaicin[34]. The ability of CGRP to enhance the resistance of the gastric mucosa to injury remains likewise unaltered by indomethacin[40]. Some investigators, though, have found that the gastroprotective action of capsaicin is reduced by indomethacin[37-42], but it has not been ascertained in these studies whether the effect of indomethacin is indeed due to inhibition of prostaglandin synthesis.

4

STIMULATION OF SENSORY NEURONS INCREASES GASTRIC MUCOSAL BLOOD FLOW

Another prominent effect of capsaicin-induced stimulation' of afferent nerve fibres in the stomach is a marked increase in gastric mucosal blood flow (GMBF). This effect is brought about by dilatation of submucosal arterioles[61] and has been demonstrated by both indirect[62] and direct[34,35,37,39,63-67] measurement of GMBF. The increase in GMBF is also seen when capsaicin is administered together with an injurious concentration of ethanol (Figure 2, right panel), and there is a remarkable correlation between the hyperaemic response to capsaicin and the concomitant reduction of gross damage (Figure 2) and deep haemorrhagic erosions[35,68]. This parallelism signifies that the gastric hyperaemia is the principal mechanism which is responsible for afferent nerve-mediated defence against injury. It should not be disregarded, however, that other mechanisms have also been suggested to play a significant part in the gastroprotective action of capsaicin[69] and CGRP[51].

A close relationship between sensory nerve-mediated increase in gastric blood flow and mucosal resistance to injury is further indicated by the identity of the mediators which are involved in the two processes. The capsaicin-induced increase in GMBF, like that in mucosal resistance to injury, is mediated by release of CGRP from afferent nerve fibres. Defunctionalization of afferent neurons by a neurotoxic dose of capsaicin abolishes the hyperaemic response to sensory nerve stimulation[35,37,61-65], and nerve-selective ablation of afferent nerve fibres has shown that only spinal, but not vagal, afferent neurons participate in the capsaicin-evoked increment of GMBF[63]. Conclusive pharmacological evidence for an involvement of CGRP comes from the finding that the CGRP antagonist $CGRP_{8-37}$ prevents the gastric hyperaemia caused by intragastric capsaicin[61,63]. This implication of CGRP is consistent with its presence[8,9] in and release[12,34,41,46,47] from perivascular afferent nerves in the stomach. CGRP is very potent in enhancing blood flow through the stomach[50,63,66,70-73], and this hyperaemic effect of the peptide is in keeping with the abundant presence of CGRP receptors on the muscle and endothelium of gastric arteries and arterioles[9,74]. Like the gastroprotective effect, the vasodilator action of CGRP released by capsaicin in the stomach appears to be brought about via formation of NO, since blockers of the NO synthase suppress the gastric hyperaemic reaction to both capsaicin[39,66] and CGRP[66,72]. It should be noticed, though, that the interaction between NO and CGRP may be more complex than considered here, as NO seems to be able to facilitate the release of CGRP from afferent nerve fibres[75-77]. Indomethacin has been reported to diminish the capsaicin-induced increase in GMBF[37,39] but the validity of this finding in terms of an implication of vasodilator prostaglandins has not yet been established.

ADDITIONAL EFFECTS OF SENSORY NERVE STIMULATION IN THE STOMACH

Capsaicin-induced stimulation of afferent nerve fibres can influence a number of functions of the rat stomach, which might be of relevance for sensory nerve-mediated gastroprotection. Although capsaicin fails to affect basal acid secretion, it is able to increase the elimination of acid from the gastric lumen[62,69,78], which results most likely from an enhanced output of bicarbonate[79]. In addition, capsaicin has been reported to stimulate the release of somatostatin and to inhibit the release of acetylcholine and gastrin in the rat isolated antrum, effects which are at least in part mediated by CGRP[47]. Both stimulant and inhibitory effects of capsaicin on the motility of the stomach have been encountered[35,37,80,81] but the relevance of these motor activity changes for gastroprotection remains to be determined. Other functions of the stomach are not directly influenced by acute intragastric administration of capsaicin. The spectrum of observed effects clearly indicates that intragastric capsaicin does not irritate the gastric mucosa and does not in any way weaken the gastric mucosal barrier as has been assumed for some red pepper spices which contain many substances other than capsaicin. This substance has no effect on the gastric content of mucus determined by an Alcian Blue technique (Lippe and Holzer, unpublished experiments), fails to enhance backdiffusion of gastric acid[62,69] does not enhance vascular permeability in the stomach[33,82], leaves the transmucosal potential difference unchanged[31,33,37,62,69,83] and fails to cause any macroscopical or histological damage by its own[31,33-35,69,78,79]. It can be ruled out, therefore, that capsaicin increases GMBF and mucosal resistance to injury by virtue of an irritant action on the mucosa. This conclusion is in keeping with the findings that the process of "adaptive cytoprotection", which is initiated by mild gastric irritants, is independent of the sensory innervation of the stomach[43,84].

INFLUENCE OF SENSORY NEURONS ON HEALING OF GASTRIC LESIONS

Capsaicin-sensitive afferent neurons play a pathophysiological role in the defence of the gastric mucosa against injurious factors as shown by the protective effects of sensory nerve stimulation and the deleterious effects of sensory nerve ablation. In contrast, sensory neurons seem to play little role in the morphological process of rapid restitution of the superficially injured gastric mucosa, a process that depends primarily on the migration of mucous cells over the denuded areas of the lamina propria[85]. There is evidence, however, that the healing of deep erosions is facilitated by the sensory innervation of the stomach, as the rate of healing of gastric ulcers is delayed by sensory nerve ablation[86,87]. The mechanisms by which afferent neurons promote healing of the injured gastric mucosa remain to be determined, but it is worth noting in this respect that the sensory nerve-derived peptides CGRP, substance P and neurokinin A are capable of stimulating the proliferation of fibroblasts, vascular smooth muscle and endothelial cells[88-90].

SENSORY NEURONS SIGNAL FOR AN INCREASE IN GASTRIC MUCOSAL BLOOD FLOW IN THE FACE OF PENDING ACID INJURY

The data described thus far clearly indicate that stimulation of sensory neurons in the stomach strengthens the resistance of the gastric mucosa against injury, a mechanism which seems closely related to a marked increase in GMBF. To further evaluate the functional potential of this neural emergency system in the gastric mucosa, a possible involvement of peptidergic afferent neurons in the gastric hyperaemic response to gastric acid backdiffusion was examined (Figures 1 and 3). This increment of blood flow is an immediate response to acid challenge of the mucosa as shown by a prompt rise of blood flow through the left gastric artery[73]. The involvement of neurons in the acid-evoked hyperaemia of the stomach was first demonstrated by the ability of tetrodotoxin to depress the response[23]. Further experiments established that the increase in GMBF due to acid backdiffusion involves capsaicin-sensitive afferent neurons[22,23,28] which pass through the coeliac ganglion and hence seem to originate from spinal ganglia[91].

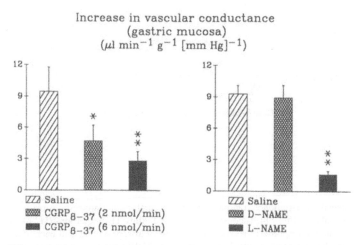

Increase in vascular conductance
(gastric mucosa)
(μl min^{-1} g^{-1} [mm Hg]$^{-1}$)

Saline
CGRP$_{8-37}$ (2 nmol/min)
CGRP$_{8-37}$ (6 nmol/min)

Saline
D−NAME
L−NAME

Figure 3. Effect of intraarterial infusion of the CGRP receptor antagonist CGRP$_{8-37}$ and of intravenous injection of the NO synthase blocker N^G-nitro-L-arginine methyl ester (L-NAME, 15 mg/kg) on the gastric mucosal hyperaemia induced by gastric acid backdiffusion. The hyperaemia is expressed as increase in vascular conductance in the gastric mucosa. D-NAME, the inactive enantiomer, was administered at the same dose as L-NAME. Means ± SEM, n = 6-8; * P < 0.05 versus saline (U test) (data taken from Holzer et al.)[73].

Inhibition of the hyperaemic reaction to gastric acid backdiffusion by tetrodotoxin or capsaicin pretreatment was associated with an aggravation of gross damage and with a significant increase in the incidence of deep erosions[23,91]. These data indicate that the acid-induced increase in mucosal blood flow is important for the protection of the mucosa from influxing acid, because it ensures that acid damage is limited to the surface of the mucosa.

The inhibitory effects of tetrodotoxin[23] and hexamethonium[92] suggest that the increment of gastric blood flow in the face of pending acid injury results from a reflex-like mechanism (Figure 4), the pathways of which are not yet completely understood. There is, however, ample evidence that CGRP plays an essential mediator role in the acid-evoked gastric hyperaemia (Figure 3), since this response is suppressed by the CGRP antagonist CGRP$_{8-37}$[71,73]. As is the case with the vasodilator response to exogenous CGRP, the acid-evoked vasodilatation depends on the formation of nitric oxide (Figure 3), since it is blocked by the NO synthase inhibitor N^G-nitro-L-arginine methyl ester in an enantiomer-selective manner[73,93]. It would appear, therefore, that the gastric hyperaemia due to acid backdiffusion is mediated by release of CGRP from perivascular afferent neurons (Figure 4). CGRP in turn acts on vascular cells, probably endothelial cells, to stimulate the formation of NO which may act as the final mediator to cause vasodilatation, increase blood flow to the stomach and strengthen the resistance of the gastric mucosa to acid injury (Figure 4). The possibility should not be neglected, though, that other mechanisms and mediators may also be involved and that hyperaemia may not be the only mechanism by which peptidergic afferent neurons contribute to defence against acid injury. Substance P acting via NK-1 receptors, histamine acting via histamine H$_1$

receptors, vasodilator prostanoids, cholinergic and noradrenergic neurons have been ruled out to take part in the gastric hyperaemic reaction to acid challenge[23,73,92,93].

Gastric mucosal hyperaemia due to acid influx

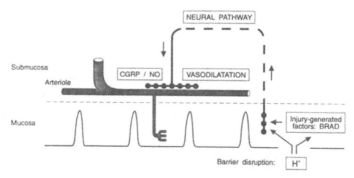

Figure 4. Schematic illustration of the pathways and mediators that are responsible for the increase in gastric mucosal blood flow in response to gastric acid influx. Hydrogen ions (H$^+$) and bradykinin (BRAD) generated in response to gastric mucosal barrier disruption and acid backdiffusion activate a reflex-like mechanism which results in the release of CGRP from nerve fibres around submucosal arterioles and the formation of NO. CGRP and NO cause vasodilatation and enhance blood flow into the mucosa.

While the effector mechanisms of the acid-evoked rise of GMBF have been delineated to some degree, it is still unclear how this response is activated by acid influx into the mucosa. One possibility is that hydrogen ions themselves activate neurons, a possibility for which there is some experimental evidence[10-13]. Another possibility is that factors generated in response to superficial mucosal injury contribute to the activation of the neural emergency system. A factor that is formed in response to injury or exposure of the tissue to low pH is bradykinin[94,95], and there is indeed experimental evidence that bradykinin, formed in response to severe acid challenge of the gastric mucosa, contributes to the protective hyperaemia associated with acid backdiffusion[96]. It remains to be examined, though, whether bradykinin causes vasodilatation by its own or by stimulating the neural vasodilator system (Figure 4).

SUMMARY AND PERSPECTIVES

Taken together the findings reviewed here have revealed the existence of a neural emergency system in the stomach, which is called into operation in the face of pending injury to the mucosa. As a result, blood flow to the stomach is greatly augmented, an effect that is likely to facilitate a variety of gastric protective mechanisms. Increased blood flow will add to the removal of injurious factors from the mucosa and promote a wide range of processes that either reduce the vulnerability, or aid the repair, of the gastric mucosa[1,4,5]. This neural emergency system is operative not only in the stomach but also in other regions of the gastrointestinal system including the oesophagus[97], small intestine[98-100], and colon[101-103].

The presence of a neural emergency system in the gastrointestinal mucosa raises the question as to its role in gastroduodenal ulcer and inflammatory bowel disease. It is conceivable that improper functioning of the system lowers mucosal defence mechanisms and predisposes to mucosal injury. Evidence in favour of this conjecture is indeed accumulating. For instance, chronic intake of nicotine by rats suppresses the increase in mucosal blood flow due to gastric acid backdiffusion, an observation that may be related to enhanced leukotriene formation in the rat stomach and delayed healing of gastroduodenal ulcers in smoking patients[104]. Other studies indicate that the gastropathy associated with experimental cirrhosis[105], portal hypertension[106] and uraemia[107] is associated with an inadequate rise of GMBF in response to acid backdiffusion through a leaky gastric mucosal barrier. Finally, the acid-evoked hyperaemia in the stomach declines with age[108,109], a change that may in part be responsible for the compromised ability of the aged gastric mucosa to defend itself against injurious factors. It will be important, therefore, to determine how the neural emergency system in the stomach and other regions of the digestive tract operates under a variety of conditions and pathological circumstances. Since CGRP has been recognized as an essential mediator of this system, it will also be important to elucidate the molecular dynamics of this peptide in order to fully understand the neural emergency system in health and disease.

ACKNOWLEDGEMENTS

Work performed in the authors' laboratories was supported by the Max Kade Foundation, the Austrian Science Foundation (grants 7845 and 9473), the Austrian National Bank (grants 4207 and 4905), the Franz Lanyar Foundation of the Medical Faculty of the University of Graz, the German Science Foundation, the National Institutes of Health and Veterans Administration Research Funds. The authors are grateful to Irmgard Russa for secretarial help.

REFERENCES

1. A. Allen, G. Flemström, A. Garner, and E. Kivilaakso, Gastroduodenal mucosal protection, *Physiol Rev.* 73: 823 (1993).
2. B.J.R. Whittle, Mechanisms underlying gastric mucosal damage induced by indomethacin and bile salts, and the actions of prostaglandins, *Br J Pharmacol.* 60: 455 (1977).
3. T.M. Bruggeman, J.G. Wood, and H.W. Davenport, Local control of blood flow in the dog's stomach: vasodilatation caused by acid back-diffusion following topical application of salicylic acid, *Gastroenterology.* 77: 736 (1979).
4. P.J. Oates, Gastric blood flow and mucosal defense, *in:* "Gastric Cytoprotection," D. Hollander and A.S. Tarnawski, eds., Plenum Press, New York (1990).
5. P.H. Guth, F.W. Leung, and G.L. Kauffman, Physiology of the gastric circulation, *in:* "The Gastrointestinal System, Handbook of Physiology," Section 6, Vol I, Part 2, S.G. Schultz, ed., American Physiological Society, Bethesda (1989).
6. P. Holzer, Capsaicin: cellular targets, mechanisms of action, and selectivity for thin sensory neurons, *Pharmacol Rev.* 43: 143 (1991).
7. J.W. Wood, "Capsaicin in the Study of Pain," Academic Press, London (1993).
8. T. Green, and G.J. Dockray, Characterization of the peptidergic afferent innervation of the stomach in the rat, mouse, and guinea-pig, *Neuroscience.* 25: 181 (1988).
9. C. Sternini, Enteric and visceral afferent CGRP neurons. Targets of innervation and differential expression patterns, *Ann New York Acad Sci.* 657: 170 (1992).

10. F. Cervero, and H.A. McRitchie, Neonatal capsaicin does not affect unmyelinated efferent fibers of the autonomic nervous system: functional evidence, *Brain Res.* 239: 283 (1982).

11. E.R. Forster, T. Green, M. Elliot, A. Bremner, and G.J. Dockray, Gastric emptying in rats: role of afferent neurons and cholecystokinin, *Am. J. Physiol.* 258: G552 (1990).

12. P. Geppetti, M. Tramontana, S. Evangelista, D. Renzi, C.A. Maggi, B.M. Fusco, and E. Del Bianco, Differential effect on neuropeptide release of different concentrations of hydrogen ions on afferent and intrinsic neurons of the rat stomach, *Gastroenterology.* 101: 1505 (1991).

13. P. Holzer, I.Th. Lippe, and R. Amann, Participation of capsaicin-sensitive afferent neurons in gastric motor inhibition caused by laparotomy and intraperitoneal acid, *Neuroscience.* 48: 715 (1992).

14. P. Holzer, Peptidergic sensory neurons in the control of vascular functions: mechanisms and significance in the cutaneous and splanchnic vascular beds, *Rev Physiol Biochem Pharmacol.* 121: 49 (1992).

15. E.D. Jacobson, Circulatory mechanisms of gastric mucosal damage and protection, *Gastroenterology* 102: 1788 (1992).

16. B.J.R. Whittle, Neuronal and endothelium-derived mediators in the modulation of the gastric microcirculation: integrity in the balance, *Br J Pharmacol.* 110: 3 (1993).

17. J. Szolcsányi, and L. Barthó, Impaired defense mechanism to peptic ulcer in the capsaicin-desensitized rat, *in:* "Gastrointestinal Defense Mechanisms," G. Mózsik, O. Hänninen and T. Jávor, eds., Pergamon Press and Akadémiai Kiadó, Oxford and Budapest (1981).

18. S. Evangelista, C.A. Maggi, and A. Meli, Evidence for a role of adrenals in the capsaicin-sensitive "gastric defence mechanism" in rats, *Proc Soc Exp Biol Med.* 182: 568 (1986).

19. P. Holzer, and W. Sametz, Gastric mucosal protection against ulcerogenic factors in the rat mediated by capsaicin-sensitive afferent neurons, *Gastroenterology.* 91: 975 (1986).

20. J.V. Esplugues, B.J.R. Whittle, and S. Moncada, Local opioid-sensitive afferent sensory neurones in the modulation of gastric damage induced by Paf, *Br J Pharmacol.* 97: 579 (1989).

21. M. Uchida, S. Yano, and K. Watanabe, Aggravation by the capsaicin treatment of gastric antral ulcer induced by the combination of 2-deoxy-D-glucose, aspirin and ammonia in rats, *Jap J Pharmacol.* 57: 377 (1991).

22. J. Matsumoto, K. Ueshima, T. Ohuchi, K. Takeuchi, and S. Okabe, Induction of gastric lesions by 2-deoxy-D-glucose in rats following chemical ablation of capsaicin-sensitive sensory neurons, *Jap J Pharmacol.* 60: 43 (1992).

23. P. Holzer, E.H. Livingston, and P.H. Guth, Sensory neurons signal for an increase in rat gastric mucosal blood flow in the face of pending acid injury, *Gastroenterology,* 101: 416 (1991).

24. S. Evangelista, C.A. Maggi, S. Giuliani, and A. Meli, Further studies on the role of the adrenals in the capsaicin-sensitive "gastric defence mechanism", *Int J Tiss React.* 10: 253 (1988).

25. B.J.R. Whittle, J. Lopez-Belmonte, and S. Moncada, Regulation of gastric mucosal integrity by endogenous nitric oxide: interactions with prostanoids and sensory neuropeptides in the rat, Br J Pharmacol. 99: 607 (1990).

26. J.V. Esplugues, and B.J.R. Whittle, Morphine potentiation of ethanol-induced gastric mucosal damage in the rat. Role of local sensory afferent neurons, *Gastroenterology.* 98: 82 (1990).

27. Y. Yonei, P. Holzer, and P.H. Guth, Laparotomy-induced gastric protection against ethanol injury is mediated by capsaicin-sensitive sensory neurons, *Gastroenterology.* 99: 3 (1990).

28. K. Takeuchi, T. Ohuchi, M. Narita, and S. Okabe, Capsaicin-sensitive sensory nerves in recovery of gastric mucosal integrity after damage by sodium taurocholate in rats, *Jap J Pharmacol.* 63: 479 (1993).

29. B.J.R. Whittle, and J. Lopez-Belmonte, Interactions between the vascular peptide endothelin-1 and sensory neuropeptides in gastric mucosal injury, *Br J Pharmacol.* 102: 950 (1991).

30. A.M. Dugani, and G.B. Glavin, Capsaicin effects on stress pathology and gastric acid secretion in rats, *Life Sci.* 39: 1531 (1986).

31. P. Holzer, M.A. Pabst, and I. Th. Lippe, Intragastric capsaicin protects against aspirin-induced lesion formation and bleeding in the rat gastric mucosa, *Gastroenterology.* 96: 1425 (1989).

32. K.G. Yeoh, J.Y. Kang, I. Yap, R. Guan, and C.C. Tan, The effect of chilli on aspirin-induced gastroduodenal mucosal injury in humans, *Abstracts II United Eur Gastroenterol Week in Barcelona.* A71 (1993).

33. P. Holzer, and I.Th. Lippe, Stimulation of afferent nerve endings by intragastric capsaicin protects against ethanol-induced damage of gastric mucosa, *Neuroscience.* 27: 981 (1988).

34. P. Holzer, M.A. Pabst, I.Th. Lippe, B.M. Peskar, B.A. Peskar, E.H. Livingston, and P.H. Guth, Afferent nerve-mediated protection against deep mucosal damage in the rat stomach, *Gastroenterology.* 98: 838 (1990).

35. P. Holzer, E.H. Livingston, A. Saria, and P.H. Guth, Sensory neurons mediate protective vasodilatation in rat gastric mucosa, *Am J Physiol.* 260: G363 (1991).

36. B.M. Peskar, M. Respondek, K.M. Müller, and B.A. Peskar, A role for nitric oxide in capsaicin-induced gastroprotection, *Eur J Pharmacol.* 198: 113 (1991).

37. K. Takeuchi, H. Niida, J. Matsumoto, K. Ueshima, and S. Okabe, Gastric motility changes in capsaicin-induced cytoprotection in the rat stomach, *Jap J Pharmacol.* 55: 147 (1991).

38. M. Uchida, S. Yano, and K. Watanabe, The role of capsaicin-sensitive afferent nerves in protective effect of capsaicin against absolute ethanol-induced gastric lesions in rats, *Jap J Pharmacol.* 55: 279 (1991).

39. T. Brzozowski, D. Drozdowicz, A. Szlachcic, J. Pytko-Polonczyk, J. Majka, and S. Konturek, Role of nitric oxide and prostaglandins in gastroprotection induced by capsaicin and papaverine, *Digestion.* 54: 24 (1993).

40. N. Lambrecht, M. Burchert, M. Respondek, K.M. Müller, and B.M. Peskar, Role of calcitonin gene-related peptide and nitric oxide in the gastroprotective effect of capsaicin in the rat, *Gastroenterology.* 104: 1371 (1993).

41. J.L. Gray, N.W. Bunnett, S.L. Orloff, S.J. Mulvihill, and H.T. Debas, A role for calcitonin gene-related peptide in protection against gastric ulceration, *Ann Surg.* 219: 58 (1994).

42. D.W. Mercer, W.P. Ritchie, and D.T. Dempsey, Sensory neuron-mediated gastric mucosal protection is blocked by cyclooxygenase inhibition, *Surgery.* 115: 156 (1994).

43. T.A. Miller, G.S. Smith, M. Stanislawska, T.M. Phan, and J.M. Henagan, Role of vagal innervation in adaptive cytoprotection, *Dig Dis Sci.* 34: 1318 (1989).

44. L. Barthó, and P. Holzer, Search for a physiological role of substance P in gastrointestinal motility, *Neuroscience.* 16: 1 (1985).

45. P. Holzer, B.M. Peskar, B.A. Peskar, and R. Amann, Release of calcitonin gene-related peptide induced by capsaicin in the vascularly perfused rat stomach, *Neurosci Lett.* 108: 195 (1990).

46. T. Inui, Y. Kinoshita, A. Yamaguchi, T. Yamatani, and T. Chiba, Linkage between capsaicin-stimulated calcitonin gene-related peptide and somatostatin release in rat stomach, *Am J Physiol.* 261: G770 (1991).

47. J.S. Ren, R.L. Young, D.C. Lassiter, and R.F. Harty, Calcitonin gene-related peptide mediates capsaicin-induced neuroendocrine responses in rat antrum, *Gastroenterology.* 104: 485 (1993).

48. Y.N. Kwok, and C.H.S. McIntosh, Release of substance P-like immunoreactivity from the vascularly perfused rat stomach, *Eur J Pharmacol.* 180: 201 (1990).

49. D. Renzi, S. Evangelista, P. Mantellini, P. Santicioli, C.A. Maggi, P. Geppetti, and C. Surrenti, Capsaicin-induced release of neurokinin A from muscle and mucosa of gastric corpus: correlation with capsaicin-evoked release of calcitonin gene-related peptide, *Neuropeptides.* 19: 137 (1991).

50. I.Th. Lippe, M. Lorbach, and P. Holzer, Close arterial infusion of calcitonin gene-related peptide into the rat stomach inhibits aspirin- and ethanol-induced hemorrhagic damage, *Regul Pept.* 26: 35 (1989).

51. J. Lopez-Belmonte, and B.J.R. Whittle, The paradoxical vascular interactions between endothelin-1 and calcitonin gene-related peptide in the rat gastric mucosal microcirculation, *Br J Pharmacol.* 110: 496 (1993).

52. G. Clementi, M. Amico-Roxas, A. Caruso, V.M.C. Cutuli, S. Maugeri, and A. Prato, Protective effects of calcitonin gene-related peptide in different experimental models of gastric ulcers, *Eur J Pharmacol.* 238: 101 (1993).

53. G. Clementi, A. Caruso, A. Prato, E. De Bernardis, C.E. Fiore, and M. Amico-Roxas, A role for nitric oxide in the anti-ulcer activity of calcitonin gene-related peptide, *Eur J Pharmacol.* 256: R7 (1994).

54. C.A. Maggi, S. Evangelista, S. Giuliani, and A. Meli, Anti-ulcer activity of calcitonin gene-related peptide in rats, *Gen Pharmacol.* 18: 33 (1987).

55. Y. Taché, Inhibition of gastric acid secretion and ulcers by calcitonin gene-related peptide, *Ann New York Acad Sci.* 657: 240 (1992).

56. S. Evangelista, I.Th. Lippe, P. Rovero, C.A. Maggi, and A. Meli, Tachykinins protect against ethanol-induced gastric lesions in rats, *Peptides.* 10: 79 (1989).

57. F. Karmeli, R. Eliakim, E. Okon, and D. Rachmilewitz, Gastric mucosal damage by ethanol is mediated by substance P and prevented by ketotifen, a mast cell stabilizer, *Gastroenterology.* 100: 1206 (1991).

58. S. Evangelista, C.A. Maggi, P. Rovero, R. Patacchini, S. Giuliani, and A. Giachetti, Analogs of neurokinin A(4-10) afford protection against gastroduodenal ulcers in rats, *Peptides.* 11: 293 (1990).

59. T. Stroff, M. Burchert, and B.M. Peskar, The gastroprotective effect of the tachykinin neurokinin A_{4-10} involves sensory neurons and calcitonin gene-related peptide, *Gastroenterology.* in press (1994).

60. B.M. Peskar, H.C. Wong, J.H. Walsh, and P. Holzer, A monoclonal antibody to calcitonin gene-related peptide abolishes capsaicin-induced gastroprotection, *Eur J Pharmacol.* 250: 201 (1993).

61. R.Y.Z. Chen, D.-S. Li, and P.H. Guth, Role of calcitonin gene-related peptide in capsaicin-induced gastric submucosal arteriolar dilation, *Am J Physiol.* 262: H1350 (1992).

62. I.Th. Lippe, M.A. Pabst, and P. Holzer, Intragastric capsaicin enhances rat gastric acid elimination and mucosal blood flow by afferent nerve stimulation, *Br J Pharmacol.* 96: 91 (1989).

63. D.-S. Li, H.E. Raybould, E. Quintero, and P.H. Guth, Role of calcitonin gene- related peptide in gastric hyperemic response to intragastric capsaicin, *Am J Physiol.* 261: G657 (1991).

64. F.W. Leung, Modulation of autoregulatory escape by capsaicin-sensitive afferent nerves in the rat stomach, *Am J Physiol.* 262: H562 (1992).

65. J.L. Wallace, G.W. McKnight, and A.D. Befus, Capsaicin-induced hyperemia in the stomach: possible contribution of mast cells, *Am J Physiol.* 263: G209 (1992).

66. B.J.R. Whittle, J. Lopez-Belmonte, and S. Moncada, Nitric oxide mediates rat mucosal vasodilatation induced by intragastric capsaicin, *Eur J Pharmacol.* 218: 339 (1992).

67. J.E. Grönbech, E.R. and Lacy, Substance P attenuates gastric mucosal hyperemia after stimulation of sensory neurons in the rat stomach, *Gastroenterology.* 106: 440 (1994).

68. P. Holzer, and M.A. Pabst, M.A. A histological and functional study of the effects of acid and capsaicin on ethanol-induced damage in the rat gastric mucosa, *Surg Res Commun.* 14: 1 (1994).

69. T.R. Sullivan, R. Milner, D.T. Dempsey, and W.P. Ritchie, Effect of capsaicin on gastric mucosal injury and blood flow following bile acid exposure, *J Surg Res.* 52: 596 (1992).

70. P. Holzer, and P.H. Guth, Neuropeptide control of rat gastric mucosal blood flow. Increase by calcitonin gene-related peptide and vasoactive intestinal polypeptide, but not substance P and neurokinin A, *Circ Res.* 68: 100 (1991).

71. D.-S. Li, H.E. Raybould, E. Quintero, and P.H. Guth, Calcitonin gene-related peptide mediates the gastric hyperemic response to acid back-diffusion, *Gastroenterology.* 102: 1124 (1992).

72. P. Holzer, I.Th. Lippe, M. Jocič, Ch. Wachter, R. Erb, and A. Heinemann, Nitric oxide dependent and -independent hyperaemia due to calcitonin gene-related peptide in the rat stomach. *Br J Pharmacol.* 110: 404 (1993).

73. P. Holzer, Ch. Wachter, M. Jocič, and A. Heinemann, Vascular bed-dependent roles of the peptide CGRP and nitric oxide in acid-evoked hyperaemia of the rat stomach, *J Physiol (London).* in press (1994).

74. T.S. Gates, R.P. Zimmerman, C.R. Mantyh, S.R. Vigna, and P.W. Mantyh, Calcitonin gene related peptide-α receptor binding sites in the gastrointestinal tract, *Neuroscience.* 31: 757 (1989).

75. E.P. Wei, M.A. Moskowitz, P. Boccalini, and H.A. Kontos, Calcitonin gene- related peptide mediates nitroglycerin and sodium nitroprusside-induced vasodilation in feline cerebral arterioles, *Circ Res.* 70: 1313 (1992).

76. P. Holzer, and M. Jocič, Cutaneous vasodilatation induced by nitric oxide-evoked stimulation of afferent nerves in the rat, *Br J Pharmacol.* in press (1994).

77. S.R. Hughes, and S.D. Brain, Nitric oxide-dependent release of vasodilator quantities of calcitonin gene-related peptide from capsaicin-sensitive nerves in rabbit skin, *Br J Pharmacol.* 111: 425 (1994).

78. J. Matsumoto, K. Takeuchi, and S. Okabe, Characterization of gastric mucosal blood flow response induced by intragastric capsaicin in rats, *Jap J Pharmacol.* 57: 205 (1991).

79. K. Takeuchi, K. Tachibana, K. Ueshima, J. Matsumoto, and S. Okabe, Stimulation by capsaicin of gastric alkaline secretion in anesthetized rats, *Jap J Pharmacol.* 59: 151 (1992).

80. U. Holzer-Petsche, H. Seitz, and F. Lembeck, Effect of capsaicin on gastric corpus smooth muscle of the rat in vitro, *Eur J Pharmacol.* 162: 29 (1989).

81. R.A. Lefebvre, F.A. De Beurme, and S. Sas, Relaxant effect of capsaicin in the rat gastric fundus, *Eur J Pharmacol.* 195: 131 (1991).

82. K. Takeuchi, H. Niida, and S. Okabe, Gastric motility changes in the cytoprotective action of orally administered N-ethylmaleimide and capsaicin in the rat stomach, *Gastroenterology.* 98: A134 (1990).

83. J. Matsumoto, K. Ueshima, K. Takeuchi, and S. Okabe, Capsaicin-sensitive afferent neurons in adaptive responses of the rat stomach induced by a mild irritant, *Jap J Pharmacol.* 55: 181 (1991).

84. S. Evangelista, C.A. Maggi, and A. Meli, Lack of influence of capsaicin-sensitive sensory fibers on adaptive cytoprotection in rat stomach, *Dig Dis Sci.* 33: 1050 (1988).

85. M.A. Pabst, E. Schöninkle, and P. Holzer, Ablation of capsaicin-sensitive afferent nerves impairs defence but not rapid repair of rat gastric mucosa, *Gut.* 34: 897 (1993).

86. B.M. Peskar, N. Lambrecht, T. Stroff, M. Respondek, and K.-M. Müller, Functional ablation of sensory neurons impairs the healing of acute gastric mucosal damage in rats, *Dig Dis Sci.* in press (1994).

87. M. Tramontana, D. Renzi, A. Calabro, C. Panerai, S. Milani, C. Surrenti, and S. Evangelista, Influence of capsaicin-sensitive afferent fibers on acetic acid-induced chronic gastric ulcers in rats, *Scand J Gastroenterol.* in press (1994).

88. J. Nilsson, A.M. von Euler, and C.-J. Dalsgaard, Stimulation of connective tissue cell growth by substance P and substance K, *Nature.* 315: 61 (1985).

89. A. Haegerstrand, C.-J. Dalsgaard, B. Jonzon, O. Larsson, and J. Nilsson, Calcitonin gene related peptide stimulates proliferation of human endothelial cells, *Proc Nat Acad Sci USA.* 87: 3299 (1990).

90. M. Ziche, L. Morbidelli, M. Pacini, P. Geppetti, G. Alessandri, and C.A. Maggi, Substance P stimulates neovascularization in vivo and proliferation of cultured endothelial cells, *Microvasc Res.* 40: 264 (1990).

91. H.E. Raybould, C. Sternini, V.E. Eysselein, M. Yoneda, and P. Holzer, Selective ablation of spinal afferent neurons containing CGRP attenuates gastric hyperemic response to acid, *Peptides.* 13: 249 (1992).

92. P. Holzer, and I.Th. Lippe, Gastric mucosal hyperemia due to acid back-diffusion depends on splanchnic nerve activity, *Am J Physiol.* 262: G505 (1992).

93. I.Th. Lippe, and P. Holzer, Participation of endothelium-derived nitric oxide but not prostacyclin in the gastric mucosal hyperaemia due to acid back-diffusion, *Br J Pharmacol.* 105: 708 (1992).

94. J.M. Hall, Bradykinin receptors: pharmacological properties and biological roles, *Pharmacol Ther.* 56: 131 (1992).

95. A. Dray, and M. Perkins, Bradykinin and inflammatory pain, *Trends Neurosci.* 16: 99 (1993).

96. P. Holzer, and G. Pethö, Bradykinin-mediated regulation of gastric blood flow in the face of acid injury, *Naunyn-Schmiedeberg's Arch Pharmacol.* 349 (Suppl.): R91 (1994).

97. B.L. Bass, K.S. Trad, J.W. Harmon, and F.Z. Hakki, Capsaicin-sensitive nerves mediate esophageal mucosal protection, *Surgery.* 110: 419 (1991).

98. Z. Rózsa, K.A. Sharkey, G. Jancsó, and V. Varró, Evidence for a role of capsaicin-sensitive mucosal afferent nerves in the regulation of mesenteric blood flow in the dog, *Gastroenterology.* 90: 906 (1986).

99. C.A. Maggi, S. Evangelista, L. Abelli, V. Somma, and A. Meli, Capsaicin- sensitive mechanisms and experimentally induced duodenal ulcers in rats, *J. Pharm Pharmacol.* 39: 559 (1987).

100. O.D. Hottenstein, G. Remak, and E.D. Jacobson, Peptidergic nerves mediate post- nerve stimulation hyperemia in rat gut, *Am J Physiol.* 263: G29 (1992).

101. S. Evangelista, and A. Meli, Influence of capsaicin-sensitive fibres on experimentally-induced colitis in rats, *J Pharmac Pharmacol.* 41: 574 (1989).

102. F.W. Leung, Role of capsaicin-sensitive afferent nerves in mucosal injury and injury-induced hyperemia in rat colon, *Am J Physiol.* 262: G332 (1992).

103. V.E. Eysselein, M. Reinshagen, A. Patel, W. Davis, C. Nast, and C. Sternini, Calcitonin gene related peptide in inflammatory bowel disease and experimentally induced colitis, *Ann New York Acad Sci.* 657: 319 (1992).

104. M. Battistel, M. Plebani, F. Di Mario, M. Jocič, I.Th. Lippe, and P. Holzer Chronic nicotine intake causes vascular dysregulation in the rat gastric mucosa, *Gut.* 34: 1688 (1993).

105. Y. Nishizaki, P.H. Guth, and J.D. Kaunitz, Pathogenesis of increased susceptibility to acid back-diffusion injury in the cirrhotic rat stomach, *Gastroenterology.* in press (1994).

106. J.L. Wallace, G.W. McKnight, P.L. Beck, S.S. Lee, and K.A. Sharkey, Impaired gastric mucosal defence in cirrhosis: a defect in sensory afferent innervation, *Gastroenterology.* 104: A288 (1993).

107. E. Quintero, J. Kaunitz, Y. Nishizaki, R. De Giorgio, C. Sternini, and P.H. Guth, Uremia increases gastric mucosal permeability and acid back-diffusion injury in the rat, *Gastroenterology.* 103: 1762 (1992).

108. J.E. Grönbech, and E.R. Lacy, Impaired gastric defense mechanisms in aged rats: role of sensory neurons, blood flow, restitution, and prostaglandins, *Gastroenterology.* in press (1994).

109. H. Miyake, K. Takeuchi, and S. Okabe, Derangement of gastric mucosal blood flow responses in aged rats: relation to capsaicin-sensitive sensory neurons, *Gastroenterology.* in press (1994).

SYMPATHO-ADRENERGIC REGULATION OF DUODENAL ALKALINE SECRETION

Lars Fändriks and Claes Jönson

Acid secreted by the gastric parietal cells is an important factor in the digestive process. The acidity is, however, a potential threat to the organism itself and several protective mechanisms are utilized to prevent autodigestion. Today, it is well established that there exists a bicarbonate secretion by the gastroduodenal surface epithelium which neutralizes gastric acid[1,2]. Micro-pH-electrodes have been used to demonstrate a pH-gradient immediately above the epithelial cells; pH at the cell surface being neutral despite highly acidic luminal contents[3,4]. In the duodenum, this transport of bicarbonate is probably the most important factor for protection against gastric acid, whereas it plays a more subordinate role in the stomach[1,5].

The physiological control of duodenal alkaline secretion is complex and involves several regulatory mechanisms including neural and hormonal (incl. local prostaglandins) mechanisms. Wright and co-workers reported in 1940 on the effects of direct electrical stimulation of the extrinsic nerves to the small intestine in cats. Vagal stimulation increased duodenal secretion and the acid-neutralizing capacity, whereas activation of the splanchnic nerves reduced the secretion[6]. During recent years additional data have emerged concerning the autonomic neural influences on the rate of alkaline secretion by the duodenal surface epithelium. Enteric nerves are important for maintaining basal secretion[7,8] and partly mediate the locally induced increase in secretion in response to mucosal exposure to acid[9,10]. The extrinsic parts of the autonomic nervous system act mainly in a classical antagonistic fashion. The parasympathetic vagal innervation is stimulatory, whereas the sympathetic splanchnic nerves are mainly inhibitory, on the duodenal alkaline secretion[10-12]. This report summarizes data concerning sympatho-adrenergic influences on duodenal bicarbonate secretion.

EFFECTS OF THE SYMPATHO-ADRENERGIC SYSTEM ON THE DUODENAL ALKALINE SECRETION

Direct Electrical Stimulation of the Splanchnic Nerves

The peripheral ends of the bilaterally cut splanchnic nerves in chloralose-anesthetized cats and rats have been electrically stimulated at 10 Hz over 10 to 15

Neuroendocrinology of Gastrointestinal Ulceration
Edited by S. Szabo and Y. Taché, Plenum Press, New York, 1995

min. Stimulation of this type resulted in an approximately 50% decrease in duodenal alkaline secretion (Fig. 1)[13,14]. The secretory decrease was associated with an increase in mean arterial pressure and a reflexly decreased heart rate. Furthermore, the secretory increment in response to vagal stimulation or to mucosal exposure to acid was inhibited by electrical splanchnic nerve stimulation[13,14].

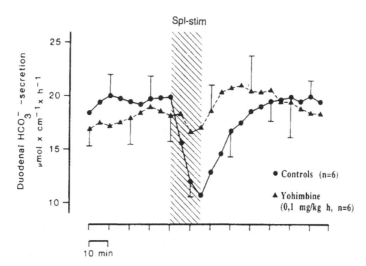

Figure 1. Effect of bilateral electrical stimulation of the splanchnic nerves (10 Hz for 15 min) on duodenal HCO_3^- secretion in a control group, compared to animals treated with yohimbine. Values are means $\pm$ SEM

Reflex Activation of the Splanchnic Nerves

The Intestino-Intestinal Reflex. Pearcy and Van Liere showed in 1926 that the distention of one part of the intestine reduced the motility in another part[15]. This effect was conveyed in the splanchnic nerves and reflexly transmitted via the spinal cord. It was later to be called the intestino-intestinal reflex[16]. Indirect evidence has been reported suggesting an influence of the intestino-intestinal reflex on the duodenal alkaline secretion. Splanchnectomy raised basal and vagally-induced alkaline in anesthetized cats, indicating on ongoing inhibitory effect by these nerves, possibly induced by the surgical trauma[11,17]. To further study such a reflex, the nerves surrounding two mesenteric vessels at the jejuno-ileal level were afferently stimulated (3Hz) in rats[18]. This procedure decreased duodenal alkaline secretion by 20 to 25%. The response was unaffected by cervical cord transection, whereas bilateral splanchnectomy markedly reduced the response[18]. These results indicate that the observed effect does not involve higher brain centers. Furthermore, the decrease in duodenal alkaline secretion during mesenteric nerve stimulation is mainly mediated via a spinal reflex and not via the decentralized pre-vertebral ganglia, a pathway originally described by Kuntz and Saccomano[19].

Effects of Hemorrhage. Hypovolemia, induced for example by hemorrhage, unloads cardiovascular volume- and baroreceptors which, in turn, decrease the afferent nerve discharge in the vagal and glossopharyngeal (carotid sinus) nerves to the vasomotor center. The vasomotor center increases the activity in the sympathoadrenal system[20]. Blood loss has been shown to increase the discharge rate in the splanchnic nerves of rats[21]. Arterial bleedings were studied in chloralose-anesthetized rats to investigate if this type of splanchnic nerve activation would influence duodenal alkaline secretion. Blood loss of approximately 10% of the total blood volume reduced duodenal alkaline secretion by about 30%. The blood loss was associated with a minor (≈20%) and transient decrease in blood pressure. Heart rate was not significantly changed by the bleeding. A smaller bleeding (5% of the total blood volume) induced a smaller secretory decrease compared to the 10% hemorrhage, suggesting a "dose-response" relationship[22]. Vagotomy lowers basal duodenal alkaline secretion and blood loss did not influence the secretion during this condition. However, when the secretion was raised by electrical stimulation of the cut vagal nerves, blood loss could again exert an inhibitory action. Apparently, the bleeding-induced inhibition of the secretion is exerted when a tonic activity in the peripheral vagi is present[22]. Blood loss also inhibited the increase in duodenal alkaline secretion due to exposure of the mucosa to acid[23].

The bleeding-induced decrease in duodenal alkaline secretion was markedly reduced by a thoraco-lumbar epidural blockade or bilateral splanchnectomy, showing that the response was mediated via spinal route, peripherally conveyed in the splanchnic nerves. Animals with ligated adrenal glands exhibited a higher basal duodenal alkaline secretion than animals with intact adrenals[24]. A 10% bleeding reduced the duodenal alkaline output even more after adrenal-ligation, suggesting that the adrenal glands are not involved in this response. Furthermore, treatment with the adrenolytic agent guanethidine (4 mg/kg) attenuated the bleeding-induced reduction of duodenal alkaline secretion. This drug has only minor effects on the release of catecholamines from the adrenals in response to splanchnic nerve activation[25]. Taken together, the data indicate that the bleeding-induced inhibition of alkaline secretion is mediated via postganglionic adrenergic nerve fibres, rather than by a humoral release from the adrenal glands.

Effects of Stimulations Within the Central Nervous System. Some structures in the hypothalamus are well-known for activating the sympatho-adrenal system and to decrease splanchnic blood flow[26,27]. In rat experiments, stimulation points within the perifornical region of the hypothalamus were chosen which raised mean arterial pressure, indicating an increased sympatho-adrenal activity. Electrical stimulation within these hypothalamic locations decreased the duodenal alkaline secretion in the majority of experiments (19 out of 25 experiments). In a few of the experiments either no change in secretion occurred (2 out of 25) or only a small secretory increase was registered (four out of 25)[28].

Hypothalamically-induced inhibition of duodenal alkaline secretion could be blocked by applying a thoraco-lumbar epidural anesthesia or by administration of the adrenolytic agent guanethidine (Fig. 2), suggesting involvement of a spinal route presumably conveyed peripherally in the splanchnic nerves[28]. Recently Lenz and collaborators demonstrated that intracerebroventricular infusions of calcitonin gene related peptide (CGRP) reduce duodenal alkaline secretion. Furthermore, this effect was inhibited by blockade of alpha-adrenoceptors with phentolamine[29].

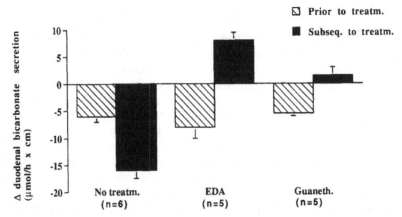

Figure 2. Effects on duodenal HCO$_3^-$ secretion of stereotaxic electrical unipolar stimulations (50Hz, 1ms pulse duration and 60-100 μA amplitude for 10 min) in the perifornical region of the hypothalamus in rats. Experiments were performed in three different groups. Two consecutive stimulations were performed in each group. Induced net changes (means ± SEM) are shown in the figure. Note that the second stimulation in the untreated control group induced a larger secretory decrease (n=6). The next group (n=5) was given an epidural anesthesia (EDA) approximately 30 min after the first stimulation. This treatment reversed the response of hypothalamic stimulation to an increase. Guanethidine, an adrenolytic agent, (8 mg/kg) was injected i.v. after the first stimulation in a third group (n=5). This compound blocked the hypothalamically induced inhibition of duodenal HCO$_3^-$ secretion (*p<0.05 (Newman-Keuls test) compared to the group without intervention).

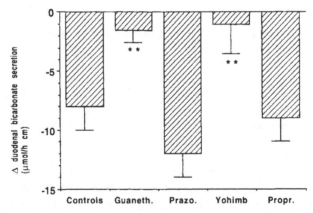

Figure 3. Effects of electrical splanchnic nerve stimulation on duodenal HCO$_3^-$ secretion in rats with different pharmacological treatments. Values are presented as net changes (lowest value during stimulation compared to the basal level) and are given as means ± SEM (**=p<0.01).

Involvement of Adrenergic Transmission

Pretreatment of the rats with the adrenolytic agent guanethidine (3 to 8 mg/kg i.v.) markedly reduced the decrease in duodenal alkaline secretion, which was induced by either electrical splanchnic nerve stimulation, afferent mesenteric nerve stimulation, hemorrhage or by electrical hypothalamic stimulations[14,18,22,28,30].

The alpha-2-adrenoceptor antagonist yohimbine, administered in doses of 0.1 to 1 mg/kg i.v., did not change basal secretion but markedly attenuated the inhibition of both basal and acid-stimulated alkaline output induced by various kinds of activation of the splanchnic nerves. Despite high doses, neither the alpha-1 blocker prazosin, nor the beta-adrenoceptor antagonist propranolol interfered with basal alkaline output or with the inhibition of duodenal alkaline secretion due to any kind of splanchnic nerve activation (Fig. 3)[13,14,18,30-32]. Administration of the alpha-2 adrenoceptor agonist clonidine inhibits duodenal alkaline output in rats, cats and man[13,33,34]. Altogether, the available data suggest that splanchnic neural inhibition of duodenal alkaline secretion to a major part is mediated by adrenergic neurons and alpha-2 adrenoceptors.

Sympatho-Adrenergic Influences on the Secretion in Relation to Effects on Blood Flow

It is well known that activation of the splanchnic nerves reduces intestinal blood flow[35-41]. Duodenal alkaline secretion in rabbits has been proposed to be determined by blood flow and arterial $[HCO_3^-]$[42,43]. Consequently, the splanchnic neural inhibition of duodenal alkaline secretion could be secondary to a reduction in blood flow and/or arterial $[HCO_3^-]$. Therefore, alkaline secretion and blood flow (measured with radioactive microspheres) were measured in the same duodenal segment in the rat. In addition, arterial $[HCO_3^-]$ was analyzed in blood from the right femoral artery. Data from ref. 43 are summarized in Table 1a and 1b. A modest arterial bleeding (10% of the total blood volume) lowered alkaline secretion by 44%, duodenal blood flow by 31% and arterial $[HCO_3^-]$ by 11%. In a group of rats with bilaterally cut vagal nerves, basal duodenal alkaline output was approximately 50% lower than that in controls (Table 1a). However, basal duodenal blood flow and basal arterial $[HCO_3^-]$ were not significantly altered by vagotomy. The 10% bleeding was not associated with changes in duodenal secretion, confirming previous the findings (see above), whereas duodenal blood flow and arterial $[HCO_3^-]$ were reduced similarly as in animals with intact vagi (Table 1b). Furthermore, pretreatment with the alpha-2 adrenoceptor antagonist yohimbine attenuated the bleeding-induced reduction in duodenal alkaline output but did not significantly change the decreases of duodenal blood flow and arterial $[HCO_3^-]$ during the bleeding period (Table 1b)[44].

When considering the data from the vagotomized and yohimbine-treated groups, it seems unlikely that changes in duodenal alkaline secretion are strictly secondary to alterations in duodenal blood flow and/or arterial $[HCO_3^-]$. Thus, during these experimental conditions, the correlation between duodenal alkaline secretion and the mucosal availability of HCO_3^- seems to reflect a co-variation rather than a causal dependence. Consequently, the present investigations in rats suggest the existence of two separate sympathetic pathways, one influencing the duodenal blood vessels and another influencing the secretion. These two pathways can apparently act independently of each other.

Table 1a. Basal Duodenal HCO_3^- Secretion, Duodenal Blood Flow and Arterial $[HCO_3^-]$

	HCO_3^- secretion (μmol/cm x h)		Blood Flow (ml/100g x min)	Art. $[HCO_3^-]$ (mmol/l)
Control	18.6 ± 2.5		339 ± 20	22.6 ± 1.4
Vagotomy	8.0 ± 1.6	**	389 ± 58	21.4 ± 0.8
Yohimbine	23.1 ± 2.9		320 ± 29	20.3 ± 1.3

One group was without any pretreatment (Controls, n=6), one group was subjected to bilateral vagotomy (n=6) and one group was treated with yohimbine 0.1-0.2 mg/kg^{-1} (n=5). Values are means ± SEM. Significant difference from the control group is indicated with asterisks (**,p<0.01, Newman-Keul's test).

Table 1b. Net Changes in Duodenal HCO_3^- Secretion, Duodenal Blood Flow and Arterial $[HCO_3^-]$ Induced by a 10% Arterial Bleeding

	HCO_3^- secretion (μmol/cm x h)		Blood Flow (ml/100g x min)	Art. $[HCO_3^-]$ (mmol/l)
Control (=10% bleeding)	-8.2 ± 2.3		-105 ± 14	-2.5 ± 0.6
Vagotomy + bleeding	-0.7 ± 0.8	**	-175 ± 37	-2.0 ± 0.5
Yohimbine + bleeding	-1.2 ± 1.0	**	-88 ± 16	-1.4 ± 0.5

The three groups are the same animals as in Table 1a. Data represent the maximal change during the bleeding period compared to the basal level (see Table 1a). Value are means ± SEM. Significant difference from the control group is indicated with asterisks (**,p<0.01, Newman-Keul's test).

FUNCTIONAL CONSIDERATIONS

Available data suggest the existence of a complex central neural control of the duodenal alkaline secretion[1,10,11,28,29]. The splanchnic nerves exert mainly an inhibitory effect by use of adrenergic neurons and, thereby, act antagonistic to the vagal neural effects. The sympatho-adrenergic neural modulation of the secretion can be a result of influencing the epithelium directly or, indirectly, via enteric secretomotor neurons. It seems less possible that the neurogenic effects are secondary to alterations of the mucosal blood circulation.

The physiological role of the sympatho-inhibitory arrangement is obscure. The sympatho-adrenal system is often being regarded as an emergency system which can be activated in response to various stimuli which are, or may be, hazardous to the organism. Example of such stressful stimuli are trauma or blood loss, or during fight or flight behavior. Such stimuli are often associated with a risk for metabolic acidosis and loss of body fluids. It has previously been proposed that sympatho-inhibitory effects on intestinal fluid secretion may contribute to the maintenance of fluid volumes in the body[45]. It may be speculated that the organism, among several homeostatic adjustments, also inhibits the alkaline secretion in order to counteract the risk for metabolic acidosis.

As mentioned in introduction, the maintenance of the mucosal integrity is dependent on the balance between aggressive digestive factors, e.g. gastric acid, and protective factors such as neutralizing bicarbonate[1,2]. An imbalance between these counterparts, for example by a dominance for sympatho-adrenergic inhibition of gastroduodenal bicarbonate secretion, may predispose for mucosal injury and disease.

ACKNOWLEDGEMENTS

Experiments in our laboratory were financed by the Swedish Medical Research Council (grants 0016, 2855, 8429, 8663), Stockholm; the Göteborg Medical Society, Göteborg; and the Swedish Medical Society, Stockholm.

REFERENCES

1. A. Allen, G. Flemström, A. Garner, and E. Kivilaakso, Gastroduodenal mucosal protection, *Physiol Rev.* 73:823 (1993).
2. J. Crampton, and W.D.W. Rees, Gastroduodenal bicarbonate secretion: its role in protecting the stomach and duodenum, *in:* "Recent Advances in Gastroenterology 6," R.E. Pounder, ed., Churchill Livingstone, Edinburgh, London, Melbourne and New York (1986).
3. G. Flemström, and E. Kivilaakso, Demonstration of a pH gradient at the luminal vivo and its dependence on mucosal alkaline secretion, *Gastroenterology.* 84:787 (1983).
4. E.M.M. Quigley, and L.A. Turnberg, pH of the microclimate lining human gastric and duodenal mucosa in vivo, *Gastroenterology.* 92:1876 (1987).
5. G. Flemström, and A. Garner, Gastroduodenal HCO_3^- transport: characteristics and proposed role in acidity regulation and mucosal protection, *Am J Physiol.* 242:G183 (1982).
6. M.A. Wright, H.W. Jennings, and R. Lium, The influence of nerves and drugs on secretion by the small intestine and an investigation of the enzymes in the intestinal juice, *Quart J Exp Physiol.* 30:73 (1940).
7. O. Nylander, G. Flemström, D. Delbro and L. Fändriks, Vagal influence on gastroduodenal HCO_3^- secretion in the cat in vivo, *Am J Physiol.* 252:G522 (1987).
8. J.R. Crampton, L.G. Gibbons, and W.D.W. Rees, Neural regulation of duodenal alkali secretion: effects of electrical field stimulation, *Am J Physiol.* 254:G162 (1988).
9. B. Smedfors, and C. Johansson, Cholinergic influence on duodenal bicarbonate response to hydrochloric acid perfusion in the conscious rat, *Scand J Gastroenterol.* 21:809 (1986).
10. L. Fändriks, C. Jönson, O. Nylander, and G. Flemström, Neural influences on gastroduodenal secretion, *in:* "Ulcer Disease, New Aspects of Pathogenesis and Pharmacology," S. Szabo, C.J. Pfeiffer, eds., CRC Press, Boca Raton (1989).
11. L. Fändriks, Vagal and splanchnic neural influences on gastric and duodenal bicarbonate secretions. An experimental study in the cat, *Acta Physiol Scand.* 128: Suppl. 555 (1986).
12. L. Fändriks, and C. Jönson, Influences of the sympatho-adrenal system on gastric motility and acid secretion and on gastroduodenal bicarbonate secretion, *Acta Physiol Scand.* 135:285 (1987).

13. L. Fändriks, C. Jönson, and O. Nylander, Effects of splanchnic nerve stimulation and of clonidine on gastric and duodenal HCO_3^--secretion in the anesthetized cat, *Acta Physiol Scand.* 130:251 (1987).

14. C. Jönson, and L. Fändriks, Splanchnic nerve stimulation inhibits duodenal HCO_3^- secretion in the rat, *Am J Physiol.* 255:G709 (1989).

15. J.F. Pearcy, and E.J. van Liere, Studies on the visceral nervous system, *Am J Physiol.* 78:64 (1926).

16. H. Hermann, and G. Morin, Mise en evidence d'un reflexe inhibiteur intestino-intestinal, *C R Soc Biol.* 115:529 (1934).

17. L. Fändriks, Sympatho-adrenergic inhibition of vagally induced gastric motility and gastroduodenal HCO_3^--secretions in the cat, *Acta Physiol Scand.* 128:555 (1986).

18. C. Jönson, and L. Fändriks, Afferent electrical stimulation of mesenteric nerves inhibits duodenal HCO_3^- secretion via a spinal reflex activation of the splanchnic nerves in the rat, *Acta Physiol Scand.* 133:545 (1988).

19. A. Kuntz, and J. Saccomano, Reflex inhibition of intestinal motility mediated through decentralized prevertebral ganglia, *J Neurophysiol.* 105:251 (1944).

20. S. Chien, Role of the sympathetic nervous system in hemorrhage, *Physiol Rev.* 47:214 (1967).

21. K. Ito, A. Sato, K. Shimamura, and R.S. Swenson, Reflex changes in sympatho-adrenal medullary functions in response to baroreceptor stimulation in anesthetized rats, *J Auton Nerv Syst.* 10:295 (1984).

22. C. Jönson, and L. Fändriks, Bleeding inhibits vagally-induced duodenal HCO_3^- secretion via activation of the splanchnic nerves in anesthetized rats, *Acta Physiol Scand.* 130:259 (1987).

23. C. Jönson, P. Tunbäck-Hansson and L. Fändriks, Splanchnic nerve activation inhibits HCO_3^- secretion from the duodenal mucosa induced by luminal acid in the rat, *Gastroenterology.* 96:45 (1989).

24. C. Jönson, O. Nylander, G. Flemström, and L Fändriks, Vagal stimulation of duodenal HCO_3^--secretion in anaesthetized rats, *Acta Physiol Scand.* 128:65 (1986).

25. A.R. Wakade, R.K. Malhotra, T.D. Wakade, and W.R. Dixon, Simultaneous secretion of catecholamines from the adrenal medulla and of [^3H] norepinephrine from sympathetic nerves from a single test preparation: different effects of agents on the secretion, *Neuroscience.* 18:877 (1986).

26. A. Cobbold, B. Folkow, O. Lundgren, and I. Wallentin, Blood flow, capillary filtration coefficients and regional blood volume responses in the intestine of the cat during stimulation of the hypothalamic "defence area", *Acta Physiol Scand.* 61:467 (1964).

27. C.P. Yardley, and S.M. Hilton, The hypothalamic and brainstem areas from which the cardiovascular and behavioral components of the defence reaction are elicited in the rat, *J Auton Nerv Syst.* 15:227 (1986).

28. L. Fändriks, C. Jönson, and B. Lisander, Hypothalamic inhibition of duodenal alkaline secretion via a sympatho-adrenergic mechanism in the rat, *Acta Physiol Scand.* 137:357 (1989).

29. H.J. Lenz, and M.R. Brown, Cerebroventricular calcitonin gene-related peptide inhibits rat duodenal bicarbonate secretion by release of norepinephrine and vasopressin, *J Clin Invest.* 85:25 (1990).

30. C. Jönson, and L. Fändriks, Bleeding-induced decrease in duodenal HCO_3^- secretion in the rat is mediated via alpha-2 adrenoceptors, *Acta Physiol Scand.* 130:387 (1987).

31. L. Fändriks, and C. Jönson, Effects of adrenoceptor antagonists on vagally induced gastric and duodenal HCO_3^--secretion in the cat, *Acta Physiol Scand.* 130:243 (1987).

32. C. Jönson, A. Hamlet, and L. Fändriks, Hypovolemia inhibits acid-induced alkaline transport in the rat duodenum via an alpha-2 adrenergic mechanism, *Acta Physiol Scand.* 142:367 (1991).

33. O. Nylander, and G. Flemström, Effects of alpha-adrenoceptor agonists and antagonists on duodenal surface epithelial HCO_3^--secretion in vivo, *Acta Physiol Scand.* 126:433 (1986).

34. L. Knutson, and G. Flemström, Duodenal mucosal bicarbonate secretion in man. Stimulation by acid and inhibition by the alpha-2 adrenoceptor agonist clonidine, *Gut.* 30:1707 (1989).

35. N.G. Kock, An experimental analysis of mechanisms engaged in reflex inhibition of intestinal motility, *Acta Physiol Scand.* 47:suppl 164 (1959).

36. B. Folkow, D.H. Lewis, O. Lundgren, S. Mellander, and I. Wallentin, Effect of graded vasoconstrictor fibre stimulation on the intestinal resistance and capacitance vessels, *Acta Physiol Scand.* 61:445 (1964).

37. B. Folkow, D.H. Lewis, O. Lundgren, S. Mellander, and I. Wallentin, The effect of the sympathetic vasoconstrictor fibres on the distribution of capillary blood flow in the intestine, *Acta Physiol Scand.* 61:458 (1964).

38. S. Sjövall, Redfors, D. Hallbäck, S. Eklund, M. Jodal, and O. Lundgren, The effect of splanchnic nerve stimulation on blood flow distribution, villous tissue osmolality and fluid and electrolyte transport in the small intestine in the cat, *Acta Physiol Scand.* 117:359 (1983).

39. O. Lundgren, Microcirculation of the gastrointestinal tract and pancreas, *in:* "Handbook of Physiology - The Cardiovascular System IV," E.M. Renkin, C.C. Michel, S.R. Geiger, eds., American Physiological Society: Bethesda (1984).

40. F.W. Leung, M. Itoh, K. Hirabayashi, and P.H. Guth, Role of duodenal blood flow in gastric and duodenal mucosal injury in the rat, *Gastroenterology.* 86:281 (1985).

41. A.P. Shepard, and G.L. Riedel, Intramural distribution of intestinal blood flow during sympathetic stimulation, *Am J Physiol.* 255:H1091 (1988).

42. R. Schiessel, M. Starlinger, E. Kovats, W. Appel, W. Feil, and A. Simon, Alkaline secretion of rabbit duodenum in vivo: its dependence of acid base balance and mucosal blood flow, *in:* "Mechanisms of Mucosal Protection in the Upper Gastrointestinal Tract," A. Allen, G. Flemström, A. Garner, W. Silen, L.A. Turnberg, eds., Raven Press, New York (1984).

43. M. Starlinger, and R. Schiessel, Bicarbonate (HCO_3) delivery to the gastroduodenal mucosa by the blood: its importance for mucosal integrity, *Gut.* 29:647 (1988).

44. C. Jönson, L. Holm, T. Jansson, and L. Fändriks, Effects of hypovolemia on duodenal blood flow, arterial [HCO_3^-] and HCO_3^- output in the rat duodenum, *Am J Physiol.* 259:G179 (1990).

45. H. Sjövall, Sympathetic control of jejunal fluid and electrolyte transport, *Acta Physiol Scand.* Suppl. 535 (1984).

POTENTIATION OF INTESTINAL SECRETORY RESPONSES
TO HISTAMINE: PATHOPHYSIOLOGIC IMPLICATIONS

P.K. Rangachari, T. Prior, and R.A. Bell

Discrimination is the essence of existence, whatever the puerile pundits of political correctness may wish to believe. Mercifully for life in general, Nature in her infinite wisdom exhibits a fine discriminant sense for the welfare of the organism. This capacity is evident in the functioning of the intestinal lining which by its selective absorption of nutrients plays such a crucial role in the body's economy. This lining is subject to a barrage of influences that impinge on it from both luminal and contraluminal aspects and the fine tuning of its transport properties by these factors keeps the organism in balance. Any disturbance can lead to alterations in gut function that are expressed as either constipation or diarrhea. Thus the release of potent endogenous chemicals during the course of inflammation may profoundly alter gut function[2].

Since its discovery by Dale and Laidlaw[1], histamine is amongst the best known of inflammatory mediators. It has marked effects on a variety of gastrointestinal functions. With specific reference to intestinal epithelia, it has been shown to alter transport functions in both large and small intestine[2-6]. If these effects of histamine were to be potentiated, the consequences for transport functions of the gut would be profound.

We have recently described the marked potentiation of colonic responses to histamine by hydroxylamines[7,8]. Since these compounds could be produced endogenously by mammalian cells as well as by colonic bacteria[9,10], the observations made may have relevance for a number of pathological conditions. The objective of this brief report is to review those observations and speculate on the possible implications. Such speculations are particularly appropriate in a volume designed to honor Hans Selye, one of the more adventurous investigators in our time. He was never averse to controversy and often used laboratory data imaginatively to speculate on larger issues. Although we cannot compete with Selye in that regard, we hope that we could at least follow his example.

MATERIALS AND METHODS

The procedures used for obtaining the colonic epithelial preparations have been described in detail earlier[7,8]. Briefly, adult dogs of either sex were euthanized with

Neuroendocrinology of Gastrointestinal Ulceration
Edited by S. Szabo and Y. Taché, Plenum Press, New York, 1995

sodium pentobarbital (100mg/kg) and the proximal colon quickly excised and immersed in oxygenated Krebs solution warmed to 37° C. Removal of the circular and longitudinal muscles was followed by a careful dissection to remove the muscularis mucosae as well. The resulting "nerve-free" preparation was set-up in conventional Ussing chambers for recording short-circuit currents. Tissues were bathed on both sides with warm oxygenated Krebs solution having the following composition (mM): NaCl 116, KCl 4.6, $MgCl_2$ 1.2, $CaCl_2$ 1.5, $NaHCO_3$ 22, NaH_2PO_4 1.2, Glucose 10. Short-circuit currents were measured using a WPI Dual Voltage Clamp.

Phosphodiesterase activity was measured in a soluble fraction prepared from the canine colon. Colonic epithelial pieces were homogenized in sucrose-MOPS (pH 7.4), the supernatant fraction obtained following homogenization and centrifugation being stored at -70° C. The enzyme in the soluble fraction was assayed by incubating aliquots in glycine buffer at pH 9.0 with 4 mM 3'5' cyclic AMP. The reaction was stopped by boiling and the 5' AMP formed converted to adenosine and inorganic phosphate by incubation with 5' nucleotidase[11]. The phosphate produced was measured colorimetrically[12]. Similar experiments were done using commercially obtained beef heart phosphodiesterase.

Histamine catabolism was measured by monitoring the disappearance of histamine from the bathing solutions, using a fluorometric method[8]. To assess the effects of hydroxylamines on diamine oxidase activities, a soluble fraction was prepared from colonic tissue homogenates. Enzyme activities in the presence and absence of the compounds were measured using the method of Kusche et al.[13]. This involved the incubation of the enzyme fraction with [1,4-^{14}C] putrescine dihydrochloride. The labelled product, [^{14}C]-Δ_1-pyrolline was extracted and measured. Details of all methods are given in Rangachari et al.[8].

Most chemicals were bought from Sigma, including all the hydroxylamine derivatives used except O-ethyl hydroxylamine which was purchased from Aldrich Chemical Co. Others (O-(2-propyl) hydroxylamine, O-butylhydroxylamine, and O-(2-hydroxyethyl) hydroxylamine) were synthesized by a modification of the Gabriel procedure for the synthesis of primary amines[14]. Beef heart phosphodiesterase was bought from Boehringer-Mannheim. The hydroxylamines (all hydrochlorides) were made up in stock solutions of 10^{-1} M in distilled de-ionised water.

RESULTS AND DISCUSSION

The colonic epithelium responded to serosal additions of histamine with sharp increases in short-circuit currents. These responses were markedly potentiated by hydroxylamine and methoxyamine which had no significant effects on their own. These potentiating effects were specific for histamine, since no effects were noted with other agonists, such as carbachol or serotonin, which also increased the short-circuit currents.

Potentiation could be demonstrated in several ways: *1)* Tissues pretreated with hydroxylamine showed exaggerated responses to histamine (Fig. 1). *2)* Pretreatment with hydroxylamine produced a marked leftward shift in the cumulative concentration-response curve to histamine (ED_{50} shifting from 9×10^{-5} M to 2.1×10^{-6} M). *3)* Following the peak in the response to histamine, the addition of hydroxylamines produced a sharp secondary increase in the short-circuit current (Fig. 2). Using these secondary increases a number of compounds were rapidly screened. Active compounds required an NH_2-O-R sequence. For instance, O-methyl hydroxylamine was active, whereas the N-methyl derivative was not. The list of active and inactive compounds is given in Rangachari et al.[8].

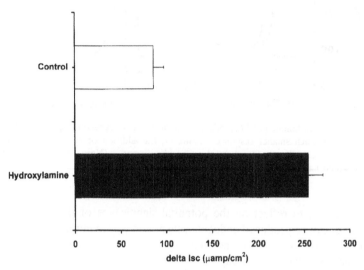

Figure 1. Tissues pretreated with 10^{-4} M serosal hydroxylamine show exaggerated responses to 10^{-5} M serosally added histamine.

Several possible mechanisms could explain the potentiations observed. These could include alterations at the receptor level, alterations in signal transduction mechanisms or alterations in the degradation/metabolism of added histamine.

The phosphodiesterase inhibitor, IBMX, also potentiated responses to histamine[7]. However, these effects were not as dramatic as those seen with the hydroxylamines. The possibility that inhibition of phosphodiesterase could explain the potentiation produced by the hydroxylamines was tested. Rao[15] reported that hydroxylamines increased both cyclic AMP and GMP levels in rabbit ileal mucosa and suggested that its effects may be mediated by inhibition of phosphodiesterase. IBMX produced a dose-dependent inhibition of soluble cAMP-dependent phosphodiesterase obtained from the colonic epithelium. However, the hydroxylamines had no significant effects. Similar results were obtained with commercially obtained beef heart phosphodiesterase (Fig. 3). Clearly then, inhibition of phosphodiesterase could not explain the effects noted.

We then tested the possibility that hydroxylamines could inhibit catabolism of histamine. We found that, in the presence of active hydroxylamines, the disappearance of histamine from the bathing solutions was considerably slower. A similar effect was seen with a standard inhibitor of diamine oxidase, amino-guanidine (see Fig. 4). A soluble fraction containing the enzyme was prepared and the effects of the hydroxylamines tested. Active hydroxylamines as well as standard inhibitors such as aminoguanidine and semicarbazide produced significant inhibition of enzyme activity (Fig. 5). An inactive hydroxylamine (N-methyl hydroxylamine) did not[8].

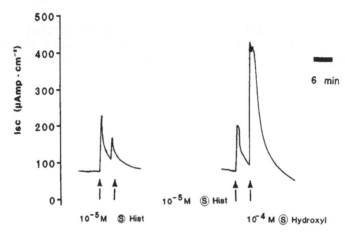

Figure 2. Responses to histamine (10^{-5} M) fade after an initial peak. A second addition of histamine (10^{-5} M) produces a much smaller response. However, the addition of hydroxylamine (10^{-4} M) produces a marked secondary increase in short-circuit current. (Note, this concentration of hydroxylamine would have elicited marginal effects on its own. See Fig. 1.)

It is interesting to reflect on the potential significance of these observations. Hydroxylamines can be produced endogenously not only by mammalian cells but also by intestinal bacteria[9,10]. For instance leukocytes produce reactive hydroxylamines from drugs such as sulfadiazine or dapsone[16,17]. These hydroxylamines may elicit toxic effects including hypersensitivity reactions[17]. In combination with endogenously produced histamine, these toxic effects can be amplified.

Intestinal bacteria are another potent source of reactive metabolites including hydroxylamines[9]. The metabolic capacity of the gut flora is diverse; the reactions catalyzed include hydrolysis, dehydroxylation, decarboxylation, dehalogenation, demethylation and a variety of reduction reactions[9,18,19]. It has been noted that "any compound taken orally, any substance entering the intestine via the biliary tract or the blood stream, or any substance secreted directly into the lumen is a potential substrate for bacterial transformation"[19]. More importantly, the metabolic capacities of the gut flora can be markedly altered by changes in diet, use of drugs etc. Goldin[19] comments that "diet can, indeed, alter the metabolic activity of the flora, and these changes may be more relevant to the host than the Latin or Greek names of the microorganisms." The potential thus exists for complex interactions. Not only can bacterial flora metabolize drugs to produce potentially toxic compounds but the drugs themselves can induce different metabolic capacities. Tannock[20], reviewing the effects of dietary and environmental stress on the gastrointestinal flora, notes that two bacterial populations may be sensitive indicators of dietary stress. Lactobacilli, occupying the proximal part of the gut, usually decrease, whereas coliforms, occupying the distal parts, usually increase under such stress. Thus these alterations in the microbial ecosystem produced by stresses could lead to the presence of microbial metabolites in concentrations that are toxic to the host.

Compounds containing nitro groups can thus be transformed to produce reactive nitroso ($-N=O$) and hydroxylamino ($-NH_2$) intermediates. These derivatives may be responsible for a number of toxic effects e.g. carcinogenesis, methemoglobinemia, aplastic anemia etc.[9]. In the gut, these compounds may not be active on their own, but they could certainly potentiate the responses to endogenously produced histamine. Thus in the presence of a background inflammatory condition, unexpected bouts of

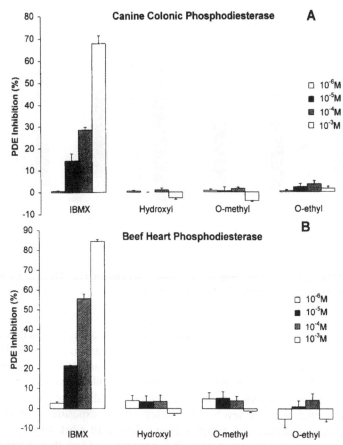

Figure 3. IBMX produces a dose-dependent inhibition of both canine colonic phosphodiesterase as well as beef heart phosphodiesterase. None of the hydroxylamines tested showed any effect.

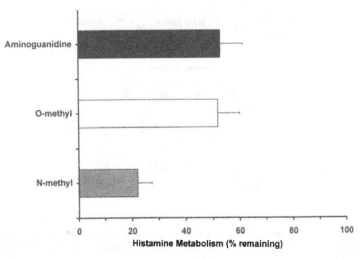

Figure 4. The effects of hydroxylamines and aminoguanidine on the disappearance of histamine from the serosal solutions bathing the isolated canine colon. Tissues were set up in Ussing chambers and a specific amount of histamine (120 μg) added to the bathing solutions. Samples were collected periodically for the fluorometric estimation of histamine. The values shown are the amounts remaining after 120 min.

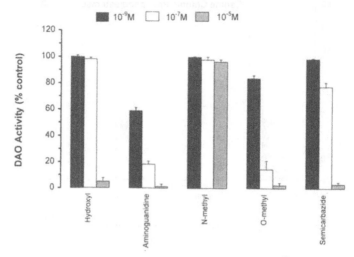

Figure 5. Inhibition of DAO activity produced by a fixed concentration (10^{-5} M) of added inhibitors. The values are expressed as percentages of control values.

diarrhea may occur. Further, as emphasized above, subtle alterations in diet can alter metabolic capacities of gut flora which can compound the problem. With specific reference to nitrate reductase activity, it has been shown that rats fed a purified fibre-free diet showed no increase in methemoglobin levels following oral administration of sodium nitrate whereas rats fed a diet supplemented with 5% pectin had significant increase in methemoglobin levels 2.5 h after nitrate exposure[21]. Such studies underscore the possibilities of complex interactions with what may appear to be innocuous manoeuvres. In recent years, it is becoming evident that individuals may exhibit adverse reactions to foods that may not be explicable on the basis of classical

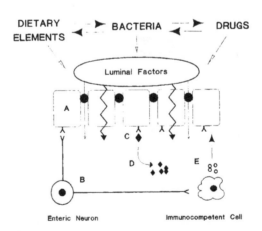

Figure 6. A schematic diagram to show interactions between luminal and contraluminal factors on the intestinal lining. Luminal factors that arise from complex interactions between bacteria, dietary elements and drugs can enter the serosal aspect either through a transcellular (⬳) or a paracellular (⬳) route. Potential loci for interaction are shown. (A) Intracellular effects (either on signal transduction or cellular metabolism) (B) Enteric neuronal activity (C) At neuro-transmitter/mediator receptors (D) Metabolism/degradation of neurotransmitters/mediators (E) Release of mediators from immunocompetent cells (mast cells etc.)

antibody mediated mechanisms. The mechanisms discussed above may be responsible in part for such pseudo-allergic food reactions[22].

The concepts discussed in this brief paper are summarized in Fig. 6. The intestinal lining is subjected to a barrage of influences from both luminal and contraluminal surfaces. Luminal factors which could arise from complex interactions between intestinal bacteria, ingested dietary elements and drugs can modulate cellular function in diverse ways. Some potential loci for interactions are shown in the figures. It is hoped that the scheme would interest pharmacologists, microbiologists and biochemists in possible collaborative ventures to unravel the complexities of the intestinal lining. Much of what has been discussed is admittedly speculative, based on limited data.

ACKNOWLEDGEMENTS

These studies were supported by a grant from the Medical Research Council of Canada.

REFERENCES

1. H.H. Dale, and P.P. Laidlaw, The physiological action of β-iminazolyl-ethylamine. *J. Physiol.* 41:318 (1910).
2. P.K. Rangachari, Six-pack balancing act: a conceptual model for the intestinal lining. *Can. J. Gastroenterol.* 4:201 (1990).
3. R.D. McCabe, and P.L. Smith, Effects of histamine and histamine receptor antagonists on ion transport in rabbit descending colon. *Am. J. Physiol.* 247:G41 (1984).
4. J. Hardcastle, and P.T. Hardcastle, The secretory actions of histamine in rat small intestine. *J. Physiol.* 388:521 (1987).
5. S.I. Wasserman, K.E. Barret, P.A. Huott *et al.*, Immune related intestinal Cl⁻ secretion. 1. Effect of histamine on the T84 cell line. *Am. J. Physiol.* 254 (Cell Physiol. 23):C53 (1988).
6. J.D. Wood, Communication between minibrain in gut and enteric immune system. *News Physiol. Sci.* 6:64 (1991).
7. P.K. Rangachari, and D. McWade, Histamine stimulation of canine colonic epithelium: potentiation by hydroxylamines. *Eur. J. Pharmacol.* 135:331 (1987).
8. P.K. Rangachari, T. Prior, R.A. Bell, and T. Huynh, Histamine potentiation by hydroxylamines: structure-activity relations; inhibition of diamine oxidase. *Am. J. Physiol.* 263:G632 (1992).
9. P. Goldman, Biochemical pharmacology of intestinal flora. *Ann. Rev. Pharmacol. Toxicol.* 18:523 (1985).
10. P. Gross, Biologic activity of hydroxylamine. A review, *CRC Crit. Rev. Toxicol.* 14:87 (1985).
11. K.G. Nair, Purification and properties of 3'5' cyclic nucleotide phosphodiesterase from dog heart. *Biochemistry* 5:150 (1966).
12. S.L. Bonting, K.A. Simon., and N.M. Hawkins, Studies on sodium-potassium activated adenosine triphosphatase. 1. Quantitative distribution in several tissues of the cat. *Arch. Biochem. Biophys.* 95:416 (1961).
13. J. Kusche, W. Lorenz, and J. Schmidt, Oxidative deamination of biogenic amines by intestinal amine oxidases: histamine is specifically inactivated by diamine oxidase. *Z. Phys. Chem.* 356:1485 (1975).
14. M.A. Gibson, M.S., and R.W. Bradshaw, The Gabriel synthesis of primary amines. *Angew. Chem. Int. Ed. Engl.* 7:919 (1968).
15. M.C. Rao, Toxins which activate guanylate cyclase: heat-stable enterotoxins, *in*: "Microbial Toxins and Diarrheal Disease," CIBA Fdn Symposium 112, Pitman, London (1985).
16. J.P. Uetrecht, N. Shear, and W. Biggar, Dapsone is metabolised by human neutrophils to a hydroxylamine. *Pharmacologist* 128:239 (1986).
17. J. Uetrecht, N. Zahid, N.H. Shear *et al.*, Metabolism of dapsone to a hydroxylamine by human neutrophils and mononuclear cells. *J. Pharmacol. Exp. Ther.* 245:274 (1986).

18. R.V. Smith, Metabolism of drugs and other foreign compounds by intestinal microorganisms. *Wld. Rev. Nutr. Diet.* 29:60 (1978).

19. B.R. Goldin, In situ bacterial metabolism and colon mutagens. *Ann. Rev. Microbiol.* 40:367 (1986).

20. G.W. Tannock, Effect of dietary and environmental stress on the gastrointestinal microbiota, *in*: "Human Intestinal Microflora in Health and Disease," D.J. Hentges, ed., Academic Press, New York (1983).

21. I.R. Rowland, A.K. Mallet, Dietary fibre and gut microflora- their effects on toxicity, *in*: "New Concepts and Developments in Toxicology." P.I. Chambers, P. Gehring, and F. Sakai, eds., Elsevier Science Publishers, Amsterdam (1986).

22. D.R. Stanworth, The scope for pseudo-allergic responses to food, *in*: "Food and the Gut," J.O. Hunter, and V. Alun Jones, eds., Bailliere Tindall, London (1985).

ACIDOSIS PLAYS AN IMPORTANT ROLE IN STRESS-INDUCED GASTRIC ULCERATION IN RATS

C.H. Cho, M.W.L. Koo and C.W. Ogle[*]

It has been shown that acidosis can decrease calcium entry and inhibit neurotransmitter release from nerve cells[1]; these effects may, therefore, interfere with the coupling action of calcium in cellular secretion and muscular contraction processes. Oral administration of ammonium chloride, an acidifying agent, induces gastric ulceration in rats[2]. Thus, acidosis appears to have a marked influence on gastric function. Cold-restraint stress is known to produce gastric ulceration[3], however, its effects on the blood acid-base balance have not been studied. Thus, it is conceivable that the occurrence of acidosis may play a role in stress-induced gastric ulceration. This chapter reports the results of a study to determine whether or not the changes in acid-base balance by cold-restraint stress are causally related to gastric ulcer formation in rats. The effects of some of antiulcer drugs on these parameters are also examined.

MATERIALS AND METHODS

General. Female Sprague-Dawley rats weighing 170-200 g were used. They were kept in an air-conditioned room in which the temperature and relative humidity were kept at $22 \pm 1°C$ and 65-70%, respectively, and exposed to a constant light (0700 - 1900 h) and dark (1900 - 0700) cycle. The animals were fed a balanced pellet diet (Ralston Purina Co., U.S.A.) and allowed to drink tap water *ad libitum*.

All rats were fasted for 48 h before use; they were kept in cages with raised floors of wide wire mesh to prevent coprophagy, and had free access to a solution of sucrose (BDH) 8% w/v in NaCl (BDH) 0.2% w/v[4] which was removed 1 h before experimentation. Animals were placed in individual close-fitting tubular wire mesh cages and exposed to $4°C$ (cold-restraint stress) or to $22 \pm 1°C$ (restraint only) for a period of 2 h[5]. The procedure of restraint only was required to immobilize the animals for i.v. infusions and blood sample collections.

---

[*]Author for correspondence

Neuroendocrinology of Gastrointestinal Ulceration
Edited by S. Szabo and Y. Taché, Plenum Press, New York, 1995

Blood vessel cannulation and blood acid-base determination. Rats were operated 24 h before experimentation. A small mid-line incision was made in the front of the neck of the ether-anesthetized animal. The right carotid artery and the left jugular vein were identified. A polyethylene tube (0.7 mm i.d., 1 mm o.d.), prefilled with a heparinized (50 IU/ml) solution of 0.9% NaCl w/v (saline), was used to cannulate the right carotid artery; the free end was made to emerge dorsally on the left side of the neck. Cannulation of the left jugular vein was carried out in the same manner. The neck incision was then closed with a suture and the open ends of the cannulae sealed before being tied in place onto the skin at the back of the neck. Drugs or saline were given by i.v. infusion and blood samples were collected from the arterial cannula.

During experiments, rats were allowed a stabilization period of 1 h inside their restraint cages before blood samples were taken for blood acid-base measurements. Samples were collected into heparinized capillary tubes and then immediately introduced into a blood-gas analyzer (Gas Check, 938 AVL). Blood acid and base were measured at 0, 30, 60, 90 and 120 min during cold restraint. A solution of $NaHCO_3$ (BDH) 8.4% w/v was infused i.v. with a Harvard infusion pump at a rate of 2.2 ml/h during the 2-h period of cold-restraint stress; HCl (BDH) 0.5 M was infused at the same rate to rats which were only restrained at $22°C$ for 2 h. Control animals were infused with saline at the same rate and period of time. Arterial blood samples were taken for blood acid-base analyses. All animals were killed 2 h after infusions. The gastric luminal content was collected and the titratable acid determined by titration with NaOH (BDH) 0.01 M to pH 7.4 using an autotitration system (Radiometer, Model TTT 80). Stomachs were examined for lesions, using an illuminated magnifying lens (3X). Ulcer size was determined by measuring each lesion along its greatest length. In the case of petechiae, five such lesions were taken as the equivalent of a 1-mm ulcer. The total lesion lengths divided by the number of rats in each group was expressed as the mean ulcer index[5].

Measurement of respiratory rate and blood lactate concentration. Each rat was restrained in a tubular wire-mesh cage which was painted with an insulating material (white paint). Two wires were attached to the left and right sides of the abdomen, just beneath the diaphragm, by two safety pins inserted under the skin. The free ends of these wires were then connected to the positive and negative poles of a transducer which measured the impedance induced by breathing; the electrical changes were recorded on a physiograph. The respiratory rate was monitored for a period of 2 h when the rats were restrained and exposed to $4°C$ in a cold room or to room temperature of $22°C$.

The method of Varley[6] was used to determine the blood lactate levels. At the end of the 2-h experiments, the rats were killed by stunning and decapitation. One ml of whole blood was collected and deproteinized with ice-cold trichloroacetic acid (E. Merck), and filtered using a Whatman no. 1 filter paper. The supernatant was removed and added to p-hydroxydiphenyl (Sigma). The final concentration of blood lactate was determined by measuring its absorbance at 560 nm using a spectrophotometer (Cary 219).

Atropine (Lacroix Laboratories), ranitidine (Glaxo) and verapamil (Knoll) were dissolved in saline, whereas astemizole (Janssen) was prepared in tartaric acid (Sigma) 0.1 M. They were given i.p. 30 min before cold-restraint stress. The arterial blood acid-base was analyzed and the stomach glandular mucosa examined for ulceration 2 h later.

The results were expressed as means ±SEM. Data were analyzed for statistical signficance by the unpaired two-tailed Student *t*-test.

RESULTS

Effects of Cold-Restraint Stress on the Respiratory Rate and Blood Lactate Level

The respiratory rate of the rats, when compared with their own value at 0 h (Table 1B) and with the corresponding rate of their controls restrained at 22°C (Table 1A), was depressed 1 h after cold-restraint stress and persisted until the end of the 2-h experiment. Stress did not significantly induce any changes in the blood lactate concentration (Table 2).

Table 1. The Effect of Cold-Restraint Stress on the Respiratory Rate (Number of Inhalations/min)

		Time (h) after stress		
0	0.5	1	1.5	2
A.	Restrained at 22°C for 2 h			
161.3 ± 5.3	148.5 ± 6.4	139.3 ± 6.8	141.5 ± 8.4	146.7 ± 9.3
B.	Restrained at 4°C for 2 h			
165.2 ± 4.5	153.3 ± 7.1	117.4 ± 7.3*+	108.7 ± 7.7***+	102.7 ± 8.5***+

Values are the means ± SEM of 9 rats in each group.
*$P < 0.05$, **$P < 0.02$, ***$P < 0.01$, compared with the corresponding value in A.
+$P < 0.001$, compared with its own control at 0 h.

Table 2. The Effects of Cold-Restraint Stress on Changes in Arterial Blood Lactate Levels

		Lactic acid (mmol/l)
A.	Restrained at 22°C for 2 h	0.65 ± 0.17
B.	Restrained at 4°C for 2 h	0.51 ± 0.07

Values are the means ± SEM of 10 rats in each group.

Effects of cold-restraint stress, NaHCO$_3$ or HCl infusion on acid-base balance

Rats restrained at 22°C for 2 h did not show any significant changes in pCO$_2$, pH and HCO$_3^-$ levels in the blood (Table 3A). Cold-restraint significantly elevated the pCO_2 at 1.5 h, whereas the blood pH was decreased 30 min after the onset of stress; the HCO_3^- level was not altered (Table 4A). HCl (0.5 M) i.v. infusion at the rate of 2.2 ml/h decreased the pH and HCO_3^- level, but pCO_2 was unaffected (Table 3B). $NaHCO_3$ infusion (8.4% w/v) at the same rate (Table 4B) significantly reversed the cold-restraint stress-induced decrease in blood pH; the HCO_3^- level was elevated when compared to the stressed saline-infused control. The increased pCO_2 by stress was unaffected by $NaHCO_3$ infusion.

Table 3. The Effects of HCl Treatment (Given as an i.v. Infusion of 2.2 ml/h at 0 h) on Arterial pCO_2, pH and HCO_3-Concentrations

Time (h) after i.v. infusion	pCO_2 (mmHg)	pH	HCO_3^- (mmol/l)
A. Saline 0.9% w/v (restrained at 22·C for 2 h)			
0	24.9 ± 1.2	7.437 ± 0.007	17.6 ± 0.5
0.5	27.6 ± 1.7	7.433 ± 0.006	17.5 ± 0.9
1.0	26.8 ± 1.8	7.427 ± 0.019	17.1 ± 1.1
1.5	27.6 ± 2.7	7.411 ± 0.025	18.9 ± 0.9
2.0	24.2 ± 2.7	7.417 ± 0.016	17.8 ± 1.0
B. HCl 0.5 M (restrained at 22·C for 2 h)			
0	25.4 ± 0.6	7.449 ± 0.007	17.4 ± 0.3
0.5	25.8 ± 0.6	7.334 ± 0.014**+	13.6 ± 0.6*+
1.0	25.7 ± 0.8	7.234 ± 0.018***+	10.6 ± 0.5***+
1.5	23.6 ± 1.5	7.087 ± 0.052***+	7.7 ± 0.9***+
2.0	23.4 ± 1.1	7.072 ± 0.033***+	6.6 ± 0.7***+

Values are the means ± SEM of 12 rats in each group.
*P < 0.01, **P < 0.001, compared with the corresponding control in A.
+P < 0.001, compared with its own control at 0 h.

Effects of Cold-Restraint Stress, $NaHCO_3$ or HCl Infusion on Luminal Titratable Acid and on Gastric Glandular Ulceration

Cold-restraint stress did not affect the luminal acid (Table 5B). The titratable acid in the gastric lumen was not significantly different between the saline-, HCl (Table 5A)- or stressed $NaHCO_3$ (Table 5B)-treated groups. Cold-restraint for 2 h produced severe hemorrhagic ulcers in the gastric glandular mucosa (Table 5B); the severity of stress-induced ulceration was markedly decreased by $NaHCO_3$ treatment. Infusion of HCl for 2 h produced gastric ulcers, with an ulcer index which was significantly higher than that of the saline-infused controls (Table 5A).

Table 4. The Effects of $NaHCO_3$ Treatment (Given as an i.v. Infusion of 2.2 ml/h at 0 h) on Cold-Restraint Stress-Induced Changes in Arterial pCO_2, pH and HCO_3^- Concentrations

Time (h) after i.v. infusion	pCO_2 (mmHg)	pH	HCO_3^- (mmol/l)
A. Saline 0.9% w/v (restrained at 4·C for 2 h)			
0	25.5 ± 1.9	7.423 ± 0.013	16.5 ± 0.5
0.5	27.3 ± 1.8	$7.307 \pm 0.015^{+++}$	15.8 ± 0.7
1.0	30.8 ± 2.2	$7.258 \pm 0.012^{+++}$	15.2 ± 0.8
1.5	$37.6 \pm 4.7^+$	$7.155 \pm 0.024^{+++}$	14.8 ± 1.5
2.0	$44.5 \pm 6.5^{++}$	$7.119 \pm 0.034^{+++}$	16.3 ± 0.9
B. $NaHCO_3$ 8.4% w/v (restrained at 4·C for 2 h)			
0	24.9 ± 1.2	7.446 ± 0.011	17.4 ± 0.5
0.5	25.8 ± 1.3	$7.482 \pm 0.018^{**}$	$19.4 \pm 1.3^*$
1.0	26.5 ± 3.6	$7.496 \pm 0.022^{**}$	$25.0 \pm 1.2^{**++}$
1.5	34.4 ± 6.6	$7.486 \pm 0.030^{**}$	$35.2 \pm 2.6^{**+++}$
2.0	41.5 ± 6.6	$7.494 \pm 0.032^{**}$	$45.5 \pm 6.3^{**+++}$

Values are the means ± SEM of 12 rats in each group.
*P < 0.05, **P < 0.001, compared with the corresponding control in A.
+P < 0.05, ++P < 0.02, +++P < 0.001, compared with its own control at 0 h.

Table 5. The Effects of HCl or $NaHCO_3$ Treatment on Gastric Glandular Ulceration and Luminal Acidity

Treatment groups (i.v. infusion, 2.2 ml/h for 2 h)	Luminal titratable acid (μequiv/100 g body weight)	Glandular ulcer index (mm)
A. Restrained at 22·C for 2 h		
Saline	4.45 ± 0.56	0.08 ± 0.05
0.5 M HCl	5.22 ± 0.98	$1.18 \pm 0.35^+$
B. Restrained at 4·C for 2 h		
Saline	4.84 ± 0.49	5.47 ± 1.82
8.4% wv $NaHCO_3$	4.13 ± 0.58	$0.29 \pm 0.06^+$

Values are the means ± SEM of 8 rats in each group.
+P < 0.01, compared with its own saline-treated group.

Effects of drug treatment on acid-base balance and gastric ulceration

Pretreatment with atropine, ranitidine, astemizole or verapamil did not affect the HCO_3^- levels, when compared to their respective vehicle-injected control level (Table 6). Only astemizole suppressed the elevation in pCO_2 at 1 and 1.5 h after stress (Table 7); this histamine H_1-receptor antagonist also lessened the degree of acidosis during the 2-h stress period (Table 8). Despite their different effects on the blood acid-base balance, all the four drugs significantly reduced the severity of gastric ulceration (Table 9).

Table 6. The Effects of Drug Pretreatment (Given i.p. 30 min Before Stress) on Cold-Restraint Stress (4°C for 2 h)-Induced Changes in Arterial Blood HCO_3^- (mmol/l) Levels

Pretreatment group	Dose	Time (h) after stress				
		0	0.5	1	1.5	2
Tartaric acid	2.0 ml/kg	14.9 ± 0.6	13.7 ± 1.3	13.6 ± 1.3	14.3 ± 1.4	15.4 ± 1.5
Saline	2.0 ml/kg	15.1 ± 0.6	14.3 ± 0.9	13.2 ± 0.8	13.4 ± 0.9	15.6 ± 0.9
Atropine	0.3 mg/kg	16.3 ± 0.8	13.0 ± 1.4	13.7 ± 0.9	14.7 ± 0.6	17.3 ± 1.3
Ranitidine	25.0 mg/kg	14.3 ± 0.6	12.1 ± 0.9	13.6 ± 0.8	14.8 ± 1.3	16.3 ± 0.8
Astemizole	1.5 mg/kg	14.6 ± 0.6	13.6 ± 0.9	11.1 ± 0.7	12.4 ± 0.5	13.2 ± 1.1
Verapamil	8.0 mg/kg	15.3 ± 0.6	12.8 ± 1.2	16.3 ± 1.4	16.9 ± 1.5	17.8 ± 1.2

Values are the means ± SEM of 12 rats in each group.

DISCUSSION

The increased blood pCO_2 during cold-restraint stress is probably due to disturbances in the vascular and respiratory systems, because cold exposure has been shown to increase pulmonary vascular resistence[7] and blood viscosity[8]. The rate and the amount of blood flow through the lung could, therefore, be lowered. These effects, coupled with a decrease in respiratory rate, may lead to accumulation of CO_2 in the blood.

Decreased blood pH indicates that acidosis occurred during cold-restraint stress. This acidosis could be either respiratory or metabolic. The present study shows that neither the blood bicarbonate nor the lactate level was affected, implying that the acidosis may not be metabolic in origin but could well be largely caused by respiratory depression. The elevation of pCO_2 during cold-restraint (Table 4A) substantiates this hypothesis.

The pCO_2 level remained at its basal value during acid infusion (Table 3B), indicating that alveolar ventilation was unaffected[9]. As the infusion of acid decreased the HCO_3^- value but did not affect the pCO_2 level, this strongly suggests that the acidosis produced by HCl infusion is largely metabolic.

Gastric ulceration was seen in the acid-infused animals (Table 5A) and this finding is in accord with the observation of Mullane and others[2, 10]; however, the

Table 7. The Effects of Drug Pretreatment (Given i.p. 30 min Before Stress) on Cold-Restraint Stress (4°C for 2 h)-Induced Changes in Arterial Blood pCO_2 (mmHg) Levels.

Pretreatment group	Dose	Time (h) after stress				
		0	0.5	1	1.5	2
Tartaric acid	2.0 ml/kg	25.2 ± 1.2	26.8 ± 2.4	32.8 ± 3.5	35.4 ± 3.7	46.3 ± 5.8
Saline	2.0 ml/kg	24.9 ± 0.9	26.6 ± 1.9	32.0 ± 2.6	35.9 ± 3.3	41.1 ± 5.7
Atropine	0.3 mg/kg	24.8 ± 2.1	23.5 ± 2.2	29.3 ± 2.2	37.5 ± 3.9	48.5 ± 9.1
Ranitidine	25.0 mg/kg	23.7 ± 0.8	23.2 ± 2.7	30.3 ± 2.2	38.3 ± 3.2	54.5 ± 4.7
Astemizole	1.5 mg/kg	25.6 ± 0.9	23.8 ± 2.0	21.2 ± 1.1[++]	26.7 ± 2.5[+]	33.8 ± 3.7
Verapamil	0.8 mg/kg	25.1 ± 1.0	27.6 ± 1.8	32.1 ± 3.4	38.1 ± 3.6	44.1 ± 5.0

Values are the means ± SEM of 12 rats in each group.
[+] $P < 0.05$, [++] $P < 0.01$, compared with the corresponding saline-pretreated group.

40

Table 8. Effects of Drug Pretreatment (Given i.p. 30 min Before Stress) on Cold-Restraint Stress (4°C for 2 h)-Induced Changes in Arterial Blood pH

Pretreatment group	Dose	Time (h) after stress				
		0	0.5	1	1.5	2
Tartaric acid	2.0 ml/kg	7.438 ±0.017	7.324 ±0.028	7.264 ±0.042	7.187 ±0.037	7.127 ±0.056
Saline	2.0 ml/kg	7.416 ±0.017	7.292 ±0.014	7.243 ±0.019	7.161 ±0.018	7.085 ±0.037
Atropine	0.3 mg/kg	7.430 ±0.019	7.308 ±0.033	7.252 ±0.024	7.201 ±0.041	7.130 ±0.070
Ranitidine	25.0 mg/kg	7.456 ±0.012	7.275 ±0.028	7.184 ±0.031	7.123 ±0.031	7.010 ±0.042
Astemizole	1.5 mg/kg	7.451 ±0.012	7.352 ++ ±0.010	7.322 ++ ±0.010	7.268 +++ ±0.015	7.199+ ±0.018
Verapamil	8.0 mg/kg	7.432 ±0.047	7.303 ±0.032	7.289 ±0.028	7.198 ±0.032	7.116 ±0.051

Values are the means ± SEM of 12 rats in each group.
+P < 0.02, ++P < 0.01, +++P < 0.001, compared with the corresponding saline-pretreated group.

Table 9. Effects of Drug Pretreatment (Given i.p. 30 min Before Stress) on Cold-Restraint Stress (4 °C for 2 h)-Induced Gastric Glandular Ulceration

Pretreatment group	Dose	Glandular ulcer index (mm)
Tartaric acid	2.0 ml/kg	7.14 ± 1.07
Saline	2.0 ml/kg	7.53 ± 0.98
Atropine	0.3 mg/kg	1.08 ± 0.66[+++]
Ranitidine	25.0 mg/kg	3.43 ± 1.14[+]
Astemizole	1.5 mg/kg	2.27 ± 1.02[++]
Verapamil	8.0 mg/kg	2.23 ± 0.54[+++]

Values are the means ± SEM of 10 rats in each group.
[+]$P < 0.02$, [++]$P < 0.01$, [+++]$P < 0.001$, compared with the saline-pretreated group.

severity of ulceration was not as great as that which occurred with cold-restraint stress (Table 5B). Although both experimental methods produce acidosis, the type of acidosis is different, as shown in the present study. In acid-infused animals, the cause of acidosis is mainly metabolic, whereas in those restrained at 4 °C it is largely respiratory. It has been found that for a similar decrease in extracellular pH, the fall in intracellular pH is greater in respiratory acidosis than in metabolic acidosis[11]. The reason for this could be due to differences in the permeability between CO_2 and ionic substances like hydrogen ions. Carbon dioxide can easily diffuse into cells, to lower intracellular pH. Cold-restraint stress induces mainly respiratory acidosis and, thus, could result in an intracellular pH which is lower than that obtained by acid infusion. Lowering of intracellular pH through decreased extracellular pH could disrupt cellular metabolism and enzymatic reactions, which may lead to greater damage of the rat stomach mucosal cells.

The observation that maintenance of the arterial pH within the normal range, by i.v. infusion of $NaHCO_3$ (Table 4B), can prevent ulcer formation in cold-restraint stressed rats (Table 5B) confirms the findings of Cheung & Porterfield (1979)[12]. Neutralization of blood acidity by $NaHCO_3$ during cold-restraint stress is due to the infusion increasing blood HCO_3^- levels (Table 4B). Neither acid nor bicarbonate infusion altered the stomach luminal acidity (Table 5), thus the production or suppression of gastric ulcers could only be related to a change in blood pH. These results indicate the maintenance of blood pH is important in preventing cold-restraint lesion formation in the rat stomach.

Except for astemizole, the various antagonists studied did not have a marked influence on acid-base balance. The ability of astemizole to lessen CO_2 retention (Table 7) could be due to its antagonistic action on the effect of histamine in the pulmonary vessels because the amine constricts bronchial smooth muscles and causes vascular congestion, leading to increased capillary permeability and edema[13]. Astemizole, a specific peripheral-acting histamine H_1-receptor antagonist[14], may, therefore, alleviate the actions of histamine on the pulmonary vessels to improve lung function during cold-restraint stress. Astemizole was found also to reduce the severity of acidosis, and reinforces the idea that the fall in pH due to cold-restraint stress may be mainly respiratory in origin. Verapamil, atropine and ranitidine did not prevent the changes in blood gases induced by stress but suppressed ulcer formation. It is, therefore, suggested that changes in acid-base balance are not an immediate cause of cold-restraint ulceration, but could contribute to lesion formation. There are other factors, e.g. vagal overactivity, disturbance of calcium homeostasis and histamine release, which may mainly be involved in the formation of gastric ulcers in stressed rats. The ulcerogenic effects of these mechanisms would indeed be attenuated by muscarinic or calcium channel blockade as well as by histamine H_1- or H_2-receptor antagonism.

REFERENCES

1. T.J. Verbeuren, W.J. Janssens and P.M. Vanhoutte, Effects of moderate acidosis on adrenergic neurotransmission in canine saphenous veins, *J Pharmacol Exp Ther.* 206:105 (1978).
2. J.F. Mullane, R.G. Wilfong, T.O. Phelps and R.P. Fischer, Metabolic acidosis, stress and gastric lesions in the rat, *Arch Surg.* 107:456 (1973).
3. E.C. Senay and R.J. Levine, Synergisim between cold and restraint for rapid production of stress ulcer in rats, *Pro Soc Exp Biol Med.* 124:1221 (1967).
4. C.H. Cho and C.W. Ogle, Cholinergic-mediated gastric mast cell degranulation with subsequent histamine H_1- and H_2-receptor activation in stress ulceration in rats. *Eur J Pharmacol.* 55:23 (1979).
5. C.H. Cho and C.W. Ogle, A correlative study of the antiulcer effects of zinc sulphate in stress rats, *Eur J Pharmacol.* 48:97 (1978).
6. H. Varley, "Practical Clinical Biochemistry," William Heineman Medical Book Ltd., London (1967).
7. K.J. Greenlees, A. Tucker, D. Robertshaw, *et al.*, Pulmonary vascular responsiveness in cold-exposed calves. *Can J Physiol Pharmacol.* 63:131 (1985).
8. M. Murkami, S.K. Lam, M. Inada, *et al.*, Pathophysiology and pathogenesis of acute gastric mucosal lesion after hypothermic restraint stress in rats, *Gastroenterology.* 88:660 (1985).
9. D.C. Flenley, Arterial blood gas tensions and pH. *Br J Clin Pharmacol.* 9:129 (1980).
10. M. Bushell and P. O'Brien, Acid-base imbalance and ulceration in the cold restrained rat, *Surgery.* 91:318 (1982).
11. G.E. Levin, P. Collinson, and D.N. Baron, The intracellular pH of human leukocytes in response to acid-base changes *in vitro. Clin Sci Mol Med.* 50:293 (1976).
12. L.Y. Cheung and G. Porterfield, Protection of gastric mucosa against acute ulceration by intravenous infusion of sodium bicarbonate, *Am J Surg.* 137:106 (1979).
13. W.W. Douglas, Histamine and 5-hydroxytryptamine (Serotonin) and their antagonists, *in:* "The Pharmacological basis of Therapeutics," A.G. Gilman, L.S. Goodman and A. Gilman, eds., MacMillan, New York (1980).
14. D.M. Richards, R.N. Brogden, R.C. Heel, *et al.*, Astemizole: A review of its pharmacodynamic properties and therapeutic efficacy. *Drugs.* 28:38 (1984).

THE DEVELOPMENT OF ACUTE GASTRIC MUCOSAL LESIONS FOLLOWING BURN STRESS

Masaki Kitajima, Masashi Yoshida, Yoshihide Otani
and Kouichiro Kumai

The stomach has many physiological functions, such as acid and pepsin secretion, mucus production and motility, which are supposed to be regulated by gastric mucosal blood flow. Therefore, blood flow is an extremely important defensive factor for the gastric mucosa. The present study was designed to elucidate the role of mucosal blood flow in the development of acute gastric mucosal lesions (AGML) after burn stress. Gastric mucosal blood flow and microcirculatory disturbance were investigated. Also, we examined the influence of gastric mucosal blood flow on the alternation of acid and pepsin activity, energy metabolism and mucus production (Hexosamine content).

MATERIALS AND METHODS

Male Wistar rats (n=64) weighing approximately 300 g, were given water ad libitum and studied at 2 (n=16), 5 (n=16) and 24 (n=16) h after burn injury. These results were compared with those obtained in sham-burned controls (n=16). Burn injury of 30% of the body surface area was created by immersing the shaved backs of the rats in boiling water for 15 seconds. This resulted in a third degree burn.

(1) Endoscopical Observation

Rat stomachs were irrigated with physiological saline mixed with gascon before observation. Under general anesthesia with pentobarbital sodium, 30-40 mg/kg intravenously, rats were fixed on the specific table (CFK-1Se CFK-1SLD) and tracheal intubation was performed by arthroscope (Olympus SED-1711k). After intubation, stomachs were observed with a fiberscope (Machida 3.5 mm in diameter) or pharyngoscope (Olympus ENF-P 4.0 mm in diameter).

(2) Measurement of Acid and Pepsin Activity

The gastric infusion method was used to measure the activity of acid and pepsin in the stomach. With care taken not to disrupt the vagal nerve or blood vessels, the esophagus was exposed and cannulated with a short length of polyethylene tubing (PE-50) through an incision to the esophagus and the duodenum. Physiological saline was constantly perfused at the rate of 10 ml/h.

Acidity was determined in accordance with the criteria established by the Committee on Assay of Gastric Juice of the Japanese Society of Gastroenterology. Using a pH meter, acid output per hour was measured by titration against 0.01N NaOH and expressed in mEq/h. Pepsin activity was also determined, using a specimen sampled by a similar procedure. In accordance with the preliminary proposal of the Committee on Assay of Gastric Juice of the Japanese Society of Gastroenterology, using a modification of Anson's method, gastric juice was added to Hb-denatured solution and the mixture was allowed to stand for 30 min at 37.0°C. At the end of this period, 5% trichloroacetic acid was added to stop the reaction and the L-tyrosine content of the filtrate was determined colorimetrically at a wavelength of 640 nm to obtain pepsin activity.

Parietal cells from the rat glandular stomach were examined. Gastric mucosal specimens were fixed in a compound solution of 2.5% glutaraldehyde and 4% paraformaldehyde for electron microscopic examination, to allow examination of the secretory condition of the parietal cells, with particular reference to the intracellular canaliculi and the tubulovesicular structure.

(3) Determination of Gastric Mucosal Blood Flow

Gastric corpus mucosal blood flow in the rat was determined by the hydrogen gas electrolytic method using a platinum electrode.

(4) Observation of Blood Vessel Damage Using Monastral Blue

From the mucosal aspect, blood vessel damage was observed using monastral blue dye (Sigma Chemical Co.). A 3% suspension of monastral blue was administered (0.5 ml/100 g of body weight) intravenously, and 10 min after the administration, the stomach was removed and fixed in 10% formalin. It was observed under stereoscopic microscope after clearing in glycerin for 48 h. Administration of monastral blue was also performed on a control group, at 15 min, 2 h, 5 h and 24 h after thermal injury.

(5) Visualization of the Microvascular Structure

A polyethylene luer end catheter (I.D. 0.034 in., O.D. 0.050 in.) was inserted into the abdominal aorta placing its tip just below the branching of the celiac artery. A second catheter was also inserted into the main mesenteric vein and advanced almost to the bifurcation at the left gastric vein and portal vein. The portal vein was ligated at the porta hepatis.

At first, orange silicon rubber, MV-117 (Canton Biomedical Products, Inc) was infused by means of a constant infusion pump (Razel Scientific Instrument Inc) and observed by two dimensional microscopy until venous blood vessels were filled with silicon rubber. Next, yellow silicon rubber, MV-122 was infused into the abdominal aorta. After infusion of the blood vessels, rats were placed in the refrigerator (4°C) overnight to complete the polymerization. Stomachs were removed the following day and opened along the greater curvature in order to inspect the gastric lesions. After gross inspection, stomachs were put between two slide glasses and soaked in 98% Hexen solution in dry iced acetone. Gastric wall separated the submucosal layer from muscular layer.

Cross sections were then cut at 50-200 microns by a freezing microtome to observe the casts. These sections were put into ethyl alcohol followed by immersion in methyl salicylate for 12 to 24 h in order to make them transparent. These sections were observed by a three dimensional viewing microscope (Olympus XT PM10-AD) with reflected light.

(6) Determination of Mucus Secretion

Hexosamine content in the gastric mucus and mucosal layer was measured as an indicator of mucosubstance.

According to Dekanski's method[1], gastric mucosal tissue was homogenized with 1 ml of 5 mM EDTA (pH 7.4) after which the homogenates were centrifugally separated at 10000 rpm for 10 min. The supernatant was then separated, 10% trichloroacetic acid and 1% phosphotungstic acid were added and kept at 4°C for 24 h. The sediment obtained was washed with 99% ethanol and kept at 4°C for 2 h, after which it was centrifuged for 10 min at 3000 rpm. This sediment was then dried in a desiccator and taken to be the soluble glycoprotein from the gastric mucosa. The dry weight of this sediment and its hexosamine content were measured. The measurement of glycoproteins in the superficial mucus layer was also performed by measuring the amounts of hexosamine in the group in which the mucus layer was carefully removed with gauze, and that group in which it was not, and estimating the amount of glycoprotein by the difference. The Neuhaus-Letzring[2] isoamyl alcohol extraction method was used to measure the amount of hexosamine.

RESULTS

Endoscopic Observation

Gastric mucosal lesions were not observed in the stomachs of normal control rats. However, after induction of burn stress, the gastric mucosa immediately changed into a pale color, which was suggested as ischemia. Multiple erosions with hemorrhage developed in the rat stomach studied at 2 or 5 h, in a manner similar to that seen in humans (Figure 1).

Acid and Pepsin Activity

The acid and pepsin activity was detected for 3 h during infusion of physiological saline. H_2 receptor antagonist and truncal vagotomy significantly reduced the acid and pepsin output as compared with that in the control group ($p < 0.05$).

In rats stressed by burn injury, acid output was lower than that seen in controls ($p < 0.05$). These results suggested that defensive factors played a more important role in the development of acute gastric mucosal lesions than offensive factors. In contrast, pepsin activity in the burn-stressed animals was significantly higher than that in controls ($p < 0.05$). These results indicated that back diffusion of pepsin did not occur in the gastric mucosal because of high molecular weight (Figure 2).

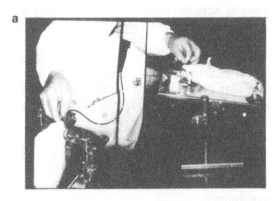

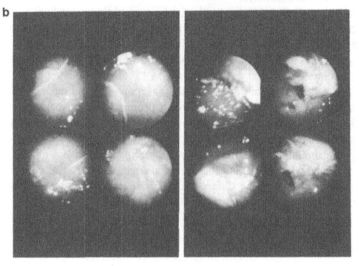

Figure 1. (a) Intratracheal intubation and endoscopic examination of the rat. (b) Endoscopic findings in normal control (left) and burned rat stomach (right), acute gastric mucosal lesion was observed in corpus of the stomach.

Gastric parietal cells in control rats revealed normal intracellular canaliculi in association with microvilli and tubulovesicular structure. These electron microscopic findings in rats stressed by burn injury changed to the hyposecretory state, consisting of a decrease in the number of intracellular canaliculi and microvilli and an increase in the tubulovesicular structure. Moreover, similar findings were demonstrated in parietal cells from rats which were treated with H_2 receptor antagonist or truncal vagotomy (Figure 3-a,b).

Microvascular Structure After Burn Injury

Normally, arteries and veins penetrate the serosa and muscularis propria and form a network in the submucosa, in which blood flow volume is more than that in the other layers. Submucosal arterioles divide into numerous metarterioles that penetrate the muscularis mucosae and branch into true capillaries at the base of the

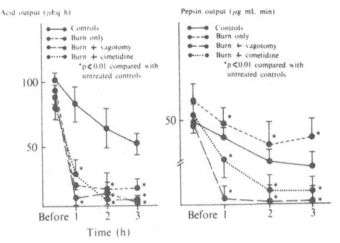

Figure 2. Acid and pepsin output in vagotomized, cimetidine-treated and untreated rats subjected to stress by burn injury.

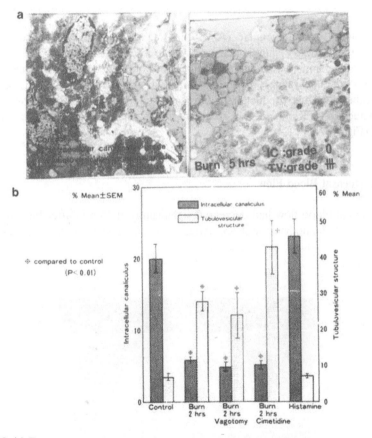

Figure 3. (a) Transmission electron micrographs showing secretory condition of a parietal cell. The intracellular canaliculi in association with the microvilli and tubulovesicular structure reveals the normal secretory condition in the gastric mucosa from a control rat (left), and hyposecretion in a burned rat (right). (b) Quantitative analysis for secretory condition of a parietal cell was estimated with a color image processor.

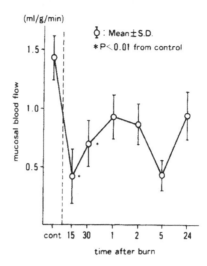

Figure 4. Gastric mucosal blood flow after burn injury detected by electrically generated hydrogen gas clearance method.

gastric glands. Arteriovenular (A-V) shunting channels normally exist in the submucosa, but no A-V shunting channels were observed in control rats. However, A-V shunting channels were prominent at 2 and 5 h after burn injury (Figure 6).

These findings suggest the impairment of gastric microcirculation as ischemia. On the other hand, early reduction of mucosal blood flow showed congestion, namely, the submucosal arterioles showed spastic narrowing and the venules were markedly dilated (Figure 7).

Gastric Mucosal Blood Flow

Mucosal blood flow had decreased significantly at 15 min after burn injury in association with hemodynamic shock. After 15 min, it gradually increased, but did not recover to control level at 2 and 5 h (Figure 4).

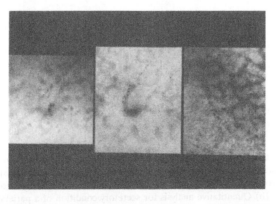

Figure 5. Observation of blood vessel damage using monastral blue. Deposition of monastral blue was observed mainly centering on collecting venules but it was also observed in capillaries.

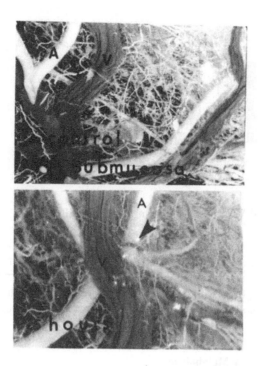

Figure 6. Microangiographic findings of the submucosa in a normal control (upper) and in a burned rat at 5 hours after stress, showing the opening of arteriovenular shunting channel (lower).

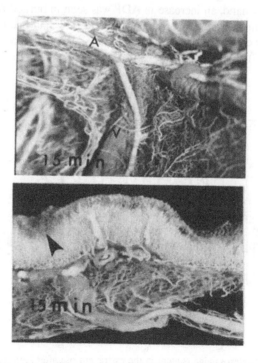

Figure 7. Microangiographic findings in a burned rat at 15 min. Submucosal venules were markedly dilated and arterioles were constricted (upper) and a shallow mucosal lesion had developed (lower).

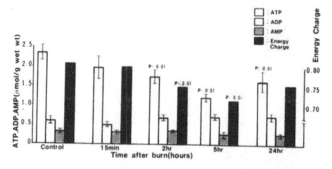

Figure 8. Adenine nucleotide and energy charge of the gastric mucosa after burn injury.

Observation of Blood Vessel Damage Using Monastral Blue

Deposition of monastral blue was observed in no rats in the control group (n=6), three rats out of seven (42.9%) in 15 the min group, four rats out of seven (57.1%) in the 3 h group, nine rats out of ten (90.0%) in the 5 h group and two rats out of six (33.3%) in the 24 h group. Deposition of monastral blue was observed mainly centering on collecting venules but it was also observed in capillaries (Figure 5).

Gastric Mucosal Energy Metabolism

ATP synthesis in the gastric mucosa decreased significantly with time after burn injury. On the other hand, an increase in ADP was seen in burned rats studied at 2 and 5 h, at which time, reduction of mucosal blood flow had been confirmed. In addition, the energy charge which was calculated by Atkinson's[3] formula, decreased significantly (p<0.05) (Figure 8).

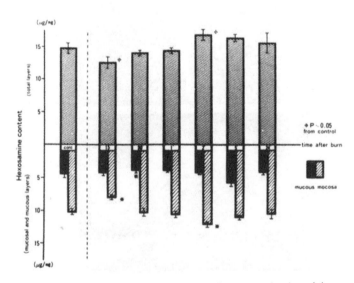

Figure 9. Hexosamine content in the gastric mucosa after burn injury.

Mucus Secretion

Hexosamine level in the mucus and mucosal layer increased gradually after burn injury. There was a significant increase measured at 2 and 5 h after burn (p<0.01) (Figure 9).

DISCUSSION

The development of peptic ulcer, including acute gastric mucosal lesions, has been conventionally explained by an imbalance[4] between offensive and defensive factors. Recently, most of the attention concerning the pathogenesis has been chiefly directed to the elucidation of gastric mucosal defensive factors, particularly mucosal blood flow[5,6]. In these studies, the influence of gastric mucosal blood flow on the defensive factors was investigated in order to elucidate the pathogenesis of stress-induced acute gastric mucosal lesions.

To demonstrate the change of acid and pepsin output in stress, electron microscopic studies were performed, that is, histochemical results were compared with electron micrographs of parietal cells and chief cells. There was a high relationship between secretory state of the parietal cells and acid secretion[7].

Not only was there an increase in back diffusion of H^+, but the parietal cells also revealed a low acid output in stress. On the other hand, pepsin activity increased in the gastric juice in stressed rats.

These results suggested that pepsin could not invade into the gastric mucosa due to its high molecular weight. Gastric mucosal blood flow was determined employing the hydrogen gas electrolytic method. After burn stress, gastric mucosal blood flow decreased significantly.

It was inferred that the impairment of gastric microcirculation was chiefly caused by ischemia and congestion. This result was supported by the findings of microvascular structure. Deposition of monastral blue was also observed after burn stress. Monastral blue was thought to deposit mainly in the vessels in which permeability had increased[8]. Monastral blue staining was described as being an early and sensitive indicator of the gastric mucosal injury[9]. Corresponding to the reduction in mucosal blood flow, the energy metabolism of the mucosa was disturbed and mitochondrial function changed from an aerobic to an anaerobic condition[10,11]. Generally speaking, mucus production decreased in coincidence with reduction of mucosal blood flow. However, on the contrary, mucus in this experimental model increased after burn stress. These results suggested the existence of a cytoprotective reaction of the gastric mucosa against the ischemic insult.

It is assumed that the destruction of the above-mentioned defensive factors, induced by the impairment of gastric mucosal blood flow, caused a secondary increase in acid back diffusion, leading ultimately to the development of acute gastric mucosal lesions[12]. Thus, it was confirmed that maintenance of gastric mucosal blood flow is an important factor in prevention of the development of gastric injury. From these experimental results, we would like to suggest that maintenance of gastric mucosal blood flow plays an important role as a key factor in the defense of the gastric mucosa.

REFERENCES

1. J.B. Dekanski, A. MacDonald, and P. Sacra, Effect of fasting stress and drugs on glycoprotein synthesis in the rat. *Brit J Pharmacol.* 55:387 (1975).
2. O.W. Neuhaus, and M. Letzring, Determination of hexosamines in conjunction with electrophoresis on starch, *Anal Chem.* 29:1230 (1957).
3. D.E. Atkinson, The energy charge of the adenylate pool as a regulatory parameter, interaction with feedback modifiers, *Biochemistry.* 7:4030 (1968).
4. H. Shay, The pathologic physiology of gastric and duodenal ulcer, *Bull N Y Acad Med.* 20:264 (1944).
5. M. Kitajima, R.D. Wolfe, R.L. Trelstad *et al.*, Gastric mucosal lesions after burn injury. Relationship to H^+ back diffusion and the microcirculation, *J Trauma.* 18:644 (1978).
6. P.H. Guth, Gastric blood flow in restraint stress, *Amer J Dig Dis.* 17:807 (1972).
7. M. Kitajima, Y. Ikeda, and H. Torii, Acute gastric mucosal lesions following burn injury: relationship to acid and pepsin output and gastric microcirculation, *in:* "New Trends in Peptic Ulcer and Chronic Hepatitis," Japanese Society of Gastroenterology, ed., Exerpta Medica, Tokyo (1987).
8. I. Joris, U. DeGirolami, K. Wortham *et al.*, Vascular labelling with monastral blue B, *Stain Technology.* 57:177 (1982).
9. S. Szabo, J.S. Trier, A. Brown *et al.*, Early vascular injury and increased vascular permeability in gastric mucosal injury caused by ethanol in the rat, *Gastroenterology.* 88:228 (1985).
10. M. Kitajima, and H. Ito, Gastric mucosal blood flow and energy metabolism in stress, *in:* "Microcirculation Annual 1985," M. Tsuchiya, ed., Excerpta Medica, Tokyo (1985).
11. R. Menguy, and Y.E. Masters, Gastric mucosal energy metabolism and stress ulceration, *Ann Surg.* 180:538 (1974).
12. M. Kitajima, J.R. Allsop, R.L. Trelstadt *et al.*, Experimental studies on stress ulcer of the stomach following thermal injury with special reference to H^+ back diffusion and microcirculation, *Gastroenterol Jpn.* 13:175 (1978).

EXPERIMENTALLY-INDUCED DUODENAL ULCERS IN RATS ARE ASSOCIATED WITH A REDUCTION OF GASTRIC AND DUODENAL CALCITONIN GENE-RELATED PEPTIDE-LIKE IMMUNOREACTIVITY

Stefano Evangelista, Paola Mantellini
and Daniela Renzi

Calcitonin gene-related peptide (CGRP) is a polypeptide produced by alternative processing of calcitonin gene transcripts[1]. CGRP-like immunoreactivity (li) is widely distributed in the rat gastrointestinal tract[2] and high concentrations of CGRP-li have been found in the rat stomach and duodenum[2]. The absence of CGRP-li in endocrine cells[3] at these levels suggests that it is more important as a neuropeptide than as a circulating hormone. Exogenous CGRP was reported to inhibit gastric acid secretion[4], increase gastric blood flow[5,6], induce a relaxation of duodenal smooth muscle[7], act as a potent antiulcer against experimentally-induced gastroduodenal ulcers[8]. All these effects have been shown to be protective in duodenal ulcers.

The aim of this study is to determine the relationships between endogenous CGRP and duodenal ulcers induced by several ulcerogens. Changes in gastric CGRP-li were also examined in view of the causative role of the increase in gastric acid secretion in the duodenal ulcerogenesis[9,10].

Since a major portion of gastroduodenal CGRP is contained in capsaicin-sensitive fibers[2], similar studies were carried out in rats treated as newborn with the sensory neurotoxin capsaicin to assess the origin of CGRP-li by cysteamine.

MATERIALS AND METHODS

General. Female albino rats, Sprague-Dawley Nossan strain, weighing 180-200 g, were housed at constant room temperature ($21 \pm 1°C$), relative humidity (60%) and with 12 h light-dark cycle (light on 0600 h).

Induction of gastric and duodenal ulcers. Duodenal ulcers were induced in 24 h-fasted rats by gavage administration of dulcerozine (300 mg/kg) or mepirizole (200 mg/kg) and in unfasted rats by cysteamine (900 mg/kg). Controls received saline. The animals were sacrificed at 0.5, 6, 16 or 24 h after cysteamine- or at 24 h after dulcerozine- or mepirizole-challenge and their duodenal ulcers scored, by an observer unaware of the treatments, according to an arbitrary scale between 0 and 4: 0 = no ulcers, 0.5 = redness, 1 or 1.5 = superficial mucosal erosion covering a small or large duodenal area, 2 or 2.5 = deep ulcer, 4 = death[11]. Gastric ulcers were scored as described above according to an arbitrary scale between 0 and 3 in relation to size[12]. In other experiments histamine (40 mg/kg s.c.) was given three times (at 2.5 h intervals) and the animals sacrificed 7.5 h after the first treatment.

Gastric and duodenal CGRP-li determination. Gastric mucosal samples were scraped from the oxyntic region, immediately frozen and weighed. Duodenal samples (the first 2 cm caudal to the pylorus) were excised, blotted in filter paper, weighed and immediately frozen at -20°C. All samples were extracted with 2N acetic acid (1/10, w/v) at 95°C, freeze-dried and reconstituted in phosphate buffer (0.1 M, pH 7.4) for CGRP-li-radioimmunoassay (RIA)[11]. Briefly 100 μl of α-rat CGRP standard or the sample were incubated with 100 μl of an appropriate dilution of rabbit anti-human CGRP II serum (Peninsula) for 48 h at 4°C. 100 μl (3000 cpm) of ^{125}I-iodohistidyl-human CGRP (Amersham) were added and further incubated for 48 h at 4°C. The separation of free from bound antigen was achieved using a second antibody immunoprecipitation. Coefficient of % variation was less than 10% for values between 25 and 300 fmol/l. The lower detection limit was 2 fmol/tube. Crossreactivity of the antiserum was 100% for both rat and human α and β CGRP and less than 0.001% for salmon calcitonin.

Sensory denervation. On the second day of life, rats received 50 mg/kg s.c. of capsaicin in a volume of 10 μl/g body weight. This treatment has been reported to cause a permanent degeneration of unmyelinated afferent neurons[13]. Controls received equal volumes of vehicle (10% ethanol, 10% Tween 80 and 80% saline v/v/v). All injections were performed under ether anaesthesia. The rats were then grown to adulthood and used for the experiments at the age of 2-3 months. In order to check the effectiveness of the treatment, one day before the experiments, a drop of a 0.33 mM solution of capsaicin was instilled into one eye of the rats and the wiping movements were counted. Capsaicin-pretreated rats that showed any wiping movements were excluded from the study[14].

Statistical analysis. Statistical analysis was performed by means of analysis of variance (ANOVA) followed by Dunnett's test and regression analysis.

RESULTS AND DISCUSSION

A transient (1-6 h) increase in gastric acid secretion is the causative factor in the cysteamine-induced duodenal ulcers[9,10]. This phenomenon is concomitant to or followed by a series of local effects which contribute to the pathogenesis of ulcer such as decrease in bicarbonate secretion[15], alteration in duodenal motility[16] and blood flow[17] and delayed gastroduodenal emptying[16,18].

A single or double ("kissing") and often perforating ulcer 2-5 mm distal to pyloroduodenal junction develops only at 16-24 h after the cysteamine challenge and is concomitant with gastric lesions.

Our findings indicate that duodenal ulcers induced by cysteamine, dulcerozine or mepirizole are accompanied with decrease in CGRP-li (Table 1). Taking into account that exogenous CGRP exerts antiulcer activity at the duodenal level[7], a protective effect of the endogenous peptide may be hypothesized in this type of ulceration.

Although (Figure 1) gastric-(panel A) and duodenal-(panel B) ulcers were present 16 h after cysteamine challenge, a significant decrease in either gastric or duodenal CGRP-li was observed only after 24 h, irrespective of the fact that exogenous cysteamine exerted a marked increase in gastric acid secretion (4-6 h after a single cystamine dose[9,10]). This is substantiated by the observation that repeated hypersecretory doses of histamine[19] were unable to produce lesions in the duodenum and did not affect gastric and duodenal CGRP-li (data not shown).

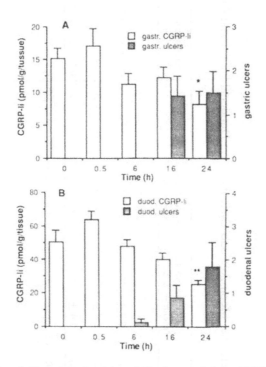

Figure 1. Gastric (panel A) and duodenal (panel B) ulcers and tissue CGRP-li at various times following the administration of a single dose of cysteamine (900 mg/kg p.o.). Statistical significance from the control group (time 0) is shown as * = P < 0.05 or ** = P < 0.01. n = 8-10 for each group.

It should be mentioned that, unlike CGRP-li, endogenous somatostatin, a gut peptide with powerful antisecretory and antiulcer activity, is markedly depleted in gastric and duodenal tissues during gastric hypersecretory phases induced by a single dose of cysteamine[20]. Thus, somatostatin and CGRP have different mechanisms and temporal involvement in duodenal ulcerogenesis.

Dissociation between acid load and blood flow in the duodenum has been recently shown[21]. The authors reported the hypothesis that decline in blood flow at the time of the development of duodenal ulcers could be related to a transmitter depletion after a prolonged stimulation of blood flow[21]. If it is the case, CGRP may be one of the likely candidates.

Table 1. Effect of Dulcerozine, Mepirizole and Cysteamine on Duodenal Ulcers and CGRP-Like Immunoreactivity (li)

Treatments:	pathologic score (mean ± S.E.M.)	duodenal CGRP-li (pmol/g/tissue)	correlation
Saline	-	55.8 ± 3.2	
Dulcerozine (300 mg/kg p.o.)	2.7 ± 0.2	23.2 ± 3.5**	y = 3.72-0.06x r = 0.79 P < 0.001
Mepirizole (200 mg/kg p.o.)	1.9 ± 0.4	40.3 ± 2.9*	y = 3.92-0.06x r = 0.88 P < 0.01
Cysteamine (900 mg/kg p.o.)	1.4 ± 0.34	26.0 ± 2.7**	y = 2.57-0.05x r = 0.64 P < 0.001

* and ** = P < 0.01 and 0.001 as compared to saline-treated group. n=8 for each group.

On the other hand, a major portion of CGRP-li contained in the stomach and duodenum is contained in capsaicin-sensitive afferent neurons[2]. In fact systemic administration of high doses of capsaicin in newborn rats produced a marked depletion in duodenal CGRP-li (78% as compared to controls), which was not further decreased by an ulcerogenic dose of cysteamine (Figure 2), indicating that cysteamine induced release of CGRP-li from a capsaicin-sensitive pool. It appears likely that the cysteamine- and capsaicin-resistant content of duodenal CGRP-li might involve the intrinsic CGRP neurons described in the rat myenteric plexus[2]. Peripheral release of CGRP and other sensory peptides has been hypothesized to participate in a capsaicin-sensitive "defense mechanism" against the ulcerogenic stimuli[6,12,22]. In fact in capsaicin pretreated rats there is a worsening of gastric ulcers induced by several stimuli[12,22] and duodenal ulcers induced by cysteamine and dulcerozine[23]. Furthermore, intragastric administration of capsaicin or capsaicin-like substances affords gastric protection against damage induced by ethanol[24] and acidified aspirin[25] and this effect seems brought about by a local release of peptides from afferent nerve endings[6,26].

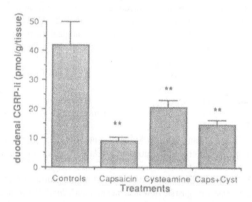

Figure 2. Effect of neonatal capsaicin pretreatment (50 mg/kg s.c.), cysteamine (900 mg/kg p.o., sacrifice 24 h later) administration or both on duodenal CGRP-li. Statistical significance from the control group is shown as ** = $P < 0.01$. n = 8-10 for each group.

The observation that a) CGRP-li is widely distributed around blood vessels in stomach and duodenum[2] and b) the specific mechanisms of sensory nerve-mediated protection involve changes in local mucosal blood flow[24,25], may suggest that this peptide is mainly involved in control of local vascular tone during cysteamine-induced ulcers[10].

In conclusion our results show that the endogenous duodenal CGRP-li depleted in animals with duodenal ulcers is sensitive to capsaicin. Furthermore, CGRP might be one of the local mediators which affords protection against ulcerogenic stimuli.

ACKNOWLEDGEMENTS

We wish to thank Drs. C. Surrenti and C.A. Maggi for helpful advice and suggestions and Miss M. Ricci for her technical assistance. Istituto Farmacobiologico Malesci S.p.A. is a company related to Industrie Farmaceutiche Menarini Firenze, Italy.

REFERENCES

1. M.G. Rosenfeld, J.J. Mermod, S.G. Amara, L.W. Swanson, P.E. Sawchenko, J. Rivier, W.W. Vale, and R.M. Evans, Production of a novel neuropeptide encoded by the calcitonin gene via tissue-specific RNA processing, *Nature.* 304:129 (1983).
2. C. Sternini, J.R. Reeve, and N. Brecha, Distribution and characterization of calcitonin gene-related peptide immunoreactivity in the digestive system of normal capsaicin-treated rats, *Gastroenterology.* 93:852 (1987).
3. G.J. Dockray, Regulatory peptides and the neuroendocrinology of gut-brain relations, *Quart J Exp Physiol.* 73:703 (1988).
4. Y. Taché, T. Pappas, M. Lauffenburger, Y. Goto, J.H. Walsh, and H. Debas, Calcitonin gene-related peptide: potent peripheral inhibitor of gastric acid secretion in rats and dogs, *Gastroenterology.* 87:344 (1984).
5. P. Holzer, and P.H. Guth, Neuropeptide control of rat gastric mucosal blood flow, *Circ Res.* 68:100 (1991).
6. I. Th. Lippe, M. Lorbach, and P. Holzer, Close arterial infusion of calcitonin gene-related peptide (CGRP) into the rat stomach inhibits aspirin- and ethanol-induced haemorrhagic damage, *Regul Peptides.* 26:35 (1989).
7. C.A. Maggi, S. Manzini, S. Giuliani, P. Santicioli, and A. Meli, Extrinsic origin of the capsaicin-sensitive innervation of rat duodenum: possible involvement of calcitonin gene-related peptide (CGRP) in the capsaicin-induced activation of intramural non-adrenergic, non-cholinergic neurons, *Naunyn-Schmiedeberg's Arch Pharmacol.* 334:172 (1986).
8. S. Evangelista, M. Tramontana, and C.A. Maggi, Pharmacological evidence for the involvement of multiple calcitonin gene-related peptide (CGRP) receptors in the antisecretory and antiulcer effect of CGRP in rat stomach, *Life Sci.* 50:PL13 (1992).
9. Y. Ishi, Y. Fuji, and M. Homma, Gastric acid stimulating action of cysteamine in the rat, *Eur J Pharmacol.* 36:331 (1976).
10. S. Szabo, Mechanisms of mucosal injury in the stomach duodenum: time-sequence analysis of morphologic, functional, biochemical and histochemical studies, *Scand J Gastroenterol.* 22(Suppl. 127):21 (1987).
11. S. Evangelista, D. Renzi, P. Mantellini, C. Surrenti, and A. Meli, Duodenal ulcers are associated with a depletion of duodenal calcitonin gene-related peptide-like immunoreactivity in rats, *Eur J Pharmacol.* 164:389 (1989).
12. S. Evangelista, C.A. Maggi, S. Giuliani, and A. Meli, Further studies on the role of the adrenals in the capsaicin-sensitive "gastric defence mechanism", *Int J Tissue React.* 10:253 (1988).
13. P. Holzer, Capsaicin: cellular targets, mechanisms of action, and selectivity for thin sensory neurons, *Pharmacol Rev.* 43:143 (1991).
14. I. Th. Lippe, M.A. Pabst, and P. Holzer, Intragastric capsaicin enhances rat gastric acid elimination and mucosal blood flow by afferent nerve stimulation, *Br J Pharmacol.* 96:91 (1989).
15. K. Ohe, Y. Miura, Y. Taoka, Y. Osakada, and A. Miyoshi, Cysteamine-induced inhibition of mucosal and pancreatic alkaline secretion in rat duodenum, *Dig Dis Sci.* 33:330 (1988).
16. G. Pihan, T.J. Kline, N.K. Hollenberg, and S. Szabo, Duodenal ulcerogens cysteamine and propionitrile induce gastroduodenal motility alterations in the rat, *Gastroenterology.* 88:989 (1985).
17. Y. Kurebayashi, M. Asano, T. Hashizume, and A. Akashi, Changes in duodenal mucosal blood flow and mucus glycoprotein content during cysteamine-induced duodenal ulceration in rats, *Arch Int Pharmacodyn Ther.* 276:152 (1985).
18. T.J. Kline, G. Pihan, and S. Szabo, Biphasic effect of duodenal ulcerogens on gastric emptying in the rat, *Dig Dis Sci.* 33:926 (1988).
19. K. Takeuchi, O. Furukawa, H. Tanaka, and S. Okabe, A new model of duodenal ulcers induced in rats by indomethacin plus histamine, *Gastroenterology.* 90:636 (1986).
20. S. Szabo, and S. Reichlin, Somatostatin in rat tissue is depleted by cysteamine, *Endocrinology.* 109:2255 (1981).
21. O.U. Scremin, F.W. Leung, and P.H. Guth, Pathogenesis of duodenal ulcer in the rat: dissociation of acid load and blood flow, *Am J Physiol.* 256:G1058 (1989).
22. P. Holzer, and W. Sametz, Gastric mucosal protection against ulcerogenic factors in the rat mediated by capsaicin-sensitive afferent neurones, *Gastroenterology.* 91:975 (1986).

23. C.A. Maggi, S. Evangelista, L. Abelli, V. Somma, and A. Meli, Capsaicin-sensitive mechanisms and experimentally induced duodenal ulcers in rats, *J Pharm Pharmacol.* 39:559 (1987).
24. M. Tramontana, D. Renzi, C. Panerai, C. Surrenti, F. Nappi, L. Abelli, and S. Evangelista, Capsaicin-like effect of resiniferatoxin in the rat stomach, *Neuropeptides.* 26:29 (1994).
25. P. Holzer, M.A. Pabst, and I. Th. Lippe, Intragastric capsaicin protects against aspirin-induced lesion formation and bleeding in the rat gastric mucosa, *Gastroenterology.* 96:1425 (1989).
26. P. Holzer, B.M. Peskar, B.A. Peskar, and R. Amann, Release of calcitonin gene-related peptide induced by capsaicin in the vascularly perfused rat stomach, *Neurosci Lett.* 108:195 (1990).

THE EFFECT OF bFGF AND PDGF ON ACUTE GASTRIC MUCOSAL LESIONS, CHRONIC GASTRITIS AND CHRONIC DUODENAL ULCER

S. Szabo, J. Folkman, S. Kusstatscher,
Zs. Sandor and M.M. Wolfe

The treatment and prevention of gastric and duodenal ulcers have been essentially limited until recently to decreasing the concentration of luminal gastric acid or inhibiting acid secretion. In these cases the erosions and ulcers were essentially left alone to heal themselves in the presence of diminished aggressive factors, but without any attempt of stimulating the healing process.

After the introduction of the concept of gastric cytoprotection[1], prostaglandins (PG) and sulfhydryls (SH) became available for the prevention of <u>acute</u> gastric mucosal lesions. Most of these agents stimulated defensive factors such as blood flow and repair processes (e.g., epithelial restitution), but were virtually inactive in stimulating the healing of <u>chronic</u> lesions. With the availability of growth factors, cellular processes related to ulcer healing could be directly stimulated, e.g., angiogenesis, granulation tissue production and cell proliferation. Thus, while superficial mucosal lesions and erosions are repaired by epithelial restitution and regeneration, the healing of deep ulcers (which involve a major loss of smooth muscle), angiogenesis, granulation tissue (e.g., fibroblast proliferation, collagen deposition, infiltration of chronic inflammatory cells), smooth muscle and neural regeneration and epithelial proliferation are also needed (Table 1).

Epidermal growth factor (EGF) stimulates the proliferation of epithelial cells, but it is weak or inactive in angiogenesis, granulation tissue production or smooth muscle regeneration. EGF, however, inhibits gastric acid secretion, and this is probably the reason that its antiulcerogenic potency in the cysteamine-induced duodenal ulcer model is similar to that of the H_2 antagonist cimetidine[2].

Heparin binding growth factors such as basic fibroblast growth factor (bFGF), on the other hand, affect several cell types with high potency (Table 2). The most potent mitogen for endothelial cells, and inducer of angiogenesis is bFGF, which also avidly stimulates the proliferation of epithelial cells, fibroblasts and smooth muscle which are essential for repairing the tissue defect in ulceration. This peptide also enhances neuronal regeneration, but it does not inhibit gastric secretion.

Platelet-derived growth factor (PDGF) is acid resistant in its natural form, and its acute or chronic administration does not modify gastric acid secretion. Unlike

Neuroendocrinology of Gastrointestinal Ulceration
Edited by S. Szabo and Y. Taché, Plenum Press, New York, 1995

bFGF, a single dose of PDGF exerts acute gastroprotection, but the effect of the peptide on cell types involved in ulcer healing has not been extensively investigated (Table 2).

In this chapter, we review our results on the affect of bFGF and PDGF on chemically induced acute gastric mucosal lesions, chronic duodenal ulcer and chronic gastritis, with reference to the pathophysiologic role of endogenous bFGF in ulcer healing.

Table 1. Cellular and Histologic Basis of Ulcer Development and Healing

Lesions	Defect	Repair
Erosion	Epithelial/mucosal	Epithelial restitution and regeneration
Ulcer	Mucosal & smooth muscle	Granulation tissue Angiogenesis Fibroblast proliferation Collagen deposition Smooth muscle proliferation Epithelial regeneration

Table 2. Comparison of bFGF and PDGF in Ulcer Pathogenesis, Prevention and Healing

	bFGF	PDGF
Gastric acid secretion	No change or stimulation	No change
Duodenal alkaline secretion	No	?
Angiogenesis stimulation	Potent	Mild or minimal
Fibroblast proliferation	Yes	Yes
Epithelial restitution	Yes	?
Epithelial proliferation	Yes	?
Smooth muscle regeneration	Yes	?
Neuronal regeneration	Yes	?
ACUTE GASTROPROTECTION	No	Mild
CHRONIC ULCER HEALING	Potent	Potent

ACUTE GASTROPROTECTION

The prevention of chemically induced acute hemorrhagic mucosal erosions by PG, SH and other compounds is referred to as "cytoprotection". This acute mucosal protection is relative since actually only hemorrhagic necrosis is absent and microscopically superficial epithelial defect can still be seen soon after administration of concentrated ethanol, acid or base, even in PG- or SH-pretreated animals[3]. Subsequently, the very superficial mucosal damage even in the "protected" animals is repaired quickly by rapid epithelial cell migration or restitution[4]. Since not all the cells in the gastric mucosa are preserved, it is more appropriate to refer to the process as gastroprotection rather than cytoprotection[3,5].

Since gastroprotection is not related to inhibition of gastric acid secretion, soon after learning about the potent ulcer healing abilities of bFGF and PDGF, which do not decrease gastric acid output[6-8], we tested the hypothesis that these growth factors might also exhibit acute gastroprotection. This was an especially relevant subject since a single dose of EGF exhibits an SH-sensitive mucosal protection against ethanol-induced acute hemorrhagic gastric erosions in the rat[9-11]. In the standard assay of acute gastroprotection, groups of fasted rats were given increasing doses, starting with the usual antiulcer dose of the acid-resistant mutein bFGF-CS23, from 100 ng/100 g or PDFG, from 500 ng/100 g, up to 1 or 2.5 μg/100 g, respectively, subcutaneously (s.c.) or by intragastric (i.g.) gavage, 30 min before the administration of 1 ml of 75% ethanol i.g. As a positive control, an additional group of rats received the SH-containing taurine, 50 mg/100 g. All the animals were killed 1 h after receiving ethanol, and the area of hemorrhagic mucosal lesions in the glandular stomach was measured by computerized stereomicroscopic planimetry (Table 3).

The results indicate that the two growth factors, even at the highest dose tested and given i.g. or s.c., were virtually ineffective in preventing the ethanol-induced acute hemorrhagic erosions (Table 3). Only 2.5 μg/100 g of PDGF administered i.g. reduced the area of acute mucosal lesions at the borderline of statistical significance ($p = 0.095$), while pretreatment with taurine resulted in about 50% reduction of gastric damage ($p - 0.027$).

Table 3. Effect of bFGF, PDGF or Taurine on Acute Hemorrhagic Gastric Mucosal Lesions Induced by 75% Ethanol in Rats

Group	Pretreatment			Mucosal lesions
	Name	Dose	Route	(% of glandular stomach)
1.	Control			10.3 $\pm$ 1.5
2.	bFGF	1 μg/100 g	s.c.	8.5 $\pm$ 2.0
3.	PDGF	2.5 μg/100 g	s.c.	7.0 $\pm$ 1.3
4.	PDGF	2.5 μg/100 g	i.g.	6.3 $\pm$ 1.7[*]
5.	Taurine	50 mg/100 g	i.g.	5.2 $\pm$ 1.8[**]

Rats of all groups were given 1 ml of 75% ethanol i.g. 30 min after the pretreatment, and were killed 1 h after the alcohol. The area of hemorrhagic mucosal lesions was measured by computerized stereomicroscopic planimetry.
Mann-Whitney U-tests: [*]$p = 0.095$, [**]$p = 0.27$

We thus conclude that despite their potency in accelerating chronic ulcer healing, bFGF and PDGF possess weak or no acute gastroprotection properties, despite the recently reported stimulation of epithelial restitution by bFGF[12]. Namely, enhanced migration of surviving epithelial cells from the edge of superficial mucosal lesions is one of the mechanisms of acute gastric cytoprotection. The discrepancy by bFGF reinforces the notion that the mechanisms of acute mucosal protection are indeed complex, e.g., prevention of microvascular injury, maintenance of blood flow and then rapid epithelial restitution or migration covers the mucosal epithelial defect[3,4,5]. Stimulation of restitution alone is not sufficient for acute gastroprotection, not even by potent antiulcer agents such as bFGF. Nevertheless, new preliminary studies on chambered gastric mucosa indicate that bFGF pretreatment does offer limited acute protection (G. Morris, personal communication). Thus, extensive new investigations are needed to define the role of bFGF and PDGF in acute mucosal protection.

CHRONIC GASTRITIS

The clinical relevance of chronic gastritis, especially because of the high prevalence of chemically induced (e.g., alcoholic or NSAID gastropathy) or H. pylori-associated gastritis, is much higher than the relevance of acute gastroprotection. Indeed, the clinical significance of gastric "cytoprotection" has been questioned from the very beginning of the introduction of the concept[1,3,4,13], and passionately attacked by some during subsequent years[14,15]. It is thus much more important to demonstrate the efficacy of new antiulcer agents in chronic gastritis than in the prevention of acute ethanol-induced hemorrhagic mucosal lesions. Unfortunately, until recently, no animal model of diffuse chronic gastritis existed, in contrast to the plethora of models of acute gastric mucosal injury and the few animal models of chronic gastric and duodenal ulcers.

We found, as a follow-up to our investigation[16] on the gastroprotective role of SH, that ingestion of low concentrations of SH alkylators (e.g., 0.1% iodoacetamide) in drinking water induced severe diffuse acute and chronic erosive gastritis in rats[17]. The lesions in the mucosa of the glandular stomach contain initially acute inflammatory cells such as leukocytes, which are mostly replaced by lymphocytes, plasma cells and macrophages in the chronic stage of gastritis, resembling the human alcoholic gastritis. We used this new animal model of gastritis to investigate the possible beneficial effect of bFGF and PDGF[18-20].

In these experiments, after induction of gastritis by the SH alkylator 0.1% iodoacetamide in drinking water for one week, rats were randomized to receive either the vehicle solvent of the growth factors or a site-specific mutated acid stable bFGF-CS23[21] at 25 or 100 ng/100 g, or PDGF-BB at 500 ng or 2.5 μg/100 g by gavage twice daily. All the rats except the absolute controls remained on drinking water containing iodoacetamide for one more week. At autopsy on the 14th day macroscopic and histologic involvement of gastric glandular mucosa was quantified, and wet and dry stomach weights were obtained.

The results revealed that vehicle-treated rats on drinking water containing the SH alkylator had severe diffuse chronic gastritis interspersed with focal erosions and superficial ulcers. The wet weight of the stomach from these rats was markedly increased due to edema and massive infiltration by acute and chronic inflammatory cells. Oral treatment with the acid stable bFGF-CS23 or PDGF dose-dependently decreased the severity of gastritis as exemplified by the dose-dependent decreases in the wet weights of the stomachs (Fig. 1). This was confirmed by macroscopic and histologic evaluations indicating less severe gastritis and more regenerated mucosa in rats treated with growth factors.

Other recent results from our laboratory indicate that oral treatment with bFGF and sucralfate, which binds endogenous bFGF[22], exerted an additive or synergistic healing effect in this animal model of chronic erosive gastritis[18,19]. Thus, growth factors administered alone or in combination with drugs such as sucralfate, which help to concentrate endogenous bFGF at the site of injury, might represent a new group of therapeutic agents that are beneficial not only for the treatment of solitary gastric or duodenal ulcers, but also for the healing of diffuse and severe chronic gastritis.

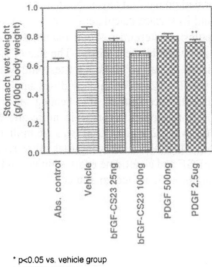

Figure 1. Effect of bFGF-CS23 and PDGF on iodoacetamide-induced chronic gastritis in rats.
*p < 0.05 vs vehicle group. **p < 0.01 vs vehicle group.

CHRONIC DUODENAL ULCER

Duodenal ulcer is still the most prevalent form of nonspecific ("peptic") ulcer worldwide, except in Japan where its incidence, nevertheless, is also rising[23]. Even in users and abusers of aspirin-like drugs, the incidence of both duodenal gastric ulcers is also increasing[24]. This is one of the main reasons that we first tested the possible ulcer healing properties of bFGF and PDGF in the animal model of duodenal ulcer[6-8,25].

In these studies we used the most widely investigated experimental duodenal ulcer induced by cysteamine or its derivatives[26,27]. This animal model resembles the human duodenal ulcer by several morpholigic and functional criteria[26,28], and like in patients, both local and central alterations contribute to the development of solitary or "kissing" duodenal ulcers on the anterior or posterior wall of rat proximal duodenum[26,28,29]. For the rapid induction of duodenal ulcers, rats were maintained on Purina Laboratory chow and tap water <u>ad libitum</u> throughout the study and received cysteamine-HCl (25 mg/100 g) i.g. three times on the first day. On the third day, laparotomy was performed under ether anesthesia, and rats were randomized to create groups with homogeneously severe ulcers (perforated or penetrated into the liver or pancreas) and treated twice a day i.g. for three weeks with vehicle solvent or bFGF-CS23 at 10 ng or 50 ng/100 g or PDGF-BB at 100 or 500 ng/100 g, or cimetidine (10 mg/100 g). These studies were preceded by dose-response experiments comparing the naturally occurring bFGF-wild and bFGF-CS23. bFGF-CS23 was found to be more potent than bFGF-w at the optimal oral dose of 100 ng/100 g twice daily for three weeks[6,7,25]. More recent dose-response studies, however, indicated that the acid-resistant mutein of bFGF-CS23 even at 50 ng.100 g could significantly accelerate the healing of cysteamine-induced chronic duodenal ulcers[30].

Autopsy was performed on the 21st day, and the ulcer size was measured in two of its largest diagonals and its surface calculated by the ellipse formula. The results were confirmed by computerized steremicroscopic planimatry, and ulcer healing was evaluated by light microscopy and histochemistry. The size of ulcers was analyzed by one-way analysis of variance, unpaired Student's t-test and Mann-Whitney U-tests. The results revealed by bFGF-CS23, even in daily doses smaller than tested before, and PDGF dose-dependently accelerated the healing of experimental chronic duodenal ulcers (Fig. 2). Both growth factors were more effective on a molar basis than cimetidine, i.e., bFGF was 7.2, while PDGF was 2.6 million, times more potent than cimetidine (Table 4). It is also interesting that other recent results from our laboratory indicate that bFGF and cimetidine exert additive, and probably synergistic, interactions in the healing of chronic duodenal ulcers[30].

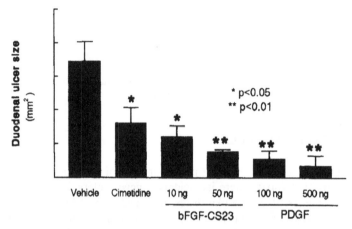

Figure 2. Effect of bFGF-CS23 and PDGF on cysteamine-induced chronic duodenal ulcer in rats.

Table 4. The Molar Comparison of Antiulcerogenic Doses of bFGF, PDGF, and Cimetidine in the Rat Model of Cysteamine-Induced Duodenal Ulcer

	bFGF	PDGF	Cimetidine
Antiulcerogenic dose:	100 ng/100 g	500 ng/100 g	10 mg/100 g
Molecular weight:	18,000	34,000	252
1 μmole:	18,000 μg	34000 μg	252 μg
Antiulcerogenic dose in μmoles:	0.00000556 μmol/100 g	0.00000147 μmol/100 g	39.69 μmol/100 g
Molar comparison	1	2.64	7.136,690.6

The histologic and histochemical evaluation of ulcers also revealed major differences. Namely, in the control vehicle-treated rats, most of the ulcers demonstrated necrotic crater, with acute and chronic inflammation, and surrounding hypovascular granulation tissue. Ulcers in cimetidine-treated rats had similar qualitative features, except the ulcers were usually smaller (see above). Duodenal sections from rats treated with bFGF-CS23 alone or in combination with cimetidine revealed dense granulation tissue with enhanced angiogenesis, minimal acute and

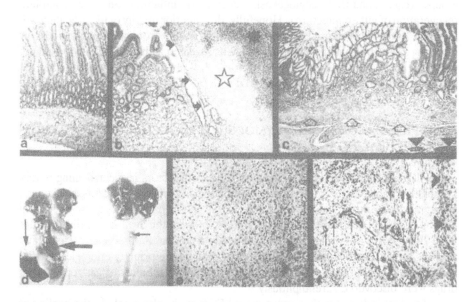

Figure 3. (A-C) Light microscopy of the rat duodenum (H&E; original magnification X 40). (A) Cross-section of the control rat duodenum shows the normal will of the mucosa (top) and the typical uninterrupted muscular layers of the duodenum. Note the free serosal surface without any cellular reaction (bottom). (B) Typical ulcer crater 21 days after cysteamine administration. Note the sharply demarcated ulcer crater (arrows) with necrotic debris (top) and the dense infiltration by acute and chronic inflammatory cells (asterisk) surrounded by granulation/connective tissue (star). (C) Healed cysteamine-induced duodenal ulcer at 21 days after treatment with bFGF-CS23 (100 ng/100 g intragastrically twice daily). The regenerated epithelium completely covers the previous ulcer crater. The muscle layer is interrupted by regenerating smooth muscle and fibrous connective tissue (open arrowheads). A serosal reaction involving attachment of the pancreas (arrowheads) indicates a previously deep, probably penetrating ulcer. (D) Gross appearance of mucosal view of two stomachs from rats that received identical doses of cysteamine to induce duodenal ulcer. On the left is the rat treated with solvent vehicle, revealing huge chronic duodenal ulcer (horizontal arrow) penetrating into the adjacent pancreas (light) and liver (dark) (vertical arrow). On the right, a barely visible, almost completely healed duodenal ulcer after 3 weeks treatment with bFGF-CS23. (E-F) Light microscopy of the base of the chronic ulcer after immunostaining for factor VIII to delineate vascular endothelial cells. The ulcer was penetrating into the liver, which is delineated by arrowheads (X100) at 3 weeks. (E) In the cysteamine-treated rats, only a few capillaries are visible in the granulation tissue. (F) Numerous blood vessels (arrows) are seen in the ulcer base after treatment with bFGF-CS23. (Reproduced with permission from *Gastroenterology.* 104:1106, 1994).

chronic inflammation if the epithelial and mucosal reconstruction were not complete. In healed ulcers, histologic signs of prominent angiogenesis and smooth muscle regeneration were also evident, often with focal serosal signs of chronic, localized peritonitis, indicating previous perforations (Fig. 3).

We conclude that bFGF-CS23 and PDGF exert not only very potent healing effects in the animal model of chronic duodenal ulcer, but the quality of ulcers healed after treatment with these growth factors is superior to that in control or cimetidine groups (e.g., enhanced angiogenesis, decreased inflammation and complete regeneration of gut architecture). Hopefully, this superior quality of ulcer healing will result in lower ulcer relapse. Indeed, the first results from an animal model of relapsing gastric ulcer suggest that rats previously treated with bFGF-CS23 to heal the ulcer had 63% lower relapse than in control[31]. It is also rewarding to know that the preliminary results of the first clinical trial with bFGF-CS23 are very encouraging[32].

FROM PHARMACOLOGY TO PHYSIOLOGY: ENDOGENOUS bFGF AND PDGF IN ULCER HEALING

Although the pharmacologic potency of growth factors in ulcer healing is very impressive (Table 4), the value of new therapeutic and preventive agents is enhanced if they have a firm physiologic basis, e.g., a role in the pathogenesis and/or natural healing of gastrointestinal (GI) ulcers. Indeed, most of the available antiulcer drugs meet these criteria only in part, and indirectly if at all. Antisecretory agents and antacids reduce gastric acidity which may not even be the primary etiologic factor in gastric and duodenal ulceration, while most of the other antiulcer or gastroprotective drugs were discovered serendipitously.

The first indications that endogenous bFGF might play a role in the healing of gastric and duodenal ulcers were derived from studies showing that the locally acting antiulcer drug sucralfate avidly binds bFGF and delivers it in large concentration to the ulcer crater[22]. Our in vitro and in vivo experiments demonstrated that sucralfate and its water-soluble component sucrose octasulfate bind bFGF more potently than heparin, and prevent the acid-peptic degradation of this peptide[22]. These data also suggest a new mechanism of action for sucralfate[22,33,34].

The contribution made by endogenous bFGF to ulcer healing was also demonstrated in our collaborative studies with H. Satoh[35]. In these studies, daily injections of neutralizing antibodies to bFGF delayed the healing of acetic acid-induced chronic gastric ulcer in rats[31]. Our preliminary results also indicate that the neutralizing bFGF antibody also retards the healing of cysteamine-induced chronic duodenal ulcers[35].

Subsequently, we investigated the role of endogenous bFGF on the natural history of ulcer disease[36,37] (Table 5). We postulated that endogenous bFGF, presumably occurring in the ulcer bed in the gut mucosa, might be critical to ulcer healing because of its angiogenic and other mitogenic properties. To test this hypothesis, rats were treated three times at 3-h intervals with the duodenal ulcerogen cysteamine-HCl, and were killed 12, 24 and 48 h after the first cysteamine administration. Mucosal scrapings from proximal duodenum from these rats and from controls were homogenized in 2 N NaCl buffered at pH 7.4 and extracted for Western blot analysis[36]. After 12 h a reduction in the low-molecular-weight (18 kD) form of bFGF (localized to the cytoplasm and extracellular space) was found, with a simultaneous increase in the high-molecular-weight (21, 24 and 25 kD) forms

Table 5. Implications for the Physiologic Role of bFGF and PDGF in Ulcer Healing

Parameters and criteria	bFGF	PDGF
Delayed ulcer healing by neutralizing antibody	Yes	Yes
Local changes in growth factors in ulcer pathogenesis and healing:		
- Western blot & ELISA	Yes	NT
- Immunocytochemistry	Yes	NT
- in situ hybridization	Yes	NT

NT = not tested

(localized to the nucleus of the cells), compared to controls. After 24 h, the high-molecular-weight forms decreased, with an increase in the low-molecular-weight form, compared to the 12-h samples. The 48-h samples showed a continuation of this trend towards the pattern seen in the controls.

The localization of bFGF in the duodenum has also been investigated by immunostaining techniques[38]. Frozen sections of proximal duodenum from control rats, and from samples taken 12, 24 and 48 h after the first of three injections with cysteamine on polylysine-coated slides, were incubated with antibodies to bFGF. Preliminary results with this technique have shown a time-dependent decrease in duodenal mucosal bFGF immunoreactivity which reached the level of statistical significance 24 h after the cysteamine injection. This reduction in submucosal bFGF was less prominent in the submucosa, and it did not involve the muscularis propria[38]. Further studies of longer duration are needed to determine the local distribution and the changes in immunoreactive bFGF in duodenal ulceration. In situ hybridization experiments to detect messenger (m)RNA for bFGF in ulcer healing indicate that bFGF synthesis, visualized mostly in endothelial cells, macrophages and fibroblasts, is enhanced in the early stages of cysteamine-induced duodenal ulceration.

Since oral treatment with PDGF also markedly accelerated the healing of cysteamine-induced chronic duodenal ulcers in rats, we recently tested the hypothesis that inhibition of endogenous PDGF delays the healing of chronic duodenal ulcer[39]. In groups of rats with cysteamine-induced duodenal ulcer daily intramuscular injections of monoclonal neutralizing anti-PDGF IgG, ulcer size increased by 84% and 392% in 7 and 14 days, respectively. Thus, growth factors might be important in initiating ulcer healing, and not only endogenous bFGF, but PDGF also seems to play a physiologic role in the healing of duodenal ulcers. Further studies, especially on the biochemical changes in the mucosal level, expression and availability of PDGF in the natural history of GI ulcers, are urgently needed.

SUMMARY

Recent studies in rats revealed that unlike EGF, oral administration of bFGF or PDGF did not inhibit gastric acid and pepsin secretion. Nevertheless, both bFGF and PDGF accelerated the healing of chronic cysteamine-induced duodenal ulcers, and their potency was 7.1 and 2.7 million times higher on a molar basis than that of cimetidine, respectively. bFGF is the most potent endothelial mitogen which also stimulates other cell types. PDGF is slightly and indirectly angiogenic and its effect on other cell types is not as well investigated as that of bFGF. Pretreatment with PDGF offers mild acute gastroprotection, while bFGF is inactive and EGF relatively potent in this respect. Very recent experiments using Western blot and ELISA revealed a biphasic change in mucosal bFGF in duodenal ulceration. The delay in healing of experimental duodenal and gastric ulcers by daily administration of neutralizing antibodies against bFGF or PDGF further confirms the pathophysiologic role of these growth factors in the natural history of ulcer disease. Thus, bFGF and PDGF are new molecules in the pathophysiology and pharmacology of gastrointestinal ulcers.

REFERENCES

1. A. Robert, Cytoprotection by prostaglandins, *Gastroenterology*. 77:761 (1979).
2. S.S. Poulsen, K.S. Olsen, and P. Kirkegaard, Healing of cysteamine-induced duodenal ulcers in the rat, *Dig Dis Sci*. 30:161 (1985).
3. S. Szabo, Critical and timely review of the concept of gastric cytoprotection, *Acta Physiol Hung*. 73:115 (1989).
4. E.R. Lacy, and S. Ito, Rapid epithelial restitution on the rat gastric mucosa after ethanol injury, *Lab Invest*. 51:573 (1984).
5. S. Szabo, Mechanisms of gastric mucosal injury and protection, *J Clin Gastroenterol*. 13:S21 (1991).
6. S. Szabo, P. Vattay, R.E. Morales, B. Johnson, K. Kato, and J. Folkman, Orally administered bFGF mutein: effect on healing of chronic duodenal ulcers in rats, *Dig Dis Sci*. 34:1323 (1989).
7. J. Folkman, S. Szabo, P. Vattay, R.E. Morales, G. Pinkus, and K. Kato, Effect of orally administered bFGF on healing of chronic duodenal ulcers, gastric secretion and acute mucosal lesions in rats, *Gastroenterology*. 98:A45 (1990).
8. P. Vattay, E. Gyömber, R.E. Morales, and S. Szabo, Effect of orally administered platelet-derived growth factor (PDGF) on healing of chronic duodenal ulcers and gastric secretion in rats, *Gastroenterology*. 100:A180 (1991).
9. S.J. Konturek, T. Brzozowski, A. Dembinski, Z. Warzecha, and J. Yamazaki, Gastric protective and ulcer-healing action of epidermal growth factor, *in:* "Advances in Drug Therapy of Gastrointestinal Ulceration," A. Garner, and B.J.R. Whittle, ed., John Wiley & Sons, New York (1989).
10. A. Garner, Strategies for the development of novel anti-ulcer drugs, *in:* "Advances in Drug Therapy of Gastrointestinal Ulceration," A. Garner, and B.J.R. Whittle, eds., John Wiley & Sons, New York (1989).
11. S.J. Konturek, T. Brzozowski, L. Nagy, E. Sheehan, D. Sacks, and S. Szabo, Sulfhydryls (SH) in the gastroprotection by epidermal growth factor (EGF), polyamines (PA) and prostaglandins (PG), *Gastroenterology*. 102:A101 (1992).
12. H. Paimela, P.J. Goddard, K. Carter, R. Khakee, P.L. McNeil, S. Ito, and W. Silen, Restitution of frog gastric mucosa in vitro: effect of basic fibroblast growth factor, *Gastroenterology*. 104:1137 (1993).
13. J.L. Wallace, "Cytoprotection" - define it or dispose it, *Dig Dis Sci*. 31:667 (1986).
14. W. Silen, What is cytoprotection of the gastric mucosa? *Gastroenterology*. 94:232 (1988).
15. C.J. Hawkey, and R.P. Walt, Prostaglandins in peptic ulcer: a promise unfulfilled, *Lancet*. 1:1084 (1986).
16. D. Dupuy, A. Raza, and S. Szabo, The role of endogenous nonprotein and protein sulfhydryls in gastric mucosal injury and protection, *in:* "Ulcer Disease: New Aspects of Pathogenesis and Pharmacology," S. Szabo, and C.J. Pfeiffer, eds., CRC Press, Boca Raton (1989).
17. S. Szabo, J.S. Trier, A. Brown, and J. Schnoor, Sulfhydryl blockers induce severe inflammatory gastritis in the rat, *Gastroenterology*. 86:1271 (1984).
18. M. Stovroff, P. Vattay, B. Marino, S. Szabo, and J. Folkman, Healing of experimental gastritis by oral fibroblast growth factor, *Surg Forum*. 452:174 (1991).
19. S. Szabo, S. Kusstatscher, and M. Stovroff, Role of bFGF and angiogenesis in ulcer healing and the treatment of gastritis, *in:* "The Stomach: Physiology, Pathophysiology and Treatment," W.

Domschke, and S.J. Konturek, eds., Springer-Verlag, Berlin, Heidelberg, New York, London, Paris, Tokyo, Hong Kong, Barcelona, Budapest (1993).

20. S. Kusstatscher, and S. Szabo, Effects of platelet-derived growth factor (PDGF) on the healing of chronic gastritis in rats, *Gastroenterology.* 104:A125 (1993).

21. K. Seno, R. Sasada, K. Iwane, K. Sudo, T. Kuroko, K. Ito, and I. Igarashi, Stabilizing basic fibroblast growth factor using protein engineering, *Biochem Biophys Res Commun.* 151:701 (1988).

22. J. Folkman, S. Szabo, M. Stovroff, P. McNeil, W. Li, and Y. Shing, Duodenal ulcer: Discovery of a new mechanism and development of angiogenic therapy that accelerates healing, *Ann Surg.* 214:414 (1991).

23. S.K. Lam, Epidemiology and genetics of peptic ulcer, *Gastroenterol Jap.* 28:145 (1993).

24. J.H. Kurata, Epidemiology of peptic ulcer disease, in: "Ulcer Disease," E.A. Swabb, and S. Szabo, eds., Marcell Dekker, Inc., New York (1991).

25. S. Szabo, J. Folkman, P. Vattay, R.E. Morales, G.S. Pinkus, and K. Kato, Accelerated healing of duodenal ulcers by oral administration of a mutein of basic fibroblast growth factor in rat, *Gastroenterology.* 106:1106 (1994).

26. S. Szabo, J. Folkman, P. Vattay, R.E. Morales, and K. Kato, Duodenal ulcerogens: effect of FGF on cysteamine-induced duodenal ulcer, in: "Mechanisms of Peptic Ulcer Healing," F. Halter, A. Garner, and G.N.J. Tytgat, eds., Kluwer Academic Publications, London (1991).

27. C. Giampaolo, A.T. Gray, R.A. Olshen, and S. Szabo, Predicting chemically induced duodenal ulcer and adrenal necrosis with classification trees, *Proc Natl Acad Sci USA.* 88:6298 (1991).

28. S. Szabo, Animal model of human disease. Duodenal ulcer disease. Animal model: cysteamine induced acute and chronic duodenal ulcer in the rat, *Am J Pathol.* 93:273 (1978).

29. S. Szabo, and C.H. Cho, From cysteamine to MPTP: structure-activity studies with duodenal ulcerogens, *Toxicol Pathol.* 16:205 (1988).

30. S. Kusstatscher, J. Folkman, L. Nagy, and S. Szabo, Additive effect of fibroblast growth factor (bFGF) and cimetidine on chronic duodenal ulcer healing in rats, *Gastroenterology.* 102:A104 (1992).

31. H. Satoh, A. Shino, I. Murakami, S. Asano, F. Sato, and N. Inatomi, New animal model for gastric ulcer relapse in the rat: bFGF may play a role in prevention of ulcer relapse, *Gastroenterology.* 102:A159 (1992).

32. M.M. Wolfe, T.E. Bynum, W.G. Parsons, K.M. Malone, S. Szabo, and J. Folkman, Safety and efficacy of an angiogenic peptide, basic fibroblast growth factor (bFGF), in the treatment of gastroduodenal ulcers: a preliminary report, *Gastroenterology.* 106:A212 (1994).

33. S. Szabo, P. Vattay, E. Scarbrough, and J. Folkman, Role of vascular factors, including angiogenesis, in the mechanisms of action of sucralfate, *Am J Med.* 91:158S (1991).

34. S. Szabo, The mode of action of sucralfate: the 1 x 1 x 1 mechanism of action, *Scand J Gastroenterol.* 26:7 (1991).

35. S. Kusstatscher, Zs. Sandor, H. Satoh, and S. Szabo, Inhibition of endogenous basic fibroblast growth factor (bFGF) delays duodenal ulcer healing in rats: implications for a physiologic role of bFGF, *Gastroenterology.* 106:A113 (1994).

36. G. Sakoulas, S. Kusstatscher, P.T. Ku, P. D'Amore, and S. Szabo, Role of endogenous basic fibroblast growth factor in duodenal ulceration, *Dig Dis Sci.* 38:A11 (1993).

37. S. Szabo, G. Sakoulas, S. Kusstatscher, Zs. Sandor

38. S. Kusstatscher, J. Bishop, L. Brown, Zs. Sandor, and S. Szabo, Basic fibroblast growth factor (bFGF) immunolocalization in duodenum of normal rats and after duodenal ulcer induction, *Gastroenterology.* 104:A125 (1993).

39. S. Kusstatscher, M. Nagata, S. Kathiresan, J. Bishop, G. Sakoulas, H. Satoh, N. Aizawa, R. Deguchi, and S. Szabo, Endogenous growth factors play a role in duodenal ulcer healing in rats, *Proc of 10th World Congr Gastroenterol.* Los Angeles, CA (1994).

ANIMAL MODELS AND PATHOGENESIS OF INFLAMMATORY BOWEL DISEASE

Gerald P. Morris, Paul L. Beck, John L. Wallace
Margaret S. Herridge and Carlo A. Fallone

Treatment of inflammatory bowel disease (IBD) is hampered by the fact that we know very little about the underlying causes of this condition. For instance, it remains unclear whether those alterations in immunologic function (such as autosensitization to epithelial cell components)[1] or in the activity or structure of the enteric nervous system (degeneration of the autonomic nerves)[2] which are characteristic of IBD represent initiating factors or are simply epiphenomena. Consequently, treatment is primarily palliative or supportive. The development of new or improved therapeutic agents and procedures, in particular, would benefit from the availability of animal models on which newly developed agents which might be of value in treatment of human disease could be tested. In particular there is a need for models which are simple to produce, which are inexpensive, and in which the induced disease is sufficiently prolonged, reproducible and severe to enable such tests to be routinely performed.

All of the available models are only approximations of the human diseases which are of interest to the clinician and the investigator and are far from Strober's ideal of an animal model in which the disease in the animal is identical in every respect to human IBD - with the same cause and the same pathophysiology[3]. What we have, as Strober so accurately pointed out, is not models of IBD but rather animal models of gastrointestinal inflammation. This observation should not, however, obscure the fact that in many studies, particularly in the assessment of novel therapeutic agents, models of gastrointestinal inflammation are precisely what is required.

NATURAL ANIMAL MODELS

Naturally occurring animal models of IBD have proven of very limited use in either furthering our understanding of the etiopathology of human IBD or in the development of therapeutic agents. Spontaneous colitic lesions have been reported from a wide variety of species, including dogs, rats, pigs, horses, and cattle[4-6], but none of these descriptions has led to the development of a useful animal model for subsequent studies.

Neuroendocrinology of Gastrointestinal Ulceration
Edited by S. Szabo and Y. Taché, Plenum Press, New York, 1995

The only naturally occurring animal disease with similarities to human IBD which has been extensively studied in recent years is the spontaneous colitis of the expensive and endangered cotton-topped tamarins or marmosets (*Saguinus oedipus*). The disease of *S. oedipus* does respond favorably to treatment with sulfasalazine[7] and it has been shown that in tamarins, as in humans, there is cross-reactivity between colonic epithelium of tamarins and epithelial cell-associated components previously characterized in the rat and the human[8]. This study provides some support for the hypothesis that ulcerative colitis has an autoimmune etiology, involving antigens derived from mucosal epithelium[1,9]. However, only very limited studies can be carried out on this interesting but impractical model, and its uses are likely to be primarily in determining whether the disease of primates is similar in various ways to ulcerative colitis of humans.

TRANSMISSION STUDIES AND BACTERIALLY INDUCED MODELS

In the classical description of Crohn's disease[10] it was noted that "regional ileitis" closely resembled intestinal tuberculosis and other bacterially induced disease states, and there is a long history of attempts to detect an etiological agent for Crohn's disease and for ulcerative colitis by animal transmission studies[11]. Numerous studies during the last two decades were stimulated by the finding that granulomas could be induced in mice by injection of foot pads with extracts of lymph node and intestinal tissues from patients with Crohn's disease[12]. Similar results were obtained after passage of homogenates through 220 nm filters and these experiments were considered to suggest a viral etiology for Crohn's disease[13]. These studies were difficult to replicate, however, and have not allowed identification of a putative causative agent. Subsequent studies by Das *et al.*[14], involving injection of cell-free filtrates into nude (athymic) mice, produced a high incidence of lymphoid hyperplasia and lymphomas that were seroreactive with Crohn's disease serum and were also described as providing support for the involvement of a transmissible "biological factor" in Crohn's disease. Recent studies using this model have, however, also demonstrated seroreactivity and hyperplastic lymph nodes when mice were pretreated with saline or homogenates from control tissues, suggesting that the seroreactivity of the Crohn's disease serum was nonspecific and could not be used as evidence for a transmissible agent[15,16].

There have been several attempts to develop bacterially-induced animal models of Crohn's disease by the administration of various strains of bacteria and bacterial components. The best known of these studies, and probably the most significant as a potential model and as an indicator of the possible infectious etiology, is the description and characterization of mycobacteria from resected tissues of Crohn's disease patients[17-19]. One of these isolates (mycobacterium strain Linda) produced granulomatous lesions in mice[17] and in young goats[20], although it failed to produce lesions in rabbits, chickens and guinea pigs. Subsequent studies indicate that *Mycobacterium* strain Linda is a strain of *M. paratuberculosis* which causes Johne's disease (ileocolitis of ruminants)[21]. Isolation of mycobacteria from human tissues is not consistently reproducible and there are studies which do not show any convincing association of mycobacteria with tissue from human Crohn's disease patients[22]. These studies, collectively, have stimulated much research on the possibility of a mycobacterial etiology for Crohn's disease[23,24], but have not, unfortunately, produced a readily available or convenient animal model.

Studies on the carrageenan model for ulcerative colitis have provided evidence for a role for gram-negative, anaerobic bacteria in the development of lesions in this animal model. The development of colitis in the cecum and colon of guinea pigs and of rabbits by feeding the animals a solution of degraded carrageenan (an extract of red seaweed) in their drinking water was described by Marcus and Watt[25,26]. The lesions appeared first in the cecum and subsequently involved the colon. In guinea pigs the cecal lesions appeared within 7 to 14 days, but could be prevented by clindamycin and metronidazole and did not develop in germ-free animals[27]. Subsequent studies showed that germ-free animals mono-associated with a strain of *Bacteroides vulgatus* developed cecal ulcerations[28] and that immunization with *B. vulgatus* before carrageenan treatment increased the severity of the lesions[29]. Adaptive transfer of immune enhancement by spleen cells was also described and the authors suggested an etiology in which carrageenan increased mucosal permeability to *B. vulgatus* antigens which might then provoke a cell-mediated mucosal inflammation[30]. There are many histological similarities between ulcerative colitis of humans and the lesions observed in rodents following carrageenan. In the animal model, however, lesions develop first in the cecum and only later involve the colon and rectum, and the maintenance of active disease depends on continual feeding of carrageenan[31].

Other studies have involved direct injection of bacterial components into the bowel wall. Sartor *et al.*[32] demonstrated that injection of streptococcal cell wall fragments containing peptidoglycan-polysaccharide complex (PG-PS) produced a long-lasting granulomatous inflammation in rats. Granulomatous inflammation was also found in mesenteric lymph nodes but mucosal ulceration was rare. Injection of PG-PS into the distal colon of rats also produced a chronic granulomatous inflammation accompanied by splenic necrosis and by arthritis[33]. Absorption of these cell wall components has also been shown to be increased, relative to controls, when radiolabelled PG-PS was injected into the cecum of anesthetized, laparotomized rats which had previously had their ascending colons damaged by 4% acetic acid[34]. In addition, PG-PS from bacterial cell walls has been shown to stimulate the systemic cellular immune response in germ-free mice and this has been cited as an example of how luminal bacteria can exert immunomodulatory effects on the gastrointestinal tract[35]. These studies have not produced models which are of significant value for screening therapeutic agents, and the procedures are complex. They do, however, demonstrate the possible consequences of a combination of products from luminal bacterial flora or from dietary constituents, and show that such macromolecules can cross the mucosal wall, that the rate of uptake is increased in the presence of acute injury, and that they are capable of initiating and perpetuating both chronic inflammation in the gastrointestinal tract and some of the extraintestinal manifestations characteristic of Crohn's disease.

CHEMICAL AND MECHANICAL INJURY

It is possible to produce long-lasting lesions in the gastrointestinal tract by relatively simple but harsh mechanical or chemical means such as the injection or serosal application of glacial acetic acid[36], or the use of scoops, or hot probes to burn holes in the stomach. The etiology of these lesions is not easily related to that of human disease although they have been useful in studying certain aspects of healing. One is studying the healing of wounds rather than the resolution of inflammation.

Similarly, intrarectal infusion of acetic acid[4,37] and of hydrogen peroxide[38], and the serosal application of capsaicin[39] have been used to injure the colonic mucosa. Variations on the acetic acid model of MacPherson and Pfeiffer[40] have been used in studies on the effects of prior acute injury on induction of colonic cancer[41], changes in arachidonic acid metabolism in acute colitis[42] and the effects of indomethacin and of a leukotriene-B$_4$ receptor antagonist on inflammation[41,43]. These studies illustrate the value of a simple and reproducible animal model, even though it has no obvious etiopathologic relevance to human IBD.

NSAID-INDUCED COLITIS

Chronic administration to dogs of indomethacin or of cinchophen (for periods of six days to about one month) produced enterocolitis in which ulcers were associated with Peyer's patches and which was accompanied by occasional granulomata in both the mucosa and adjacent mesenteric lymph nodes[44]. Despite these, and some other features characteristic of Crohn's disease, the authors reported that chronic treatment of dogs with indomethacin for as long as three years did not produce bowel wall thickening, strictures, or fistulae. In the rat, orally administered indomethacin produces lesions throughout the length of the jejunum and the ileum within 24 h[45]. The incidence of lesions is dose-dependently reduced by dexamethasone and by salicylazosulfapyridine. Two subcutaneous injections of indomethacin, spaced 24 h apart, produced a more chronic inflammation in the rat, with extensive mucosal injury and ulceration that lasted at least two weeks post-injection[46]. Inflammation was accompanied by increased mucosal permeability and by translocation of bacteria into the mesenteric lymph nodes, liver and spleen. This NSAID-induced colitis may be a very useful model of IBD.

"IMMUNE" MODELS

Many of the most widely used animal models have been developed as a result of manipulations of the immune system. The early studies of LeVeen et al.[47] indicated that dogs injected with anti-colon antibodies developed transient bloody diarrhea. Rosenberg, Bicks and coworkers developed models for ulcerative colitis in swine and in guinea pigs which were based on a cell-mediated immune response to intracolonic administration of dinitrochlorobenzene (DNCB) to animals in which repeated skin exposures had developed a delayed hypersensitivity reaction[48-52]. The DNCB model was subsequently extended to include the rabbit[53]. Histologically, the changes produced by intrarectal infusion of DNCB into sensitized animals resemble those seen in ulcerative colitis with ulceration, crypt abscesses and granulocyte infiltration of the mucosa. Healing occurred if the hapten was not repeatedly administered, but repeated exposure to the hapten did result in the production of lymphocytes reactive to a saline extract of rabbit colon. A cell-mediated model has also been described in which alloimunization of guinea pigs with a mucosal protein in complete Freund's adjuvant results, over a period of weeks or months, in the appearance of scattered mucosal ulcerations and congestion, followed by mononuclear cell infiltration and granulomatous inflammation of the ileum and colon[54]. An interesting model, which may also have an autoimmune basis, has been recently described in which two cm segments of the distal ileum of rats are excised and left attached on the mesentery[55]. The retention, on the mesentery, of the excised ileal segment corresponded with the appearance after four to six weeks of para-anastomotic ileal ulcers. The authors speculate that the time necessary for lesion development and the presence of mononuclear cells and of occasional epithelioid granulomata in the bowel wall, favored an immune etiology, perhaps involving sensitization of the immune system by microbial or cellular elements in the isolated ileal segment.

An apparently chronic, immune colitis in rabbits has been described by Mee *et al.*[56]. In a previous study, the Auer technique was used to induce experimental colitis in rabbits by intravenous injection of preformed immune complexes into animals in which the colon had been locally irritated by instillation of 1% formalin[57]. Mee *et al.*[56] immunized rabbits with the common enterobacterial antigen of Kunin and then induced immune complex colitis. Although only three animals were examined, the authors described plasma cell and lymphocyte infiltrates, crypt abscesses and shortened, branched glands present six months after initiation of damage. They speculated that the initial, severe acute injury allowed access of luminal antigens, and that the combination of mucosal permeability and prior sensitization to bacterial antigen could lead to chronic disease. This interesting model has been used in investigations of eicosanoid metabolism in colitis[58,59] and of the effects of interleukins on colitis[60-62].

DEVELOPMENT OF THE TRINITROBENZENESULFONIC ACID (TNB) MODEL

Several years ago we became interested in the development of procedures which would convert the readily produced acute gastric damage into chronic gastritis or chronic ulcer. We had found, as had others[63], that repeated administration of a "barrier breaker" such as ethanol did not result in the development of chronic ulcers. Rather, the mucosa became tolerant to ethanol-induced damage and the extent of ulceration decreased. Those ulcers which were present after four weeks of daily administration of ethanol did not extend below the muscularis mucosae and the inflammatory infiltrate in these lesions was primarily composed of neutrophils and eosinophils. It was apparent that some additional factor was necessary to effect the conversion of acute erosions into chronic gastric ulcer or gastritis. Our working hypothesis, based on the concepts developed by Ward[64] and by Shorter[65], was as follows: (1) Chronic ulceroinflammatory disease (peptic ulcer or IBD) is initiated by acute damage, with consequent development of acute inflammation and increased mucosal permeability; (2) Chronic inflammation depends on entry into the lamina propria of a luminal antigen which is not adequately cleared by the mucosal immune system; (3) Conventional therapy may produce transient healing, but recurrence of acute injury or enhanced permeability in the presence of the inciting antigen will result in relapse of peptic ulcer or IBD.

Gastric Ulcer Studies

Our first studies involved an attempt to adapt the DNCB model of ulcerative colitis. The procedure, as developed for rabbits[53], involved skin sensitization with the hapten DNCB, and then intracolonic infusion with DNCB in a vehicle (acetone) which is also a damaging agent. We reasoned that orogastric intubation of the same mixture into a suitably sensitized rat might produce a condition resembling either gastritis or gastric ulcer. We used the hapten TNB rather than DNCB, and dissolved it in 1 ml of 50% ethanol. TNB was selected for trivial reasons - we had no DNCB readily available - and ethanol was used as a vehicle because we were familiar with its effects on the gastric mucosa.

When 50 mg of TNB in 0.75 ml of either 40% or 100% ethanol was administered by orogastric intubation to fasted rats, the damage at days one and three was essentially identical to that seen following intubation with ethanol alone. By day eight after intubation, however, damage produced by ethanol alone had healed completely, while most animals intubated with ethanol/TNB contained one or more chronic-type ulcers which were preferentially located in the antral mucosa on the lesser curvature (Figure 1)[66]. These ulcers were sharply demarcated and the inflammatory cell mass extended through the muscularis mucosae and submucosa. Some chronic-type ulcers were also located in the fundic mucosa, near the fundus-forestomach junction. These fundic ulcers were often elongated, fissure ulcers rather than the circular ulcers which predominated in the antrum. All of these ulcers were surrounded by regions of gastritis. The primary difference, histologically, from chronic ulcers of humans, was the presence in the rat mucosae of numerous giant cells and associated granulomata (Figure 2).

Figure 1. Chronic-type antral ulcer 10 days after intubation with 1 ml 50% ethanol containing 50 mg TNB. Epon section stained with toluidine blue. Magnification x 64.

Chronic-type ulcers were found in both immunized (TNB-bovine gamma globulin in complete Freund's adjuvant) and non-immunized animals demonstrating that prior sensitization was not necessary for production of chronic-type gastric ulcer. Ulcers were variable in size, with typical diameters being between two and five mm. The ulcers did not persist for more than about three weeks, although inflammation and alterations in microvascular architecture persisted for longer periods[67].

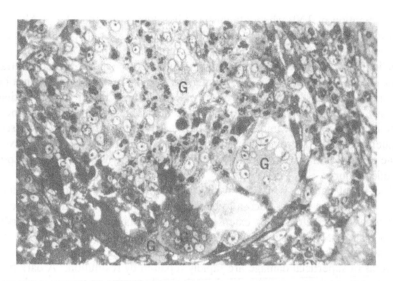

Figure 2. Granuloma and Langhan's type-giant cells (G) from antral ulcer, eight days after intubation with ethanol/TNB. Epon section stained with toluidine blue. Magnification x 380.

Induction of TNB Colitis

The presence of giant cells and granulomata and the appearance of fissure ulcers in the gastric model suggested that the chronic inflammation which developed following administration of TNB in ethanol had certain similarities to Crohn's disease. To determine whether a similar chronic inflammation could be induced in the colon we instilled various doses of TNB in ethanol into the distal colon of virgin, female Sprague-Dawley rats which were lightly anesthetized with ether. We obtained the most effective and reproducible inflammation with a dose of 30 mg TNB in 0.25 ml 50% ethanol which was delivered to a point eight cm proximal to the anus (at approximately the splenic flexure) by a rubber catheter attached to a 1 ml syringe[68]. The process takes two or three seconds and does not require fasting, enemas or any prior immunological manipulation. Others have used different doses of TNB or different concentrations and volumes of ethanol with different strains of rat[69,70].

In a separate study we sought to determine whether the damaging agent (ethanol) and the hapten had to be presented simultaneously to produce the chronic inflammation[71]. We were seeking to determine, in a model system, whether prexisting acute damage could be converted to chronic ulceroinflammatory disease by subsequent appearance of an inciting antigen or hapten. We administered TNB in saline at times ranging from 15 min to 24 h after intracolonic infusion of 0.25 ml 50% ethanol. TNB can be administered at least 24 h after acute damage was induced without any lessening in the severity of chronic inflammation assessed two weeks later.

The hypothesis that increased intestinal permeability is an etiological factor in Crohn's disease has gained support recently with convincing demonstration not only of increased intestinal permeability in patients with Crohn's disease but also in clinically unaffected relatives[72-74]. The primary role of ethanol in the TNB model may be to increased mucosal permeability. No studies have been carried out on the relative retention of TNB when administered in saline or in ethanol, but Allgayer *et al.*[70] have shown that serosal injection of TNB in saline at multiple points in the colon wall produced similar damage to intracolonic infusion of ethanol/TNB. The tissue destruction and recruitment of neutrophils which follows exposure to concentrated ethanol may be essential to the development of chronic inflammation. It has been argued that TNB causes damage by direct mucosal toxicity rather than by acting as a hapten[75]. If so, this would lessen the value of this model in understanding the immunological events that initiate human IBD. However, for reasons discussed in subsequent sections, there is substantial evidence that the model does have an immune etiology and that the progression of disease activity is not a simple response to chemical injury.

The initial damage is due primarily to the vascular congestion and necrosis induced by exposure of the mucosa to concentrated ethanol. The mucosa becomes thickened and edematous with extensive areas of inflammation and hemorrhage (Figures 3 and 4). During the first few hours after exposure to ethanol/TNB there is extensive damage to the luminal epithelium over the distal eight cm of colon, but the regions of superficial damage are rapidly repaired by restitution. A dark brown pseudomembrane forms over sites of ulceration and remains on the mucosal surface for about one week (Figure 4).

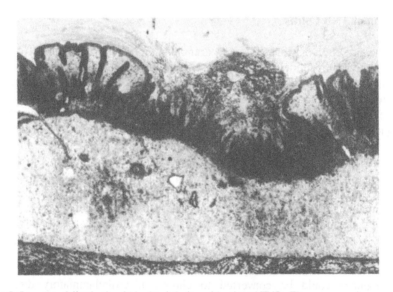

Figure 3. Segment of distal colon 24 h after infusion with ethanol/TNB. The mucosa, submucosa, and muscularis are edematous and there is extensive infiltration by neutrophils. Epon section stained with toluidine blue. Magnification x 70.

Sites of inflammation varied from the perirectal region to about seven cm proximal to the anus but no damage was detected proximal to the splenic flexure. Cobblestoning, in which islands of grossly normal mucosa were surrounded by ulcers, was common (Figures 4 and 5). Focal ulcers often circumscribed the colon and extended more than three cm along its length. At such sites the circumference of the colon was often enlarged to more than three times normal. In addition to severe transmural inflammation, localized accumulations of serosal fat and fibrinous adhesions to the small bowel and uterine horns were frequent. Incomplete bowel

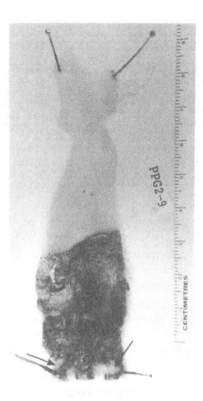

Figure 4. Severely damaged colon one week after intrarectal infusion of ethanol/TNB. Arrows indicate distal islands of grossly normal mucosa. Ulcerated regions are covered by a pseudomembrane in this specimen.

obstruction (marked narrowing of the lumen of the colon adjacent to inflamed sites)was present in more than 50% of the animals studied over an eight-week period after infusion of ethanol/TNB. Perforation did not occur although fistulae between the colon and small bowel were present in a few (less than 1%) of the animals. Body weight decreased about 8-10% during the first week after induction of disease, after which growth rates returned to normal.

Assessment of Inflammation

Damage is quantified by three methods: (1) a damage score based on the extent of gross morphologic damage when the colon is pinned out and viewed at low magnification with a dissecting microscope; (2) wet weight of the distal eight cm colon; (3) colonic myeloperoxidase (MPO) activity is determined[76,77]. There is a highly significant linear correlation between these three indices of inflammation. A lesion score can be assigned and wet weight determined within a few minutes.

Animals which received only ethanol or TNB in saline developed acute colonic damage but this was sufficiently resolved by within one week that these groups were not significantly different from controls. In the groups which received ethanol/TNB, damage scores and colon weight remained significantly above control levels for at least eight weeks while MPO activity was significantly elevated during weeks one to seven.

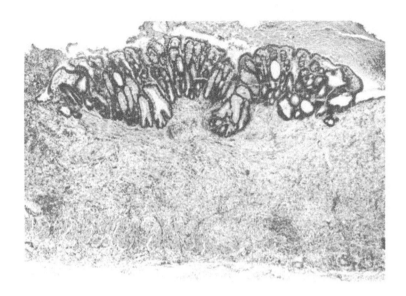

Figure 5. "Island" of distorted but grossly intact mucosa from colon four weeks after intrarectal administration of ethanol/TNB. Paraffin section stained with hematoxylin and eosin. Magnification x 45.

Changes in Inflammatory Cells During the Progression of Inflammation

Chronic inflammation of the colon characterized by accumulation of neutrophils, monocytes and macrophages, and proliferation of fibroblasts and small vessels was established by one week after ethanol/TNB infusion and persisted for at least eight weeks. A chronic, persistent inflammatory response was never established by exposure to ethanol alone. Giant cells and granulomata were much less frequent than in the gastric mucosa after ethanol/TNB but were present in histological sections from about 50% of the animals at two and three weeks after ethanol/TNB (Figures 6 and 7).

The numbers of connective tissue mast cells (CMC) changed in a characteristic fashion during the progression of inflammation in the TNB model. Intracolonic administration of ethanol/TNB caused a reduction in mast cell numbers during the first two days. A mast cell stabilizer, ketotifen, reduced the severity of acute inflammation[78], suggesting that ethanol-induced release of mast cell mediators may be important in the development of the initial inflammation. We found that the periods of most severe chronic ulceration and inflammation in this model (weeks two to four), however, were associated with a five-fold increase in the number of CMC. This is the opposite to what occurred in animals exposed only to ethanol (acute damage) where few mast cells were present during the periods of most severe acute inflammation and ulceration and maximum numbers were found after the tissue returned to a macroscopically normal appearance. In the chronic model there was a marked drop in mast cell numbers and a return to normal values during the fifth week after ethanol/TNB.

Mast cell involvement has been previously noted in chronic inflammatory states of the gastrointestinal tract. For instance, increased numbers of mast cells have been reported in Crohn's disease, ulcerative colitis, celiac sprue and diverticular disease[79-83].

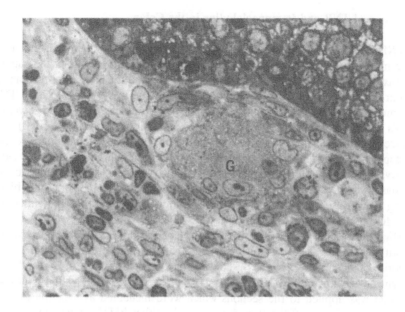

Figure 6. Multinuclear, giant cell (G) at base of mucosa, from colon two weeks after intrarectal administration of ethanol/TNB. Epon section stained with toluidine blue. Magnification x 770.

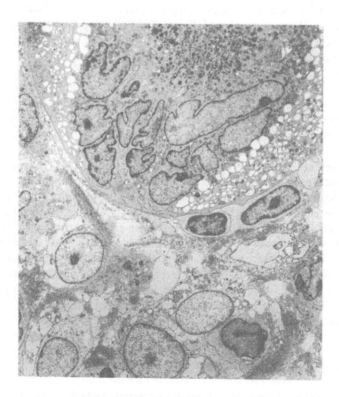

Figure 7. Transmission electron micrograph of a Langhan's-type giant cell from the submucosa of the distal colon, two weeks after administration of ethanol/TNB. Magnification x 3,000.

Localization of TNB and Immunoglobulins in the Colon

TNB was localized in tissue fixed one hour to five weeks following ethanol/TNB to determine the sites of localization and the persistence in the tissue of the hapten. Both wax- and plastic(Epon)-embedded tissues were used for immunohistochemical procedures. An indirect avidin-biotin staining technique was used to localize TNB and immunoglobulins IgM, IgG, and IgA, which were studied in tissue fixed at weekly intervals from week one to week four. Horseradish peroxidase (HRP) and fluorescein isothiocyanate (FITC) were used as labels. The plastic-embedded sections were deplasticized by immersing sections for 30 min in sodium methylate. All slides were trypsinized for 10 min and, when HRP was used as a label, endogenous peroxidase was blocked with hydrogen peroxide in methanol. An avidin-biotin blocking kit was used to block any nonspecific binding of the biotinylated antibodies and avidin-bound labels.

High titer rabbit anti-TNB serum was used as the primary antibody to localize TNB within sections. Sections were incubated with normal serum of the species of the biotinylated antibody (goat), rinsed and incubated for one h at $37°$ and then overnight at $4°$ with primary antibody (rabbit anti-TNB)[84]. Biotinylated goat anti-rabbit IgG was applied to the sections which were incubated as with the primary antibody. After rinsing off the secondary antibody, either FITC-conjugated avidin or HRP-conjugated avidin was added as a label.

At one day, TNB was located diffusely throughout the mucosa, muscularis mucosae and submucosa. The muscularis externa stained positively for TNB throughout weeks one to four but generalized staining of the mucosa and the muscularis mucosae was not present during weeks three and four. TNB was localized in lipid bodies of large macrophages (Figure 8) during the period from weeks one to five. Similar cells have been described for Dvorak et al.[85] as sites of active metabolism of arachidonic acid in tissues from IBD patients.

Plasma cell populations changed in a characteristic manner during the progression of the ulceroinflammatory response. IgA^+ plasma cell numbers were elevated by one week but did not appear to change during subsequent weeks. IgG^+ plasma cells were the most numerous type during the first two weeks but decreased in number during weeks three and four. IgM^+ plasma cells were low in number during weeks one and two but were the most numerous type during weeks three and four (followed in decreasing order by IgG and IgA^+ cells). All three types of plasma cell were present in numbers well above controls at all times studied.

The pattern of plasma cell numbers seen in tissues from rats exposed to ethanol/TNB and the elevated levels of all three types studied are similar to those described previously in tissues from patients with IBD[86,87]. Increased numbers of plasma cells in the mucosa were probably associated with increased levels of antibody production. However, there were no signs of glomerulonephritis in animals sacrificed from one day to five weeks following ethanol/TNB and immunocytochemical examination of frozen sections failed to demonstrate IgG deposition in glomerular capillaries. Therefore, from these very preliminary data, there was no evidence of immune complex involvement in the ethanol/TNB-induced disease.

Comparison of Different Haptens

We have carried out a separate study to determine whether other haptens might be equally effective at inducing chronic inflammation in the colon[88]. All of the

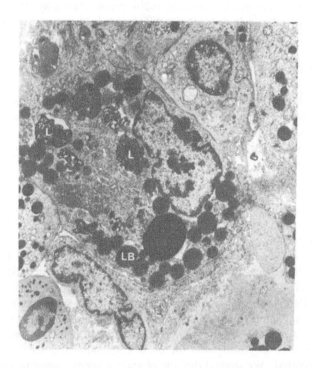

Figure 8. Transmission electron micrograph of a TNB-positive macrophage from the upper layers of the inflammatory cell mass of an ulcer from a colon fixed two weeks after intracolonic administration of ethanol/TNB. The numerous, large, electron-dense lipid bodies (LB) stained positively for TNB. L, lysosome. Magnification x 8,200.

haptens tested were administered by intrarectal infusion of haptens at total doses of 5.11×10^{-5} mol (equivalent to 15 mg of TNB) in 0.25 ml of 50% ethanol. We compared the effects of each hapten on wet weight of the colon, lesion score and MPO activity in animals killed one week after administration. TNB produced significantly more severe inflammation and ulceration, by all three criteria, than did 2,4-dinitrophenol, 2,4-dinitrobenzene sulfonic acid, picryl chloride or ethanol alone. Only TNB and DNCB consistently produced chronic inflammatory responses and only the MPO levels in the tissue exposed to TNB were significantly higher than those in tissue exposed to DNCB. TNB and DNCB were thus similar in their ability to induce chronic inflammation. Certainly, the inflammation produced by DNCB in 50% ethanol was more severe than that reported by Rabin and Rogers[53] when 100 mg DNCB in acetone was instilled into the rabbit colon. This may reflect a difference in the effect of the vehicle or, perhaps, differences in species susceptibility.

Experimental Manipulation of Disease Activity

We have tested the effects of a variety of agents currently used in therapy of IBD on the progression of disease in the TNB model. Damage was not reduced by treatment after the induction of damage by indomethacin, aspirin, sulfasalazine, 5-aminosalicyclic acid, or prednisolone[69,89]. Pretreatment with cytoprotective doses of prostaglandins (PG) does reduce the severity of disease (assessed by lesion score and colon weight but not by MPO activity) which develops subsequent to ethanol/TNB[77] but administration of PG after disease induction did not accelerate healing in our study, although others have described beneficial effects[70]. Recent studies in which recently developed agents have been tested on the TNB model are discussed in a separate chapter.

Immunosuppressants have been shown in some studies to be effective in the treatment of inflammatory bowel disease[90-92]. In addition, therapeutic agents used in treatment of IBD may act via effects on the immune system. Corticosteroids have numerous effects on the immune system, including the suppression of lymphocyte proliferation and interferon production, and the induction of mast cell degranulation. Cyclosporine A, another immunosuppressant, reduced damage scores 14 days after TNB treatment[93], suggesting that the chronic inflammation seen in this model is immune-mediated. We studied the role of antigen-specific immune processes in the pathogenesis of the TNB model by determining if induction of humoral tolerance to TNB would affect the severity of disease. When the severity of inflammation was assessed 2 weeks after infusion of ethanol/TNB, all three indices of inflammation (lesion score, colon weight, MPO activity) were significantly lowered in tolerized animals. If a causal antigen responsible for the induction of IBD in humans could be identified, similar immunological manipulation may provide an effective avenue for treatment of these diseases. Such an antigen might induce IBD subsequent to acute inflammation and increased mucosal permeability, as suggested by the positive correlation between the severity of colonic inflammation and the extent of bacterial translocation in the TNB model[94].

SUMMARY

Like all of the other animal models for IBD, the TNB model has certain advantages and weaknesses. The model provides an inexact stimulation of the chronic transmural inflammation and histopathological features observed in human Crohn's colitis. Transmural inflammation, lymphoid aggregates, the characteristic chronic inflammatory infiltrate, and the presence of occasional giant cells and granulomata are reminiscent of Crohn's disease. Similarly, at a macroscopic level, the large persistent lesion sites, thickened, edematous affected segments, fibrinous adhesions and adherent pericolic mesenteric fat are suggestive of IBD. Cobblestoning is observed in the distal bowel as a result of transverse and longitudinal fissure-ulcers, and a sharp demarcation exists between the lesion site and proximal, grossly inflamed mucosa. However, granulomas are infrequently seen in the colonic model, and there is no histological evidence of disease beyond the site exposed to ethanol/TNB. Although, as Rachmilewitz et al.[95] have pointed out, the model does not accurately mimic all the histopathological features of Crohn's disease, it is representative of severe, transmural chronic colonic inflammation.

The TNB model does not go through periodic stages of remission and relapse. We have not been able to maintain disease severity at levels above that of control groups receiving only an initial infusion of ethanol/TNB by subsequent infusion of 30 mg TNB in 0.25 saline at two-week intervals for a period of twelve weeks. Relapse can only be mimicked by reinfusion with ethanol/TNB. In the initial three days, there is a 50-80% decrease in food intake; a decrease which was prevented by indomethacin but not by a 5-lipoxygenase inhibitor[96]. The TNB model may thus be useful in the study of IBD-associated anorexia, suggesting that it is prostaglandin mediated. Weight loss following ethanol/TNB is transient, and after an initial loss of about 10% body weight, animals infused with ethanol/TNB gain weight at the same rate as controls. Diarrhea is, however, present in almost all animals for about four weeks and persists in some for at least twelve weeks.

In its present form, the TNB model offers certain significant advantages over other animal models of IBD. The animal used, the rat, is inexpensive and widely available. The procedure is simple and for the most part very reproducible. The model can be adapted to work in other animals although we have had only occasional successes in attempts to induce TNB colitis in mice. The principal advantage to the model is perhaps utilitarian. The combination of ethanol and TNB in the colon of the rat produces an IBD-like condition of sufficient severity, duration and reproducibility to make it suitable for experiments in which potential new treatments and agents are to be assessed.

REFERENCES

1. J.K. Roche, C. Fiocchi, and K. Youngman, Sensitization to epithelial antigens in chronic mucosal inflammatory disease, *J Clin Invest.* 75:522 (1985).
2. A.M. Dvorak, What is the evidence for damaged enteric autonomic nerves in Crohn's disease, in: "Inflammatory Bowel Disease: Current Status and Future Approach," R.P. MacDermott, ed., Elsevier Science Publishers B.V., Amsterdam: (1988).
3. W. Strober, Animal models of inflammatory bowel disease - an overview, *Dig Dis Sci.* 30:3S (1985).
4. B. Macpherson, and C.J. Pfeiffer, Experimental colitis, *Digestion.* 14:424 (1976).
5. J.F. Mayberry, J. Rhodes, and R.V. Heatley, Infections which cause ileocolic disease in animals: are they relevant to Crohn's disease, *Gastroenterology.* 78:1080 (1980).
6. R. Lindberg, Pathology of equine granulomatous enteritis, *J Comp Path.* 94:233 (1984).
7. J.L. Madara, D.K. Podolsky, N.W. King, P.K. Sehgal, R. Moore, and H.S. Winter, Characterization of spontaneous colitis in cotton-top tamarins (*Saguinus oedipus*) and its response to sulfasalazine, *Gastroenterology.* 88:13 (1985).
8. H.S. Winter, P.M. Crum, N.W. King, P.K. Sehgal, and J.K. Roche, Expression of immune sensitization to epithelial cell-associated components in the cotton-top tamarin: a model of chronic ulcerative colitis, *Gastroenterology.* 97:1075 (1989).

9. W. Strober, and S.P. James, The immunologic basis of inflammatory bowel disease, *J Clin Immunol.* 6:415 (1986).

10. B.B. Crohn, L. Ginsberg, and G.D. Oppenheimer, Regional ileitis: a pathological and clinical entity, *J Amer Med Assoc.* 99:1323 (1932).

11. D.B. Sachar, Animal transmission models, *Mt Sinai J Med.* 50:166 (1983).

12. W.L. Beeken, Transmissible agents in inflammatory bowel disease: 1980, *Med Clin N Am.* 64:1021 (1980).

13. W.L. Beeken, D.N. Mitchell, and D.R. Cave, Evidence for a transmissible agent in Crohn's disease, *Clin Gastroenterology.* 5:289 (1976).

14. K.M. Das, I. Valenzuela, S.E. Williams, R. Soeiro, A.S. Kadish, and S.G. Baum, Studies on the etiology of Crohn's disease using athymic nude mice, *Gastrotenterology.* 84:364 (1983).

15. J.F. Collins, R.G. Strickland, E.L. Kekahbah, M.H. Arthur, F. Naeim, and G.L. Gitnick, Histological response and antigen transmission in the lymph nodes of athymic nu/nu mice inoculated with Crohn's disease tissue filtrates, *Gut.* 29:983 (1988).

16. H.C. Walvoort, G.E. Fazzi, and A.S. Pena, Seroreactivity of patients with Crohn's disease with lymph nodes of primed nude mice is independent of the tissue used for priming, *Gastroenterology.* 97:1097 (1989).

17. R.J. Chiodini, H.J. Van Kruiningen, W.R. Thayer, R.S. Merkal, and J.A. Coutu, Possible role of mycobacteria in inflammatory bowel disease. I. An unclassified Mycobacterium species isolated from patients with Crohn's disease, *Dig Dis Sci.* 29:1073 (1984).

18. W.R. Thayer, J.A. Coutu, R.J. Chiodini, H.J. Van Kruiningen, and R.S. Merkal, Possible role of mycobacteria in inflammatory bowel disease. II. Mycobacterial antibodies in Crohn's disease, *Dig Dis Sci.* 29:1080 (1984).

19. R.J. Chiodini, H.J. Van Kruiningen, R.S. Merkal, W.R. Thayer, and J.A. Coutu, Characteristics of an unclassified Mycobacterium species isolated from patients with Crohn's disease, *J Clin Microbiol.* 20:966 (1984).

20. H.J. Van Kruiningen, R.J. Chiodini, W.R. Thayer, J.A. Coutu, R.S. Merkal, and P.L. Runnels, Experimental disease in infant goats induced by a Mycobacterium isolated from a patient with Crohn's disease. A preliminary report, *Dig Dis Sci.* 31:1351 (1986).

21. H.H. Yoshimura, D.Y. Graham, M.K. Estes, and R.S. Merkal, Investigation of association of mycobacteria with inflammatory bowel disease by nucleic acid hybridization, *J Clin Microbiol.* 25:45 (1987).

22. K. Kobayashi, M.J. Blaser, and W.R. Brown, Immunohistochemical examination for mycobacteria in intestinal tissues from patients with Crohn's disease, *Gastroenterology.* 96:1009 (1989).

23. D.H. Graham, D.C. Markesich, and H.H. Yoshimura, Mycobacteria as the cause of Crohn's disease, *Gastroenterology.* 97:1354 (1989).

24. M.J. Blaser, W.R. Brown, and K. Kobayashi, Correspondence. Reply to letter by D.Y. Graham *et al.*, *Gastroenterology.* 97:1355 (1989).

25. R. Marcus, and J. Watt, Seaweeds and ulcerative colitis in laboratory animals, *Lancet.* 2:489 (1969).

26. J. Watt, and R. Marcus, Carrageenan-induced ulceration of the large intestine in the guinea pig, *Gut.* 12:164 (1971).

27. A.B. Onderdonk, J.G. Bartlett, Bacteriological studies of experimental ulcerative colitis, *Am J Clin Nutr.* 32:258 (1979).

28. A.B. Onderdonk, M.L. Franklin, and R.L. Cisneros, Production of experimental ulcerative colitis in gnotobiotic guinea pigs with simplified microflora, *Infect Immun.* 32:225 (1981).

29. A.B. Onderdonk, R.L. Cisneros, and R.T. Bronson, Enhancement of experimental ulcerative colitis by immunization with *Bacteroides vulgatus*, *Infect Immun.* 42:783 (1983).

30. A.B. Onderdonk, R.M. Steeves, R.L. Cisneros, and R.T. Bronson, Adoptive transfer of immune enhancement of experimental ulcerative colitis, *Infect Immun.* 46:64 (1984).

31. J. Watt, and R. Marcus, Experimental ulcerative disease of the colon in animals, *Gut.* 14:506 (1973).

32. R.B. Sartor, W.J. Cromarties, D.W. Powell, and J.H. Schwab, Granulomatous enterocolitis induced in rats by purified bacterial cell wall fragments, *Gastroenterology.* 89:587 (1985).

33. T. Yamada, R.B. Sartor, S. Marshall, R.D. Specian, and M.B. Grisham, Mucosal injury and inflammation in a model of chronic granulomatous colitis in rats, *Gastroenterology.* 104:759 (1993).

34. R.B. Sartor, T.M. Bond, and J.H. Schwab, Systemic uptake and intestinal inflammatory effects of luminal bacterial cell wall polymers in rats with acute colonic injury, *Infect Immun.* 56:2101 (1988).

35. C.J. Woolverton, E.J. Chapman, P.B. Carter, and R.B. Sartor, Systemic immunoregulation in germ free mice fed bacterial cell wall polymers, *in:* "Inflammatory Bowel Disease: Current Status and Future Approach," R.P. MacDermott, ed., Elsevier Science Publishers B.V., Amsterdam (1988).

36. S. Okabe, K. Nakamura, K. Takeuchi, and K. Takagi, Effects of antacid, antipeptic and anticholinergic agents on the development and healing of duodenal ulcers in the rat, *Digestion.* 10:113 (1974).

37. R. Fabia, R. Willen, A. Ar'Rajab, R. Andersson, B. Ahren, and S. Bengmark, Acetic acid-induced colitis in the rat: A reproducible experimental model for acute ulcerative colitis, *Eur Surg Res.* 24:211 (1992).

38. D.R.E. Reeve, An electron microscopical study of chronic ulcers of the colon in rats. *Br J Exp Path.* 58:63 (1977).

39. C.H. Weischer, W. Jahn, and I. Szelenyi, A new model for inflammatory colonic disease induced by capsaicin in rats. *Agents Actions.* 26:222 (1989).

40. B.R. MacPherson, and C.J. Pfeiffer, Experimental production of diffuse colitis in rats, *Digestion.* 17:135 (1978).

41. P.G. Davey, E.E. McGuiness, S. Nichol, and K.G. Wormsley, Experimental colitis and colonic cancer in the rat, *Mt Sinai J Med.* 52:209 (1984).

42. P. Sharon, and W.F. Stenson, Metabolism of arachidonic acid in acetic acid colitis in rats. Similarity to human inflammatory bowel disease, *Gastroenterology.* 88:55 (1985).

43. D.J. Fretland, S. Levin, B.S. Tsai *et al.*, The effect of leukotriene-B$_4$ receptor antagonist, SC-41930, on acetic acid-induced colonic inflammation, *Agents Actions.* 27:395 (1989).

44. T.H.M. Stewart, C. Hetenyi, H. Rowsell, and M. Orizaga, Ulcerative enterocolitis in dogs induced by drugs, *J. Pathol.* 131:363 (1980).

45. P. Del Soldato, D. Foschi, L. Varin, and S. Daniotti, Indomethacin-induced intestinal ulcers in rats: effects of salicylazosulfapyridine and dexamethasone, *Agents Actions.* 16:393 (1985).

46. T. Yamada, E. Deitch, R.D. Specian, M.A. Perry, R.B. Sartor, and M.B. Grisham, Mechanisms of acute and chronic intestinal inflammation induced by indomethacin, *Inflammation.* 17:641 (1993).

47. H.H. LeVeen, G. Falk, and B. Schatman, Experimental ulcerative colitis produced by anticolon sera. *Ann Surg.* 154:275 (1961).

48. E.W. Rosenberg, and R.W. Fischer, DNCB allergy in the guinea pig colon, *Arch Dermatol.* 89:99 (1964).

49. R.O. Bicks, and E.W. Rosenberg, A chronic delayed hypersensitivity reaction in the guinea pig colon *Gastroenterology.* 46:543 (1964).

50. R.O. Bicks. G. Brown, H.D. Hickey, and E.W. Rosenberg, Further observations on a delayed hypersensitivity reaction in the guinea pig colon, *Gastroenterology.* 48:425 (1965).

51. R.O. Bicks, M.M. Azar, E.W. Rosenberg, W.J. Dunham, and J. Luther, Delayed hypersensitivity reactions in the intestinal tract. I. Studies of 2,4-dinitrochlorobenzene-caused guinea pig and swine colon lesions, *Gastroenterology.* 53:422 (1967).

52. R.O. Bicks, M.M. Azar, and E.W. Rosenberg, Delayed hypersensitivity reactions in the intestinal tract. II. Small intestinal lesions associated with xylose malabsorption, *Gastroenterology.* 53:437 (1967).

53. B.S. Rabin, and S.J. Rogers, A cell-mediated immune model of inflammatory bowel disease in the rabbit, *Gastroenterology.* 75:29 (1978).

54. M.S. Nemirovsky, and J.S. Hugon, Immunopathology of guinea-pig autoimmune enterocolitis induced by alloimmunisation with an intestinal protein, *Gut.* 27:1434 (1986).

55. G. Wickbom, S.-O. Cha, and P. O'Leary, Experimentally induced ileal ulcers in rats. A model to study non-specific inflammatory bowel disease, *Scand J Gastroenterol.* 23:808 (1988).

56. A.S. Mee, J.E. McLaughlin, H.J.F. Hodgson, and D.P. Jewell, Chronic immune colitis in rabbits, *Gut.* 20:1 (1979).

57. H.J.F. Hodgson, B.J. Potter, J. Skinner, and D.P. Jewell, Immune-complex mediated colitis in rabbits, *Gut.* 19:225 (1978).

58. H.W. Kao, and R.D. Zipser, Exaggerated prostaglandin production by colonic smooth muscle in rabbit colitis, *Dig Dis Sci.* 33:697 (1988).

59. N.K. Boughton-Smith, and B.J.R. Whittle, The role of eicosanoids in animal models of inflammatory bowel disease, *in:* "Inflammatory Bowel Diseases - Basic Research and Clinical Implications," H. Goebell, B.M. Peskar, H. Malchow, eds., MTP Press Ltd., Lancaster, England (1988).

60. F. Cominelli, C.C. Nast, R. Lierena, C.A. Dinarello, and R.D. Zipser, Interleukin 1 suppresses inflammation in rabbit colitis. Mediation by endogenous prostaglandins, *J Clin Invest.* 85:582 1990).

61. F. Cominelli, C.C. Nast, B.D. Clark *et al.*, Interleukin 1 (IL-1) gene expression, synthesis and effect on specific IL-1 receptor blockade in rabbit immune complex colitis, *J Clin Invest.* 86:972 (1990).

62. F. Cominelli, C.C. Nast, A. Duchini, and M. Lee, Recombinant interleukin-1 receptor antagonist blocks the proinflammatory activity of endogenous interleukin-1 in rabbit immune colitis, *Gastroenterology.* 103:65 (1992).

63. K.J. Ivey, A. Tarnawski, J. Stachura, H. Werner, T. Mach, and M. Burks, The induction of gastric mucosal tolerance to alcohol by chronic administration, *J Lab Clin Med.* 96:922 (1980).

64. M. Ward, The pathogenesis of Crohn's disease, *Lancet.* 903 (1977).

65. R.G. Shorter, K.A. Huizenga, and R.J. Spencer, A working hypothesis for the etiology and pathogenesis of nonspecific inflammatory bowel disease, *Am J. Dig Dis.* 17:1024 (1972).

66. G.P. Morris, L. Rebeiro, M.S. Herridge, M. Szewczuk, and W.T. Depew, An animal model for chronic granulomatous inflammation of the stomach and colon, *Gastroenterology.* 86:A1188 (1984).

67. T.T. Hynna, and G.P. Morris, Gastric mucosal microcirculation in a chronic ulcer model, *Gastroenterology.* 94:A199 (1988).

68. G.P. Morris, P.L. Beck, M.S. Herridge, W.T. Depew, M.R. Szewczuk, and J.L. Wallace, Hapten-induced model of chronic inflammation and ulceration in the rat colon, *Gastroenterology.* 96:795 (1989).

69. N.K. Boughton-Smith, J.L. Wallace, G.P. Morris, and B.J.R. Whittle, The effect of anti-inflammatory drugs on eicosanoid formation in a chronic model of inflammatory bowel disease in the rat. *Br J Pharmacol.* 94:65 (1988).

70. H. Allgayer, K. Deschryver, and W.F. Stenson, Treatment with 16,16'-dimethyl prostaglandin E2 before and after induction of colitis with trinitrobenzenesulfonic acid in rats decreases inflammation, *Gastroenterology.* 96:1290 (1989).

71. P.L. Beck, and G.P. Morris, Acute ulceration and inflammation is required for the induction of chronic ulceration and inflammation in the colon of the rat, *Gastroenterology.* 96:A35 (1989).

72. D. Hollander, Crohn's disease--A permeability disorder of the tight junction, *Gut.* 29:1621 (1988).

73. G. Olaison, P. Leandersson, R. Sjodahl, and C. Tagesson, Intestinal permeability to polyethyleneglycol 600 in Crohn's disease. Perioperative determination in a defined segment of the small intestine, *Gut.* 29:196 (1988).

74. D. Hollander, C.M. Vadheim, E. Brettholz, G.M. Petersen, T. Delahunty, and J.I. Rotter, Increased intestinal permeability in patients with Crohn's disease and their relatives. A possible etiologic factor, *Ann Intern Med.* 105:883 (1986).

75. T. Yamada, S. Marshall, R.D. Specian, and M.B. Grisham, A comparative analysis of two models of colitis in rats, *Gastroenterology.* 102:1524 (1992).

76. S. Ahmad, J. Rausa, E. Jang, and E.E. Daniel, Calcium channel binding in nerves and muscle of canine small intestine, *Biochem Biophys Res Commun.* 159:119 (1989).

77. J.L. Wallace, W.K. MacNaughton, G.P. Morris, and P.L. Beck, Inhibition of leukotriene synthesis markedly accelerates healing in a rat model of inflammatory bowel disease, *Gastroenterology.* 96:29 (1989).

78. R. Eliakim, F. Karmeli, E. Okon, and D. Rachmilewitz, Ketotifen effectively prevents mucosal damage in experimental colitis, *Gut.* 33:1498 (1992).

79. K. Matsueda, J.J. Rimpila, H.E. Ford, B. Levin, and S.C. Kraft, Tissue mast cells in Crohn's disease and ulcerative colitis, *in:* "Recent Advances in Crohn's Disease," A.S. Pena, E.T. Weterman, C.C. Booth, W. Strober, eds., Martinus Nijhoff, The Hague (1981).

80. S.N. Rao, Mast cells as a component of the granuloma in Crohn's disease, *J. Pathol.* 109:79 (1972).

81. A.M. Dvorak, R.A. Monahan, J.E. Osage, and G.R. Dickersin, Crohn's disease: transmission electron microscopic studies. II. Immunologic inflammatory response. Alterations of mast cells, basophils, eosinophils, and the microvasculature, *Human Pathology.* 11:606 (1980).

82. M.N. Marsh, and J. Hinde, Inflammatory component of celiac sprue mucosa. I. Mast cells, basophils, and eosinophils, *Gastroenterology.* 89:92 (1985).

83. P. Ranlov, M.H. Nielsen, and J. Wanstrup, Ultrastructure of the ileum in Crohn's disease. Immune lesions and mastocytosis, *Scand J Gastroent.* 7:471 (1972).

84. P. Brandtzaeg, Prolonged incubation time in immunohistochemistry: Effects on fluorescence staining of immunoglobulins and epithelial components in ethanol- and formaldehyde-fixed paraffin-embedded tissues, *J. Histochem Cytochem.* 29:1302 (1981).

85. A.M. Dvorak, H.F. Dvorak, S.P. Peters *et al.*, Lipid bodies: cytoplasmic organelles important to arachidonate metabolism in macrophages and mast cells, *J Immunol.* 131:2965 (1983).

86. K. Baklien, and P. Brandtzaeg, Immunohistochemical characterization of local immunoglobulin formation in Crohn's disease of the ileum *Scand J. Gastroenterol.* 11:447 (1976).

87. P.C.M. Rosekrans, C.J.L.M. Meijer, A.M. Van der Wal, C.J. Cornelisse, J. Lindeman, Immunoglobulin containing cells in inflammatory bowel disease of the colon: a morphometric and immunohistochemical study, *Gut.* 21:941 (1980).

88. P.L. Beck, A.W. Wade, and G.P. Morris, Chronic inflammation in the rat colon: a comparison of different haptens, *Gastroenterology.* 96:A35 (1989).

89. G.P. Morris, M.S. Herridge, P.L. Beck, and L.R. Rebeiro, An animal model for colonic Crohn's disease, *Dig Dis Sci.* 30:389 (1985).

90. D.J.G. White, Cyclosporin A. Clinical pharmacology and therapeutic potential *Drugs.* 24:322 (1982).

91. D.B. Sachar, and D.H. Present, Immunotherapy in inflammatory bowel disease, *Med Clin N Amer.* 62:173 (1978).

92. J. Brynskov, L. Freund, O.O. Thomsen, C.B. Andersen, S.N. Rasmussen, and V. Binder, Treatment of refractory ulcerative colitis with cyclosporin enemas, *Lancet.* 1:721 (1989).

93. H. Hoshino, S. Sugiyama, A. Ohara, H. Goto, Y. Tsukamoto, and T. Ozawa T, Mechanism and prevention of chronic colonic inflammation with trinitrobenzene sulfonic acid in rats, *Clin Exp Pharmacol Physiol.* 19:717 (1992).

94. K.R. Gardiner, P.J. Erwin, N.H. Anderson, J.G. Barr, M.I. Halliday, and B.J. Rowlands, Colonic bacteria and bacterial translocation in experimental colitis, *Br J Surg.* 80:512 (1993).

95. D. Rachmilewitz, P.L. Simon, L.W. Schwartz, D.E. Griswold, J.D. Fondacaro, and M.A. Wasserman, Inflammatory mediators of experimental colitis in rats, *Gastroenterology.* 97:326 (1989).

96. K.J. McHugh, H.P. Weingarten, C. Keenan, J.L. Wallace, and S.M. Collins, On the suppression of food intake in experimental models of colitis in the rat, *Am J. Physiol Regul Integr Comp Physiol.* 264:R871 (1993).

LEUKOTRIENES IN HUMAN AND EXPERIMENTAL ULCERATIVE COLITIS

John L. Wallace, Gerald P. Morris, and
Catherine M. Keenan

In addition to the ulcerative diseases which affect the upper gastrointestinal tract, such as peptic ulcer, "stress" ulcer, and the gastropathy associated with the use of non-steroidal anti-inflammatory drugs, ulceration of the more distal regions of the gastrointestinal tract are also an important clinical problem. Inflammatory bowel disease (IBD), which is an umbrella term for Crohn's disease and ulcerative colitis, is characterized by chronic inflammation and ulceration of the gastrointestinal tract. In both of these subsets of IBD the affected region is most commonly the distal ileum and colon. While there are many clinical and histopathological differences between Crohn's disease and ulcerative colitis, they share the features of being idiopathic and being poorly responsive to existing therapy.

There are many theories as to the initiating factor or factors of IBD, but none have yet been generally accepted. One of the more widely accepted theories is that people with IBD have an abnormally permeable intestinal epithelium[1], allowing entry into the lamina propria of microbes, endotoxin, formylated peptides or other foreign substances which are poorly cleared by the mucosal immune system (Figure 1). These factors can trigger an acute inflammatory response, which under normal circumstances would result in removal of the invading organisms or substances. For unknown reasons the inflammatory response continues to develop in IBD leading to extensive tissue damage and loss of function. Regardless of what first initiates it, the acute inflammatory response is relatively uniform. The release of various mediators by cells resident in the lamina propria and by infiltrating cells leads to changes in vascular tone and permeability in the region, the release of reactive oxygen metabolites and proteases which can damage the surrounding tissue and to the recruitment of more inflammatory cells. Thus, the response to the initiating factor is amplified. This amplification stage may represent a rational target for the development of therapeutic agents. If the inflammatory response can be inhibited or controlled, it is possible that the tissue damage which normally accompanies it can be prevented. As stated by Stenson[2]: "*as long as the initiating event remains unknown, it is likely that further advances in the medical therapy of IBD will result from pharmacological modulation of inflammatory mediators*".

Neuroendocrinology of Gastrointestinal Ulceration
Edited by S. Szabo and Y. Taché, Plenum Press, New York, 1995

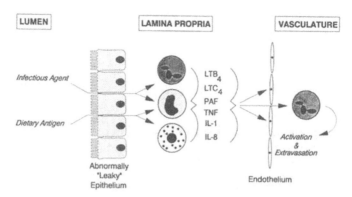

Figure 1. Schematic diagram illustrating the potential role of several inflammatory mediators, including leukotriene B_4, in amplifying the response of the colon to breach of the epithelial barrier by bacterial products. Endotoxin (LPS) and formylated peptides (such as f-methionyl-leucyl-phenylalanine; FMLP) can activate resident inflammatory cells within the lamina propria, including neutrophils and macrophages. Release of mediators from these cells results in changes in the vascular endothelium which promotes adherence and diapedesis of circulating granulocytes. Mediators such as leukotriene B_4 and interleukin-8 also have chemoattractant actions on neutrophils. Thus, the recruitment of granulocytes into the colonic tissue, with subsequent release of more inflammatory mediators, results in an amplification of the response to the bacterial products. Release of proteases and reactive oxygen metabolites from the infiltrating granulocytes may contribute to the tissue necrosis which characterizes ulcerative colitis.

In an inflammatory response such as that which accompanies a bout of colitis there is a plethora of mediators released which might contribute to the ulceration and inflammation which characterizes this condition (Figure 1). There is emerging evidence that the 5-lipoxygenase products of arachidonic acid, the leukotrienes, may play a particularly important role in the pathogenesis of human and experimental colitis. This evidence falls into the following categories 1) leukotrienes have biological actions consistent with a role in colitis; 2) leukotriene synthesis is elevated in human and experimental colitis; 3) leukotrienes can exacerbate colonic ulceration and inflammation; 4) inhibition of leukotriene synthesis or antagonism of the interaction of leukotrienes with their receptors leads to accelerated healing in colitis.

Biological Actions of Leukotrienes

The leukotrienes (LT) derived from arachidonic acid include LTB_4 and the peptido-leukotrienes, LTC_4, LTD_4 and LTE_4. Leukotriene B_4 is synthesized primarily by neutrophils and has potent chemotactic and chemokinetic effects on neutrophils. Thus, LTB_4 can play an important role in amplifying the inflammatory response since its release from activated neutrophils leads to recruitment of more neutrophils to that site. LTB_4 can also activate neutrophils to release superoxide and proteases and to aggregate. The peptido-leukotrienes have potent stimulatory effects on gastrointestinal smooth muscle, can increase mucus secretion and increase vascular permeability[3]. Furthermore, these leukotrienes also appear to be capable of increasing intestinal secretion and inhibiting absorption[4]. It is therefore possible that peptido-leukotrienes contribute to both the pathogenesis and the symptoms of colitis.

The importance of LTB_4 as a chemotactic factor in human ulcerative colitis was recently demonstrated by Lobos *et al.*[5], who showed that a lipid extractable factor in inflamed colonic tissue was capable of inducing chemotaxis in an *in vitro* Boyden chamber preparation. This lipid extract co-eluted with authentic LTB_4 when analyzed by high performance liquid chromatography. Furthermore, the chemotactic activity of the lipid extract was removed if it was pre-incubated a specific antibody to leukotriene B_4.

There are numerous cellular sources of the peptido-leukotrienes, but perhaps the most important in terms of a role in colitis are the mucosal mast cells and the endothelium. LTC_4, LTD_4 and LTE_4 are all capable of increasing the tone of vascular and non-vascular smooth muscle, and of increasing vascular permeability. This latter action may be important for promoting the movement of granulocytes from the intravascular to the extravascular compartment at a site of injury or infection. Elevated release of peptido-leukotrienes at a site of inflammation may contribute to the motor disturbances which accompany inflammation of the intestine.

Exacerbation of Colonic Ulceration by Leukotrienes

While there is some evidence to suggest that peptido-leukotrienes may play a role in the pathogenesis of acute ulceration in the upper gastrointestinal tract, there is not yet similar evidence to support such a role in the colon or ileum. There is, however, some evidence to suggest that LT B_4 can promote colonic ulceration and inflammation in the rat. Intracolonic administration of 40% ethanol resulted in widespread damage to the epithelium and an acute inflammatory response. Within three days, most of the inflammation had resolved and colonic ulcers persisted in only about a third of the animals. When two nmoles of LTB_4 was added to the 30% ethanol and the animals examined three days later, 90% exhibited macroscopically visible colonic ulceration (Figure 2) and there was extensive infiltration of peroxidase-containing granulocytes (Figure 3). In similar experiments in which two nmoles of either LTC_4, LTD_4 or LTB_5 was added to the 30% ethanol, significant augmentation of colonic ulceration was not observed. Addition of an equimolar concentration of another pro-inflammatory substance, FMLP, also did not produce colonic ulceration which differed in severity from that seen in rats receiving the ethanol alone (Figure 2).

Elevation of Leukotriene Synthesis in Colitis

There is ample evidence from both man and animals that the synthesis of leukotrienes, particularly LTB_4, is markedly elevated in colitis. For example, Sharon and Stenson[6] demonstrated that biopsies from ulcerative colitis patients could synthesize significantly more LTB_4 when incubated *in vitro* that biopsies from normal subjects. Peskar *et al.*[7] demonstrated a similar elevation in the capacity of colonic tissue from ulcerative colitis and Crohn's disease patients to synthesize peptido-leukotrienes. Using radiolabelled arachidonic acid as a substrate, Broughton-Smith *et al.*[8] demonstrated elevated formation of 12- and 15-hydroxyeicosatetraenoic acid (15-HETE) by samples of colon from IBD patients versus control samples. One problem with all studies in which tissue samples are incubated *in vitro* for measurement of eicosanoid release is that one assumes that because there is an increased capacity to synthesize leukotrienes under the conditions of the incubation (likely because of increased numbers of inflammatory cells), that increased synthesis occurs *in vivo*. This problem has been addressed by Lauritsen *et al.*[9]. Using *in vivo* equilibrium dialysis, they have demonstrated elevated production of LTB_4 in the rectum of ulcerative colitis patients and that the levels of LTB_4 correlated with severity of ulcerative colitis. Importantly, this increased production of LTB_4 was observed without any exogenous stimulus. While one must make the assumption that

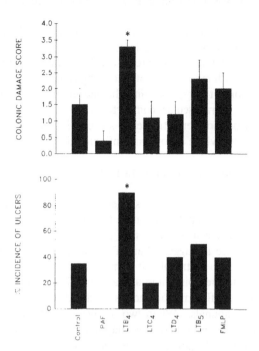

Figure 2. Effects of intracolonic administration of 30% ethanol plus an inflammatory mediator on colonic ulceration 72 h later. Groups of 10 rats each received 0.25 ml of 30% ethanol in which was dissolved two nmoles of either LTB$_4$, LTC$_4$, LTD$_4$, LTB$_5$ or the bacterially derived chemotactic peptide, FMLP. Control rats received the ethanol vehicle alone. After 72 h, the rats were killed and colonic damage was scored on a 0 (normal) to 5 (severe ulceration and inflammation) scale by an observer unaware of the treatment. The incidence of colonic ulceration in each group was also calculated. Only LTB$_4$ caused a significant increase in colonic damage score and ulcer incidence when compared to controls (*p<0.05).

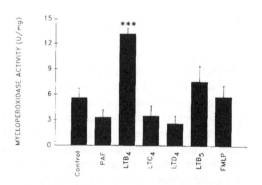

Figure 3. Same experimental protocol to that in Figure 2. Colonic myeloperoxidase activity was measured 72 h after intracolonic administration of 30% ethanol plus one of several inflammatory mediators. Only LTB$_4$ caused a significant increase in myeloperoxidase activity (an enzyme marker of granulocyte infiltration) when compared to the control rats receiving only the ethanol vehicle (***p<0.001).

the levels of LTB_4 entering the lumen are reflective of mucosal production, this technique nonetheless provides the first direct evidence on increased leukotriene production *in vivo*.

Various animal models of colitis have been developed and the changes in colonic leukotriene production which accompany the inflammation have been characterized. These studies are outlined in Table 1. Boughton-Smith and Whittle[10] demonstrated increased metabolism of arachidonic acid to lipoxygenase products (5,11,12-HETE) in an immune model of colitis in the guinea pig. Sharon and Stenson (1985) demonstrated changes in arachidonic acid similar to those observed in human colitis when they induced colitis in the rat by intracolonic administration of acetic acid. In that study, the primary lipoxygenase products detected were LTB_4 and 5-HETE. Several groups have used the trinitrobenzene sulfonic acid (TNB) model of colitis for studies of leukotriene synthesis by the colon. Boughton-Smith *et al.*[11] demonstrated increased formation of 12,15-HETE from labelled arachidonic acid, and increased formation of LTB_4 from endogenous substrate[12]. We have also measured elevated production of LTB_4 by colonic tissue from rats with colitis induced by TNB. These increases have been observed both with *in vitro* incubation of colonic tissue[13] and with *in vivo* intracolonic dialysis[14]. Zipser *et al.*[15] have used a rabbit model in which colitis is induced by immune complexes. In this model, they detected elevated production of LTB_4 and LTC_4 using the *in vivo* dialysis method.

Inhibition of Leukotriene Synthesis in Human and Experimental Colitis

While there can be little doubt that leukotriene synthesis is elevated in human and experimental colitis, there remains uncertainty regarding the possibility that drugs which are presently used in the treatment of ulcerative colitis reduce colonic inflammation via reduction of leukotriene synthesis. Several *in vitro* studies have provided evidence that the active moiety of sulfasalazine, 5-aminosalicylic acid (5-ASA), can inhibit 5-lipoxygenase at concentrations in the five to six mM range[16,17,18,19]. However, it is questionable that these concentrations are achieved within the colonic mucosa following oral administration of sulfasalazine or intracolonic administration of 5-ASA. Grisham and Granger[20] recently demonstrated, in the cat, that colonic interstitial and lymph concentrations of 5-ASA following luminal perfusion with a concentration of 10 mM only reached ~130 μM. These concentrations were therefore far below the range necessary for inhibition of 5-lipoxygenase to be achieved. Grisham and Granger[20] suggested that the anti-inflammatory effects of 5-ASA are likely attributable to its antioxidant and anti-peroxidase action, which are observed *in vitro* at concentrations in the five to twenty-five μM range.

The beneficial actions of corticosteroids in IBD have also been attributed to inhibition of leukotriene synthesis[2]. These drugs can inhibit liberation of arachidonic acid from membrane phospholipids by inducing the synthesis of a family of proteins (lipocortins) which inhibit the activity of phospholipase A_2. There are data from *in vivo* studies indicating that corticosteroids can cause a decrease of colonic leukotriene synthesis. For example, Lauritsen *et al.*[9] demonstrated that intrarectal treatment with prednisolone for two to four weeks resulted in a significant decrease in colonic LTB_4 and PGE_2 synthesis, as measured by *in vivo* rectal dialysis. They reported similar effects of 5-ASA in this system. One must be cautious, however, in drawing the conclusion that healing induced by a drug occurs as a consequence of decreased leukotriene production as opposed to the reverse situation. It is entirely possible that colonic LTB_4 synthesis was decreased as a consequence of reduced numbers of granulocytes within the colonic tissue. Using the TNB model of colitis, we have demonstrated a significant correlation between colonic LTB_4 synthesis and colonic neutrophil numbers (as measured by tissue myeloperoxidase activity)[13]. Furthermore, depletion of circulating neutrophils by treating with an anti-neutrophil serum prior to induction of colitis with TNB resulted in inhibition of the increase in <u>both</u> myeloperoxidase activity and LTB_4 synthesis (Figure 4). Similar findings have been reported by Stenson[2] using the acetic acid model of colitis. We have further demonstrated that healing of colonic ulceration in this model is clearly paralleled by a reduction of tissue neutrophil numbers and LTB_4 production[13,14]. Taken together, these findings suggest the effects of drugs such as sulfasalazine and corticosteroids on the healing of colonic ulcers could result in a decrease in colonic leukotriene production <u>independent</u> of specific effects of these drugs on 5-lipoxygenase or phospholipase A_2 activity.

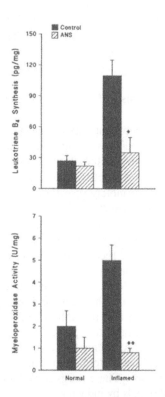

Figure 4. Leukotriene B_4 synthesis (top) and myeloperoxidase (bottom) in samples of grossly normal and inflamed colonic tissue 24 h after intracolonic administration of trinitrobenzene sulfonic acid. One group of rats was pretreated with an anti-neutrophil serum (ANS) which reduced circulating neutrophil numbers by 90%. In control rats, administration of TNB resulted in significant increases in LTB_4 synthesis and myeloperoxidase activity. However, in neutropenic rats, neither parameter was significantly increased ($^*p<0.05$) in "inflamed" samples when compared to normal samples.

To more directly assess the role of leukotrienes in the pathogenesis of colitis, we and others have assessed the effects of treatment with more specific inhibitors of 5-lipoxygenase in the TNB model (see Table 2). The first compound we tested in this model was BW755c, which is a dual cyclo-oxygenase/lipoxygenase inhibitor which also

Table 1. Evidence of Elevated Colonic Leukotriene Synthesis in Experimental Colitis

Reference	Experimental Model	Species	Method for LT Measurement	LT Elevated
10	DNCB	Guinea Pig	^{14}C-AA TLC (in vitro)	5, 11, 12-HETE
21	Acetic Acid	Rat	^{14}C-AA TLC (in vitro)	LTB$_4$, HETE
22	Ovalbumin	Guinea Pig	RIA (in vitro)	LTC$_4$
15	Immune Complex	Rabbit	HPLC/RIA (in vivo)	LTB$_4$, LTC$_4$
11	TNB	Rat	^{14}C-AA TLC (in vitro)	12, 15 HETE
12	TNB	Rat	RIA (in vitro)	LTB$_4$
13	TNB	Rat	RIA (in vitro)	LTB$_4$
14	TNB	Rat	RIA (in vitro)	LTB$_4$

Abbreviations: ^{14}C-AA: ^{14}carbon-labelled arachidonic acid; TLC: thin-layer chromatography; HPLC: high pressure liquid chromatography; RIA: radioimmunoassay; DNCB: dichloronitrobenzene; TNB: trinitrobenzene sulfonic acid.

Table 2. Effects of Leukotriene Synthesis Inhibitors in Experimental Colitis

Drug	Period of Treatment (days)	Experimental Model	Species	Effects on LT Synthesis	Effect on Colitis	References
BW755c	1-7	TNB	Rat	Decreased	Decreased MPO	12
BW755c	7-21	TNB	Rat	Decreased	Inconsistent	12
Sulfasalazine	7-21	TNB	Rat	No effect	No effect	12
Prednisolone	1	Immune Complex	Rabbit	No effect	No effect	15
L651,392	1-4	TNB	Rat	Decreased	Accelerated healing	13
5-ASA	1-4	TNB	Rat	Decreased	Accelerated healing	13
MK-886	1-7	TNB	Rat	Decreased	Accelerated healing	14
Sulfasalazine	1-7	TNB	Rat	No effect	No effect	14
MK-886	7-14	TNB	Rat	Decreased	No effect	14
Sulfasalazine	7-14	TNB	Rat	No effect	No effect	14

Abbreviations: MPO: myeloperoxidase activity; 5-ASA: 5-aminosalicylic acid possesses considerable anti-oxidant activity. Treatment with this compound on a daily basis at a dose which significantly decreased colonic LTB$_4$ synthesis resulted in inconsistent effects on healing[12]. One possible explanation for these inconsistent results is that the haemolytic properties of BW755c and/or its inhibitory effects on colonic prostaglandin synthesis may have offset any beneficial effects of 5-lipoxygenase inhibition.

More recently, we have assessed the effects of a specific 5-lipoxygenase inhibitor, L-651,392[13]. Daily intracolonic administration of this compound during the first 96 h after induction of colitis resulted in a marked acceleration of the healing of colonic ulcers and a significant reduction of colonic myeloperoxidase activity (an enzyme marker of granulocytes). This compound was found to be inactive in this model when administered orally or intraperitoneally, likely because of poor bioavailability.

The vast majority of 5-lipoxygenase inhibitors act through reduction of the enzyme to an inactive form. Recently, Gillard et al.[23] described a compound, MK-886, which is a derivative of a series of indol-2-propanoic acids and which appears to inhibit the activation of 5-lipoxygenase by preventing its translocation from the cytosol to the plasma membrane. This compound is active after oral administration and produces long-lasting inhibition of leukotriene synthesis. Indeed, in the TNB model a single dose of MK-886 (10 mg/kg) given two h prior to induction of colitis produced complete suppression of leukotriene synthesis for more than 24 h [14]. Daily oral treatment with this dose of MK-886 for one week after induction of colitis resulted in a significant acceleration of healing[14.]

Clearly, the testing of specific 5-lipoxygenase inhibitors in a clinical setting is essential before the promising results from animal models can be extended to man. Hawkey and Rampton[24] performed an open trial involving ulcerative colitis patients in which the putative 5-lipoxygenase inhibitor, benoxaprofen, was found to be of no therapeutic value. However, questions have subsequently been raised concerning the ability of this compound to inhibit 5-lipoxygenase in vivo[25]. Clinical trials using more recently developed, specific inhibitors of 5-lipoxygenase are presently underway with a number of experimental compounds. Another possible avenue for therapeutic intervention in IBD is the use of specific receptor antagonists to leukotrienes. Few studies to date have been performed using such compounds. We tested the effects of daily administration of an LTD$_4$ antagonist (L649,923) in the TNB model and found no significant effect on ulcer healing or tissue myeloperoxidase activity[26]. Recently, Fretland et al.[27] reported a significant inhibitory effect of an LTB$_4$ antagonist (SC-41930) on neutrophil infiltration into the colon of the guinea pig following intracolonic administration of acetic acid.

SUMMARY

There is overwhelming evidence that metabolism of arachidonic acid via the 5-lipoxygenase pathway is markedly elevated in both experimental colitis and in human inflammatory bowel disease. Furthermore, the biological properties of this family of mediators are consistent with a role in both the pathogenesis and the production of the symptoms of colitis. The modulation of the synthesis or actions of leukotrienes may therefore be a rational target for drug development for IBD.

REFERENCES

1. D. Hollander, C.M. Vadheim, E. Brettholz *et al.*, Increased intestinal permeability in patients with Crohn's disease and their relatives - a possible etiological factor, *Ann Intern Med.* 105:883 (1986).

2. W.F. Stenson, Leukotriene B_4 in inflammatory bowel disease, *in:* "Inflammatory Bowel Disease Basic Research and Clinical Implications," *in:* H Goebell, B.M. Peskar, and H. Malchow, eds., MTP Press Ltd., Lancaster (1988).

3. R.A. Lewis, and K.F. Austen, The biologically active leukotrienes. Biosynthesis, functions, and pharmacology, . *Clin Invest.* 73:889 (1984).

4. D.M. Montzka, P.L. Smith, and J.D. Fondacaro, Action of peptidoleukotrienes (PLTs) on electrolyte transport in rat small intestine, *Gastroenterology.* 92:1803 (1987).

5. E.A. Lobos, P. Sharon, and W.F. Stenson, Chemotactic activity in inflammatory bowel disease: role of leukotriene B_4, *Dig Dis Sci.* 32:1380 (1987).

6. P. Sharon, and W.F. Stenson, Enhanced synthesis of leukotriene B_4 by colonic mucosa in inflammatory bowel disease, *Gastroenterology.* 86:453 (1984).

7. B.M. Peskar, K.W. Dreyling, B.A. Peskar *et al.*, Enhanced formation of sulfidopeptido-leukotrienes in ulcerative colitis and Crohn's disease: inhibition by sulfasalazine and 5-aminosalicylic acid, *Agents Actions.* 18:381 (1986).

8. N.K. Boughton-Smith, C.J. Hawkey, and B.J.R. Whittle, Biosynthesis of lipoxygenase and cyclooxygenase products from ^{14}C-arachidonic acid by human colonic mucosa, *Gut.* 24:1176 (1983).

9. K. Lauritsen, L.S. Laursen, K. Bukhave *et al.* Effects of topical 5-aminosalicylic acid and prednisolone on prostaglandin E_2 and leukotriene B_4 levels determined by equilibrium *in vivo* dialysis of rectum in relapsing ulcerative colitis, *Gastroenterology.* 91:837 (1986).

10. N.K. Boughton-Smith, and B.J.R. Whittle, Increased metabolism of arachidonic acid in an immune model of colitis in guinea pig, *Br J Pharmacol.* 86:439 (1985).

11. N.K. Boughton-Smith, J.L. Wallace, and B.J.R. Whittle, Relationship between arachidonic acid metabolism, myeloperoxidase activity and leukocyte infiltration in a rat model of inflammatory bowel disease, *Agents Actions.* 25:115 (1988).

12. N.K. Boughton-Smith, J.L. Wallace, G.P. Morris, and B.J.R. Whittle, The effect of anti-inflammatory drugs on eicosanoid formation in a chronic model of inflammatory bowel disease in the rat, *Br J Pharmacol.* 94:65 (1988).

13. J.L. Wallace, W.K. MacNaughton, G.P. Morris *et al.*, Inhibition of leukotriene synthesis markedly accelerates healing in a rat model of inflammatory bowel disease, *Gastroenterology.* 96:29 (1989).

14. J.L. Wallace, and C.M. Keenan, An orally active inhibitor of leukotriene synthesis accelerates healing in a rat model of colitis, *Am J Physiol.* 258:G527 (1990).

15. R.D. Zipser, C.C. Nast, M. Lee *et al.*, *In vivo* production of leukotriene B_4 and leukotriene C_4 in rabbit colitis. Relationship to inflammation, *Gastroenterology.* 92:33 (1987).

16. W.F. Stenson, and E. Lobos, Sulfasalazine inhibits the synthesis of chemotactic lipids by neutrophils, *J Clin Invest.* 69:494 (1982).

17. B.M. Peskar, and C.H. Coersmeier, Effect of anti-inflammatory drugs on human colonic leukotriene formation, *in:* "Inflammatory Bowel Disease - Basic Research and Clinical Implications," H. Goebell, H., B.M. Peskar, and H. Malchow, eds., MTP Press Ltd., Lancaster (1988).

18. K.W. Dreyling, U. Hoppe, B.A. Peskar *et al.*, Leukotrienes in Crohn's disease: effect of sulfasalazine and 5-aminosalicylic acid, *in:* "Advances in Prostaglandin, Thromboxane, and Leukotriene Research," B. Samuelsson, P.W. Ramwell, and R. Paoletti, eds., Vol. 17, Raven Press, New York (1987).

19. H. Allgayer, and W.F. Stenson, A comparison of effects of sulfasalazine and its metabolites on the metabolism of endogenous vs exogenous arachidonic acid, *Immunopharmacology.* 15:39 (1988).

20. M.B. Grisham, and D.N. Granger, 5-aminosalicylic acid concentration in mucosal interstitium of cat small and large intestine. *Dig Dis Sci.* 34:573 (1989).

21. P. Sharon, and W.F. Stenson, Metabolism of arachidonic acid in acetic acid colitis in rats: similarity to human inflammatory bowel disease, *Gastroenterology.* 88:55 (1985).

22. R.H. Wolbling, U. Aehringhaus, B.A. Peskar *et al.*, Release of slow-reacting substance of anaphylaxis and leukotriene C_4-like immunoreactivity from guinea pig colonic tissue, *Prostaglandins.* 25:809 (1983).

23. J. Gillard, A.W. Ford-Hutchinson, C. Chan *et al.*, L-663,536 (MK-886) (3-[1-(4-chlorobenzyl)-3-t-butyl-thio-5-isopropylindol-2-yl]-2,2-dimethylpropanoic acid), a novel, orally active leukotriene biosynthesis inhibitor, *Can J Physiol Pharmacol.* 67:456 (1989).
24. C.J. Hawkey, and D.S. Rampton, Benoxaprofen in the treatment of active ulcerative colitis, *Prostaglandins Leukotrienes Med.* 10:405 (1983).
25. J.A. Salmon, L.C. Tilling, and S. Moncada, S., Benoxaprofen does not inhibit formation of leukotriene B_4 in a model of acute inflammation, *Biochem Pharmacol.* 33:2928 (1984).
26. J.L. Wallace, G.P. Morris, and W.K. MacNaughton, Evaluation of the role of leukotrienes as mediators of colonic inflammation and ulceration in an animal model, *in:* "Inflammatory Bowel Disease: Current Status and Future Approach," R.P. MacDermott, ed., Elsevier Science Publishers, New York (1988).
27. D.J. Fretland, S. Levin, B.S. Tsai *et al.*, The effect of leukotriene-B_4 receptor antagonist, SC-41930, on acetic acid-induced colonic inflammation, *Agents Actions.* 27:395 (1989).

THE BRAIN-GUT AXIS AND THE MUCOSAL
IMMUNOINFLAMMATORY RESPONSE

Fergus Shanahan

The concept of a brain-gut axis is now well established, although it is usually discussed in the context of neuroendocrine regulation of gastrointestinal secretory and motor function. That there might also be a neuroendocrine controlling effect on gastrointestinal mucosal immune function seems intuitively correct, but has only recently been investigated. While most investigators have examined systemic neuroimmune interactions[1-4], it is becoming evident that the mucosal immune system may respond to distinct neuroregulatory influences[5-8]. The intent here is to present a brief overview of the evidence for a mucosal neuroimmune axis and to discuss its potential relevance to mucosal inflammatory disease.

Maintenance of the internal milieu requires efficient homeostatic processes including a recognition system for sensing challenges by foreign or infectious agents. This requires the ability to distinguish "self" from "non-self" which is the fundamental function of the immune system. As with the handling of other forms of sensory information, the immune system has receptors or recognition structures for distinguishing self from non-self, a central mode of processing of the information, an effector response for elimination of foreign (or non-self) material, and a memory response. Thus, in many respects, the immune system may be regarded as the sixth sense. Like all other sensory/effector mechanisms, regulation must be exquisitely precise. Failure to distinguish self from non-self and to respond appropriately has the potential for development of autoimmunity or immune deficiency. Therefore, it is to be expected that the immune system is subject to regulatory signals from the neuroendocrine system. Several lines of experimental evidence support this and are presented in overview below (Table 1).

PAVLOVIAN CONDITIONING OF THE MUCOSAL IMMUNE SYSTEM

The interrelationship between the neuropsyche and immune function has been demonstrated in striking fashion by behavioral conditioning, in much the same way as Pavlov originally studied gastric secretion. Thus, the ability to enhance or suppress the systemic immune system by classical Pavlovian conditioning has been demonstrated convincingly by several investigators[9]. More recently, it has been shown

Neuroendocrinology of Gastrointestinal Ulceration
Edited by S. Szabo and Y. Taché, Plenum Press, New York, 1995

that mucosal mast cell function may also be modified by behavioral conditioning[10]. Conditioning of mucosal lymphocytes has not yet been formally demonstrated, but a remarkable study in children who were taught self-hypnosis with specific suggestions for control of salivary immunoglobulins, raised the possibility of voluntary or self-regulation of IgA levels[11].

Table 1. Evidence for Neuroendocrine Regulation of the Mucosal Immune System

Pavlovian conditioning
Innervation of gut-associated lymphoid tissue (GALT)
Neurogenic inflammation
Presence of neuropeptide receptors on mucosal immunocytes
Differential effects of neuropeptides on mucosal vs systemic immunocytes *in vitro*
Upregulation of substance P receptors in GALT in mucosal inflammation

INNERVATION OF MUCOSA-ASSOCIATED LYMPHOID TISSUE

While the innervation of the gastrointestinal mucosa has been well studied, there is limited information on the innervation of the mucosal immune system[12,13]. Several investigators have found evidence for peptidergic innervation of the mucosal lymphoid tissue[13,14]. The mucosal lymphoid tissue is organized as discrete lymphoid follicles or as aggregates of follicles (Peyer's patches) and as a diffusely scattered population of cells within the lamina propria and epithelium. Within Peyer's patches, there are nerve fibers containing substance P, somatostatin, VIP, and CGRP, and the extensive nerve plexus in the lamina propria also contains a range of neuropeptides including substance P, somatostatin, VIP and CCK[12,14-16].

Elegant morphometric studies by Stead and colleagues at the light microscopic and ultrastructural level demonstrated a close association between substance P- and CGRP-containing nerve fibers and mucosal mast cells[17]. Mucosal mast cell hyperplasia was induced in rats parasitized with a nematode (*Nippostrongylus brasiliensis*), and combined immunocytochemical techniques showed that approximately 60% of the mast cells were in contact with nerves containing substance P-like and CGRP-like immunoreactivity. Intimate membrane to membrane contact between mast cells and what appeared to be peptidergic neurons was confirmed by electron microscopy. The same investigators have also reported similar nerve-mast cell apposition in the human gastrointestinal mucosa[18]. Others have found that such an association is not limited to mast cells but also occurs with other immunocytes[19].

NEUROGENIC INFLAMMATION

Antidromic stimulation of sensory nerves results in an alteration of vascular permeability and a variable inflammatory response known as neurogenic inflammation. This is thought to be mediated by sensory neuropeptides released locally[20,21]. While substance P released from nerve terminals is generally believed to be the primary mediator of neurogenic inflammation, sensory nerves also contain several other neuropeptides such as CGRP that may contribute to or modify the inflammatory response[5]. CGRP is itself a potent vasodilator and may act synergistically with substance P, but can also limit the duration of its action by triggering release of mast cell proteases[22].

Neurogenic inflammation has been well studied in the skin, the respiratory mucosa and elsewhere[20,21,23], and there is some evidence that it also occurs in the intestinal mucosa[24]. The role of neurogenic inflammation in the pathogenesis of intestinal inflammatory disorders is not certain but there is extensive evidence to implicate neurogenic involvement in the altered intestinal physiology during acute mucosal inflammation[7,25].

DIFFERENTIAL EFFECTS OF NEUROPEPTIDES ON MUCOSAL VS SYSTEM IMMUNITY

Evidence for the presence of neuropeptide receptors has been found on several immunocytes and a wide range of peptides has been implicated in modulating the immune system in vitro[4-7]. Many of these peptide-immunocyte interactions are probably of pharmacologic curiosity, whereas others appear to have potential physiologic significance. Although most of the *in vitro* functional studies have been performed using peripheral blood mononuclear cells, there are several examples of differences in responsiveness to regulatory peptides between the mucosal and systemic immune systems.

Stanisz and colleagues found that the effects of neuropeptides on immunoglobulin production by conconavalin A-stimulated murine lymphocytes, were tissue-dependent, peptide-dependent, and immunoglobulin isotype-dependent[26]. Immunoglobulin production was enhanced by substance P and suppressed by somatostatin, but the effects of substance P was three to fourfold greater with cells from gut-associated lymphoid tissue compared with splenocytes, and the enhancement of immunoglobulin synthesis by intestinal lymphocytes favored IgA over IgM with little effect on IgG. The effects of VIP were also organ-specific in that IgA production by Peyer's patch lymphocytes was inhibited, whereas that by splenocytes was enhanced. Tissue-specific differences in receptor expression for peptides is also apparent. For example, murine T and B lymphocytes express substance P receptors, but receptor numbers are significantly greater on cells from intestinal Peyer's patches in comparison with splenocytes[27]. The same is true for somatostatin receptors[28].

Tissue-dependent secretagogue effects of certain neuropeptides on mast cells reflects the heterogeneity of mast cells. Mucosal mast cells differ from connective tissue or non-mucosal mast cells cytochemically, functionally, and pharmacological[29,30]. These tissue differences are largely due to conditioning by the local microenvironment[30]. Most basic peptides, including neuropeptides such as substance P, trigger histamine release from connective tissue mast cells, but mucosal mast cells tend to be unresponsive to such stimuli[31,32]. In addition, there is indirect evidence that somatostatin may be inhibitory to mucosal mast cells and stimulatory to non-mucosal mast cells[33]. The basis for these tissue-dependent neuropeptide/mast cell interactions in unclear. Differences in receptor expression are unlikely since there is no convincing evidence for receptors for neuropeptides on mast cells. However, there is evidence favoring a receptor-independent mechanism for the action of substance P and other neuropeptides on mast cells involving direct G-protein activation[34,35].

The differential effect of neuropeptides on mucosal and systemic immune function is also revealed by the finding that VIP alters the migration and homing patterns of immune effector cells[36]. Preincubation of radiolabelled mouse T cells *in vitro* with VIP resulted in a dose-dependent reduction in their subsequent *in vivo* localization within Peyer's patches and mesenteric lymph nodes but not within the spleen and other major organs[37].

ROLE OF PSYCHONEURO-IMMUNE INTERACTIONS IN MUCOSAL INFLAMMATION

The influence of neuropsychic factors such as stress on the immunoinflammatory response has received much attention recently[38-40]. This is difficult to study in humans, and there is no convincing evidence to suggest that stress alters the activity of inflammatory bowel disease. Despite this, there is a widely prevalent suspicion that the clinical course of these conditions may be influenced, in some individuals, by endogenous factors such as the psychoneuroendocrine system. A possible indicator of a neuropsychic influence on the course of chronic inflammatory bowel disease is the marked placebo effect in both Crohn's disease and ulcerative colitis[41]. Given the evidence for neuropeptidergic regulation of the mucosal immune system and the clear role of mucosal immunologic effector mechanisms in mediating the tissue damage in these diseases[42], it is tempting to speculate that the link between the neuropsyche and disease activity may be through the neuroendocrine-immune axis.

Inflammatory activity might be altered by increased release of neuropeptides locally or by a change in the sensitivity of immune effector cells to the effects of neuropeptides. Proinflammatory effects of substance P have been well demonstrated in vivo by injection into various tissues including the skin, eyes and joints[5]. However, the levels of substance P and other peptides within the inflamed mucosa of patients with Crohn's disease and ulcerative colitis have been inconsistent in different reports and do not appear to be altered substantially[5-7,43]. In contrast, there is convincing autoradiographic evidence for an alteration in the expression of receptors for substance P, but not for other neuropeptides, in the mucosa of patients with Crohn's disease and ulcerative colitis[44,45]. The degree of receptor upregulation was estimated to be a thousand-fold, and was most pronounced in the vasculature but also in the germinal centers of lymphoid nodules. Because the limited resolution of autoradiography, it was not possible to determine whether immunocytes within the intestinal lymphoid tissue actually express the increased substance P receptors. However, the development of biotinylated analogues of substance P and other molecular probes will now permit identification of the cellular source of substance P binding sites within the inflamed mucosa[46].

SUMMARY

Several lines of evidence provide strong circumstantial support for the notion that the brain-gut axis probably includes a link between the neuropsyche or neuroendocrine system and mucosal immune system. Neuroendocrine immune interactions are well established in relation to the systemic immune system but may have special significance within the mucosa where juxtaposition of mucosal immunocytes with nerve terminals is optimal for bi-directional communication. The relevance of neuroimmune interactions to the pathogenesis or healing of chronic mucosal inflammatory disorders such as ulcerative colitis and Crohn's disease is not fully understood but is an exciting challenge that may have therapeutic importance in the future.

REFERENCES

1. J.E. Blalock, A molecular basis for bidirectional communication between the immune system and neuroendocrine systems. *Physiol Rev.* 69:1 (1989).
2. B.S. Rabin, S. Cohen, R. Ganguli, D.T. Lysle, and J.E. Cunnick, Bidirectional interaction between the central nervous system and the immune system, *Crit Rev Immunol.* 9:279 (1989).
3. S. Reichlin, Neuroendocrine-immune interactions, *N Engl J Med.* 329:1246 (1993).
4. E.J. Goetzl, D.C. Adelman, and S.P. Sreedharan, Neuroimmunology, *Adv Immunol.* 48:161 (1990).
5. F. Shanahan, and P. Anton, Role of peptides in the regulation of the mucosal immune and inflammatory response, *in:* "Gut Peptides: Biochemistry and Physiology," J.H. Walsh, G.J. Dockray, eds., Raven Press, New York (1994).
6. F. Shanahan, and P. Anton, Neuroendocrine modulation of the immune system. Possible implications for inflammatory bowel disease, *Dig Dis Sci.* 33:41S (1988).
7. K. Croitoru, P.B. Ernst, A.M. Stanisz, R.H. Stead, and J. Bienenstock, Neuroendocrine regulation of the mucosal immune system, *in:* "Immunology and Immunopathology of the Liver and Gastrointestinal Tract," S.R. Targan, F. Shanahan, eds., Igaku-Shoin Press, New York, Tokyo (1989).
8. J. Bienenstock, M. Perdue, A. Stanisz, and R. Stead, Neurohormonal regulation of gastrointestinal immunity, *Gastroenterology.* 93:1431 (1987).
9. R. Ader, "Psychoneuroimmunology," Academic Press, London (1981).
10. G. MacQueen, J. Marshall, M. Perdue, S. Siegel, and J. Bienenstock, Conditional secretion of rat mast cell protease II by mucosal mast cells, *Science.* 243:83, (1989).
11. K. Olness, T. Culbert, and D. Uden, Self-regulation of salivary immunoglobulin A by children, *Pediatrics.* 83:66 (1989).
12. H.J. Cooke, Neurobiology of the intestinal mucosa, *Gastroenterology.* 90:1057 (1986).
13. D.L. Bellinger, D. Lorton, T.D. Romano, J.A. Olschowka, S.Y. Felten, and D.L. Felten, Neuropeptide innervation of lymphoid organs, *Ann NY Acad Sci.* 594:17 (1990).
14. C.A. Ottaway, D.L. Lewis, and S.L. Asa, Vasoactive intestinal peptide-containing nerves in Peyer's patches, *Brain Behav Immun.* 1:148 (1987).
15. E. Ekblad, C. Winther, R. Ekman, R. Hakanson, and F. Sundler, Projections of peptide-containing neurons in rat small intestine, *Neuroscience.* 20:169 (1987).
16. L. Probert, J. Demey, and J.M. Polak, Distinct subpopulations of enteric p-type neurons contain substance P and vasoactive intestinal polypeptide, *Nature.* 294:470 (1981).
17. R.H. Stead, M. Tomioka, G. Quinonez, G. Simon, S.Y. Felten, and J. Bienenstock, Intestinal mucosal mast cells in normal and nematode-infected rat intestines are in intimate contact with peptidergic nerves, *Proc Natl Acad Sci USA.* 2975 (1987).
18. R.H. Stead, M.F. Dixon, N.H. Bramwell, R.H. Ridell, and J. Bienenstock, Mast cells are closely apposed to nerves in the human gastrointestinal mucosa, *Gastroenterology.* 97:575 (1989).
19. N. Arizono, S. Natsuda, T. Hattori, Y. Kojima, T. Maeda, and S.J. Galli, Anatomical variation in mast cell nerve associations in the rat small intestine, heart, lung, and skin: similarities of distances between neural processes and mast cells, eosinophils, or plasma cells in the jejunal lamina propria, *Lab Invest.* 62:626 (1990).
20. J.C. Foreman, Peptides and neurogenic inflammation, *Br Med Bull.* 43:386 (1987).
21. B Pernow, Role of tachykinins in neurogenic inflammation, *J Immunol.* 135:812S (1985).
22. S.D. Brain, and T.J. Williams, Substance P regulates the vasodilator activity of calcitonin gene related peptide, *Nature.* 335:73 (1988).
23. M.L. Kowalski, and M.A. Kaliner, Neurogenic inflammation, vascular permeability and mast cells, *J Immunol.* 140:3905 (1988).
24. T. Bani-Sacchi, M. Barattini, S. Bianchi *et al.*, The release of histamine by parasympathetic stimulation in guinea pig auricle and rat ileum, *J Physiol.*, 371:29 (1986).
25. G.A. Castro, Immunophysiology of enteric parasitism, *Parasitology Today.* 5:11 (1989).
26. A.M. Stanisz, D. Befus, and J. Bienenstock, Differential effects of vasoactive intestinal polypeptide, substance P, and somatostatin on immunoglobulin synthesis and proliferations by lymphocytes from Peyer's Patches, mesenteric lymph nodes, and spleen, *J Immunol.* 136:152 (1986).
27. A.M. Stanisz, R. Scicchitano, P. Dazin, J. Bienenstock, D.G. Payan, Distribution of substance P receptors in murine spleen and Peyer's patch T and B cells, *J Immunol.* 139:749 (1987).
28. R. Scicchitano, P. Dazin, J. Bienenstock, D.G. Payan, and A.M. Stanisz, Distribution of somatostatin receptors on murine spleen and Peyer's patch T and B lymphocytes, *Brain Behav Immunol.* 1:173 (1987).
29. A.A. Irani, and L.B. Schwartz, Mast cell heterogeneity, *Clin Exp Allergy.* 19:143 (1989).

30. Y. Kitamura, Heterogeneity of mast cells and phenotypic change between subpopulations, *Ann Rev Immunol.* 7:59 (1989).

31. F. Shanahan, J.A. Denburg, J. Fox, J. Bienenstock, and D. Befus, Mast cell heterogeneity: effect of neuroenteric peptides on histamine release, *J Immunol.* 135:1331 (1985).

32. F. Shanahan, T.D.G. Lee, J. Bienenstock, and A.D. Befus, The influence of endorphins on peritoneal and mucosal mast cell secretion, *J Allergy Clin Immunol.* 74:499 (1984).

33. E.J. Goetzl, T. Chernov-Rogan, K. Furuichi, L.M. Goetzl, J.Y. Lee, and F. Renold, Neuromodulation of mast cell and basophil function, *in:* "Mast Cell Differentiation and Heterogeneity," A.D. Befus, J. Bienenstock, J.A. Denburg, eds., Raven Press, New York (1986).

34. H. Repke, and M. Bienert, Mast cell activation-a receptor-independent mode of substance P action, *FEBS Lett.* 221:236 (1987).

35. M. Mousli, J.L. Bueb, C. Bronner, B. Rouot, and Y. Landry, G protein activation: a receptor independent mode of action for cationic amphiphilic neuropeptides and venom peptides, *Trends Pharmacol Sci.* 11:358 (1990).

36. C.A. Ottaway, Migration of lymphocytes within the mucosal immune system, *in:* "Immunology and Immunopathology of the Liver and Gastrointestinal Tract," S.R. Targan, F. Shanahan, eds., Igaku-Shoin Press, New York, Tokyo (1989).

37. C.A. Ottaway, In vitro alteration of receptors for vasoactive intestinal peptide changes the in vivo localization of mouse T cells, *J Exp Med.* 160:1054 (1984).

38. D.N. Khansari, A.J. Murgo, and R.E. Faith. Effects of stress on the immune system, *Immunol Today.* 11:170 (1990).

39. D. Mason, Genetic variation in the stress response: susceptibility to experimental allergic encephalomyelitis and implications for human inflammatory disease, *Immunol Today.* 12:57 (1991).

40. S. Cohen, D.A.J. Tyrell, and A.P. Smith, Psychological stress and susceptibility to the common cold, *N Engl J Med.* 325:606 (1991).

41. S. Meyers, and H.D. Janowitz, "Natural history" of Crohn's disease. An analytical review of the placebo lession, *Gastroenterology.* 87:1189 (1984).

42. F. Shanahan, and S. Targan, Mechanisms of tissue injury in inflammatory bowel disease, *in:* "Inflammatory Bowel Disease: From Bench to Bedside," S. Targan, F. Shanahan, eds., Williams & Wilkins, Baltimore (1994).

43. V.E. Eysselein, and C.C. Nast, Neuropeptides and inflammatory bowel disease, *in:* "Inflammatory bowel diseases. Progress in basic research and clinical implications," H. Goebell, K. Ewe, H. Malchow, C.H. Koelbel., eds., Kluwer Academic Publishers, Dordrecht (1991).

44. C.R. Mantyh, T.S. Gates, R.P. Zimmerman, M.L. Welton, E.P. Passaro, S.R. Vigna, R. Maggio, L. Kurger, and P.W. Mantyh, Receptor binding sites for substance P, but not substance K or neuromedin K, are expressed in high concentrations by arterioles, venules, and lymph nodules in surgical specimens obtained from patients with ulcerative colitis and Crohn's disease, *Proc Natl Acad Sci USA.* 85:3235 (1988).

45. P.W. Mantyh, M.D. Catton, C.G. Boehmer, M.L. Welton, E.P. Passaro, J.E. Maggio, and S.R. Vigna, Receptors for sensory neuropeptides in human inflammatory bowel diseases: implications for the effector role of sensory neurons, *Peptides.* 10:627 (1989).

46. P.A. Anton, J.R. Reeve, A. Vidrich, E. Mayer and F. Shanahan, Development of a biotinylated analog of substance P for use as a receptor probe, *Lab Invest.* 64:703 (1991).

47. P.W. Mantyh, D.J. Johnson, C.G. Boehmer, M.D. Catton, H.V. Vinters, J.E. Maggio, H.-P. Too, and S.R. Vigna, Substance P receptor binding sites are expressed by glia *in vivo* after neuronal injury, *Proc Natl Acad Sci USA.* 86:5193 (1989).

48. J. Nilsson, A.M. von Euler, and C.-J. Dalsgaard, Stimulation of connective tissue cell growth by substance P and substance K, *Nature.* 315:61 (1985).

49. E.S. Kimball, and M.C. Fisher, Potentiation of IL-1-induced BALB/3T3 fibroblast proliferation by neuropeptides, *J Immunol.* 141:4203 (1988).

50. P. Holzer, M.A. Pabst, I. Th. Lippe, Intragastric capsaicin protects against aspirin-induced lesion formation and bleeding in the rat gastric mucosa, *Gastroenterology.* 96:1425 (1989).

51. S. Evangelista, and A. Meli, Influence of capsaicin-sensitive fibres on experimentally-induced colitis in rats, *J Pharm Pharmacol.* 41:574 (1989).

NEUROENDOCRINE CONTROL OF GASTRIC ACID SECRETION

Yoshitsugu Osumi, Yasunobu Okuma, Kunihiko Yokotani,
Mitsuhiro Nagata and Toshio Ishikawa

A wide variety of the stressors experienced by humans and induced in experimental animals leads to gastric ulceration. As gastrointestinal functions are to a great extent regulated by neurogenic mechanisms, considerable attention has been focused on the possible role of the hypothalamus and other regions of the brain in stress. Actually there was little information regarding the relationship of central neurotransmitters to gastric functions, partly because not much is known of localization and distribution of central neurotransmitters, including peptides.

For the past fifteen years, we have directed attention to the regulatory mechanisms of gastric acid secretion, with special reference to central neurotransmitters and peripheral adrenergic mechanisms. In this report, therefore, possible inhibitory mechanisms of gastric acid secretion by central neurotransmitters and transmitter candidates are briefly discussed based on our observations. The experiments were performed using urethane anesthetized gastric fistula rats.

POSSIBLE ROLES OF THE HYPOTHALAMUS AND CENTRAL EXCITATORY DESCENDING PATHWAYS IN REGULATION OF GASTRIC ACID SECRETION

In order to clarify central inhibitory mechanisms of gastric acid secretion, the scene of events related to possible roles of the hypothalamus in regulation of gastric acid secretion and central descending pathways to induce increase in this gastric parameter is briefly discussed here.

Within the hypothalamus, the ventromedial hypothalamus (VMH) and the lateral hypothalamic area (LHA) are probably the most important nuclei involved in the central regulation of gastric acid secretion. The VMH is located adjacent to the 3rd ventricle, and the LHA is located lateral to the VMH and corresponds to the region of medial forebrain bundles. These hypothalamic nuclei, the VMH and LHA are known as the so-called, "Satiety center" and "Feeding center", respectively. These hypothalamic nuclei are functionally interrelated[1], and anatomical links between these two areas were also reported[2]. In connection with peripheral autonomic functions, the VMH and LHA are thought to be integrative centers of sympathetic and

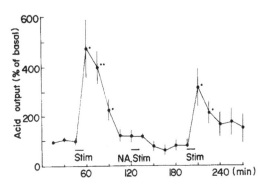

Figure 1. Repeated electrical stimulation of LHA, and the effect of NA on the gastric acid output. NA 10 μg per animal was injected into the lateral ventricle five min prior to the second electrical stimulation. Results are expressed as percent of the respective basal value (Mean ± SEM, n=5). Stim: electrical stimulation. Statistically significant from the respective value immediately before the stimulation. *p<0.05, **p<0.01. [Upper part of Fig. 5. in the original paper(7). Reprinted with permission from Life Sci 1977, Pergamon Press, Ltd.]

parasympathetic neuron systems, respectively[3]. Electrical stimulation of the VMH inhibits gastric acid secretion when the basal gastric acid level is maintained at a relatively high level[4,5], while destruction of this region induced increase[6]. On the other hand, repetitive electrical stimulation of the LHA induced a constant and reproducible increase in gastric acid secretion[7], as already reported by Misher and Brooks[4].

The question then arises as to how stimuli of the LHA are conveyed to the stomach. The increase in acid output by electrical stimulation is blocked when the animals are given atropine intravenously (i.v.), as a pretreatment[7]. Therefore, it seems that stimulation of the LHA activates the dorsal motor nucleus of the vagus (DMV) in the brain stem. Accumulating evidence shows that a number of central neurotransmitter candidates activate this excitatory neuron system and an increase in gastric acid output occurs. These central excitatory neurotransmitter or neurotransmitter candidates are; acetylcholine[8], thyrotropin releasing hormone (TRH)[9], oxytocin[10] and atrial natriuretic peptide[11]. Furthermore, an existence of central GABAergic mechanism stimulating gastric acid secretion was also suggested[12,13]. Reports on cholecystokinin-8 by various investigators are inconsistent probably due to different experimental conditions[14,15]. As related to the roles of the central cholinergic neuronal system, activation with bethanechol of muscarinic receptors in the LHA and DMV, and that with nicotine of nicotinic receptors in the posterior parts of the VMH induced significant increases in gastric acid secretion[8,16]. Furthermore, neurotransmission of the central descending pathway to the DMV in excitatory regulation of gastric functions is probably mediated through cholinergic muscarinic receptors[17]. However, TRH activated the DMV independent of the cholinergic neuronal system in this nucleus and a resultant increase in gastric acid secretion occurred[18,19].

CENTRAL INHIBITORY MECHANISMS OF GASTRIC ACID SECRETION

Noradrenaline (NA)[7], bombesin[20], corticotropin releasing factor[21], calcitonin gene-related peptide [22,23], β-endorphin[24], neuropeptide Y[25], gastrin releasing peptide[26],

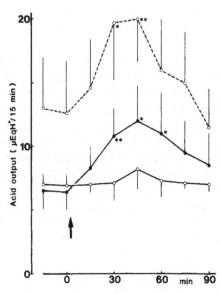

Figure 2. Effect of ACh given into the LHA on gastric acid output. In non-reserpinized rats: (O———O), Ach 30 nmoles (n=7); (•———•), Ach 300 nmoles (n=10). In reserpinized rats: (O-----O), ACh 30 nmoles (n=6). *:p<0.05, **: P<0.01 (statistically significant as compared to the respective basal values just before the application of ACh). [Reprinted with permission from Life Sci 1983, Pergamon Press, Ltd. (8)].

insulin-like growth factor II[27], prostaglandins[28,29] and interleukin-1β[30] induced central inhibition of gastric acid secretion. Reports on neurotensin by various investigators are inconsistent probably due to different experimental conditions[31,32,33]. There exist two types of central inhibitory mechanisms, as noted in our investigations.

Inhibition of Excitatory Neuronal Systems within the Brain

NA in the brain is a representative transmitter for this type of inhibition. Administration of NA intracerebroventricularly (i.c.v.) 5 min prior to the second electrical stimulation of the LHA almost completely blocked the stimulation-induced increase in acid output[7], as shown in Fig. 1. Seventy-five min after the second stimulation, gastric acid output responded well to the third stimulation.

Both the LHA and DMV have a large number of noradrenergic nerve terminals[34-38], thus the inhibitory effect of NA i.c.v. applied is probably mediated by adrenergic receptors located in these regions. Then, the effect of NA microinjected into the DMV was examined in reserpinized animals[39]. Twenty h after the intraperitoneal administration of reserpine two mg/kg, both NA and dopamine concentrations in the various brain regions were much lower than in respective controls. This pretreatment with reserpine significantly elevated the basal level of acid secretion and potentiated the inhibitory effect of NA. The administration of NA (100 - 500 ng) but not dopamine (500 ng) into the DMV, induced a dose-dependent decrease in gastric acid output.

We then asked whether this inhibition of acid output, as induced by NA would be reproducible by electrical stimulation of the central noradrenergic neuron system[39].

The locus coeruleus, located in the dorsolateral tegmentum of the pons, is the largest nucleus composed of the cell bodies which contain NA. Unilateral electrical stimulation of the locus coeruleus decreased the basal level of acid output. Furthermore, the increase in gastric acid secretion by electrical stimulation of the LHA was also blocked by the simultaneous stimulation of the locus coeruleus. Histochemical evidence shows that the DMV receives noradrenergic nerve terminals from the locus coeruleus[40]. Therefore, these results clearly show that the central noradrenergic neuron system originating from the locus coeruleus exerts inhibitory effects on gastric functions, at the level of the DMV. In another series of experiments, we found that activation of cholinergic nicotinic mechanisms in the DMV with nicotine, decreased acid secretion by its NA releasing action[41,42].

The LHA was thought to be another possible site of action of NA in inhibiting gastric acid secretion. NA 30 nmoles microinjected into the LHA in normal animals did decrease acid output[8]. Such an inhibitory effect of NA was markedly potentiated in reserpine-pretreated animals. Three nmoles of NA, which was without effect in non-reserpinized animals, significantly decreased acid output. In these reserpinized animals, acetylcholine in a dose of 30 nmoles, induced a marked increase in acid output. In the controls, it is interesting to note that the acetylcholine-induced increase was observed only with a 10 times higher dose, 300 nmoles, as shown in Fig. 2. Therefore, the central noradrenergic inhibitory mechanism of gastric acid secretion may be present at the level of the LHA as well as at the DMV in the brain stem (Fig. 3). It is likely that the cholinergic neuron system in the LHA has an excitatory effect on acid secretion, and is antagonized by a noradrenergic inhibitory mechanism[8].

Central Inhibition of Gastric Excitatory Pathway at Peripheral Sites.

The second type of central inhibition of gastric acid secretion is revealed with bombesin, and prostaglandin. In 1980, Taché et al. found that the intracisternal administration of bombesin inhibited gastric acid secretion[20]. The bombesin-induced secretion of adrenal catecholamines was also reported[44]. We, therefore, were much interested in this peptide, since our attention had been focused on the roles of peripheral adrenergic system in the regulation of gastric acid secretion.

When the left vagus nerve was stimulated continuously at three cycles/sec, 0.5 ms, one mA, gastric acid output increased and a steady level was reached with 60 min. Bombesin (one nmole)-induced inhibition of gastric acid output was also observed in these vagus nerve stimulated animals as well as in 2-DG-pretreated animals[45]. However, the same one nmole dose of bombesin did not alter the increase in this gastric parameter induced by the i.v. infusion of bethanechol, a cholinergic muscarinic agonist. These results show that bombesin-induced inhibition of vagal nerve activity is probably exerted at the level of intragastric ganglia such as the Auerbach plexus.

We reported that electrical stimulation of the greater splanchnic nerve, an activation of adrenergic receptors at the level of intragastric vagal plexus, inhibited vagally-induced gastric acid secretion[46,47]. Then, in the next series, the effect of bilateral cutting of the greater splanchnic nerves on bombesin-induced reduction of acid output was examined. This pretreatment markedly reduced the effect of bombesin[45]. Preganglionic greater splanchnic nerves ramify into two major branches; the adrenal branch and the branch terminating in the sympathetic coeliac ganglion. Therefore, the effects of chemical sympathectomy with 6-hydroxydopamine and surgical adrenalectomy on bombesin-induced inhibition of acid output were examined[45].

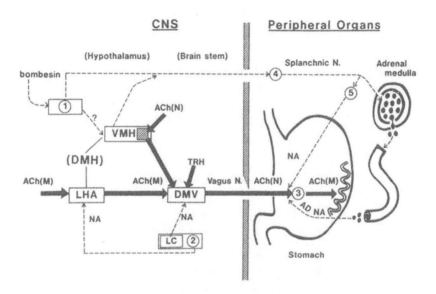

Figure 3. Schematic illustration of possible inhibitory mechanisms of gastric acid secretion with NA and bombesin in the brain ———►: Neuronal system to induce gastric acid secretion; ----►: Neuronal or humoral system to inhibit gastric acid secretion; 1: Central sites of action of bombesin; 2: Central NA neuron system; 3: Intragastric nerve plexus such as the Auerbach plexus; 4: Paravertebral ganglion chains (mainly Th 5-9); 5: Coeliac ganglion; LC: locus coeruleus; ML muscarinic receptors; N: nicotinic receptors; AD: adrenaline; TRH: thyrotropin releasing hormone. [Reprinted with permission from Folia Pharmacol Jpn 1990, Japanese Pharmacological Society (43)].

6-Hydroxydopamine 50 mg/kg i.v., markedly reduced the content of NA in the stomach, 3 days after its administration (less than 12% that of control), however, this agent was without effect on the content of either adrenaline or NA in the adrenal glands. NA content in the brain was also not affected by the treatment. The inhibitory effect of bombesin one nmole i.c.v. on the vagally-mediated increase in acid output was not modified by either pretreatment with 6-hydroxydopamine or by bilateral adrenalectomy alone. However, when chemical sympathectomy was combined with adrenalectomy, the inhibitory effect of bombesin was abolished, as shown in Fig. 4. Sites of action for this peptide are probably the paraventricular nucleus and its adjacent regions[48,49].

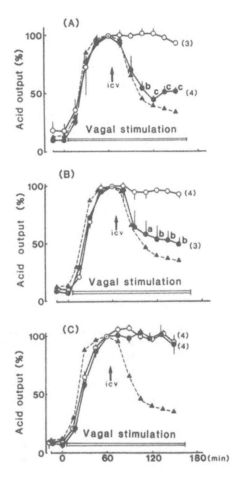

Figure 4. Effects of chemical sympathectomy with 6-OHDA(A), bilateral adrenalectomy(B) and combination of chemical sympathectomy with 6-OHDA and adrenalectomy(C) on the bombesin-induced inhibition of gastric acid secretion. (O) Saline (control), (•) bombesin one nmole. In these two groups, both saline and bombesin were applied i.c.v. during electrical stimulation of the vagus nerve in the animals pretreated with (A), (B) or (C). Number of animals in each experiment between parentheses. (▲) Bombesin one nmole was applied i.c.v. during electrical stimulation of the vagus nerve in the animals given no pretreatment (n=4). a:p<0.05, b:p<0.01, c:p<0.001 (statistically significant difference from the respective controls with saline). [Reprinted with permissions from Eur J Pharmacol 1987, Elsevier Science Publishers (45)].

Such a central inhibition of gastric acid secretion was also revealed by administration of small doses of prostaglandins (PGs)[29]. It is well known that various PGs are formed in the central nervous system[50]. PGE_2 (0.01-0.5 μg/animal) given i.c.v. induced a dose-dependent inhibition of vagally-mediated gastric acid secretion[29]. Effects of PGs such as PGD_2 and $PGF_{2\alpha}$ were also compared with that of PGE_2. The effect of PGE_2 proved to be the most potent. This inhibitory effect of PGE_2 was abolished by α-adrenergic receptor blockade using phentolamine but not by β-adrenergic blockade using propranolol, as shown in Fig. 5. This PGE_2-induced inhibition could be blocked only by combined treatment with 6-hydroxydopamine and adrenalectomy[29].

These central inhibitions of acid secretion after intracerebroventricular administration of bombesin and PGE_2 are classified as the second type. Central excitation of the sympathoadreno-medullary system by these substances reduced cholinergic vagal tones at the level of the intragastric nerve plexus and as a result, gastric acid secretion is inhibited, as shown in Fig. 3.

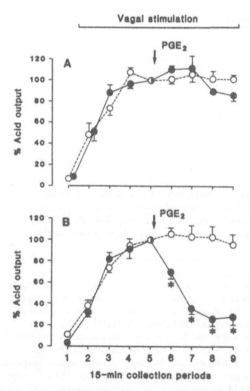

Figure 5. Effects of phentolamine and propranolol on i.c.v.-administered PGE_2 vagally stimulated gastric acid output. A, phentolamine-treated group; B, propranolol-treated group. Phentolamine (5 mg/kg) or propranolol (5 mg/kg) was administered 30 min before the start of the vagal stimulation. (O) control rats (n=4), (●) PGE_2 (0.1 μg)-administered rats (n=4). The actual values at the 5th 15-min collection period were 75.6 ± 17.1 μEq/15 min in the control rats and 79.5 ± 12.7 μEq/15 min in PGE_2-administered rats in group A; 83.9 ± 19.5 μEq/15 min in the control rats and 97.8 ± 16.7 μEq/15 min PGE_2-administered rats in group B, respectively. [Reprinted with permission from J Pharmacol Exp Ther 1988, The American Society for Pharmacology and Experimental Therapeutics (29)].

It is interesting to note that sympathetic hypofunction is well compensated for by circulating catecholamines released from the adrenal glands. Conversely, adrenomedullary dysfunction is well compensated for when sympathetic function is fully activated. We recently confirmed that intracerebroventricular administration of bombesin induced a marked increase in the plasma level of adrenaline and a slight increase in the plasma level of NA in urethane anesthetized rats[51]. It can be assumed that the sympathetic or adrenomedullary system alone sufficiently activates intragastric adrenergic receptors to decrease acid output.

SUMMARY

There exist two types of central inhibition of gastric acid secretion. One is the inhibition of gastric excitatory signals within the brain as observed with NA, and another is the inhibition of gastric excitatory signals at peripheral sites, as observed with bombesin and PGE_2 in the brain.

ACKNOWLEDGEMENTS

A part of this work was supported by a Grant-in-Aid for Special Project Research, a Grant-in-Aid for Cooperative Research, a Grant-in-Aid for Scientific Research, and a Grant-in-Aid for Encouragement of Young Scientists from the Ministry of Education, Science and Culture, Japan, and a Grant from the Smoking Research Foundation, Japan.

REFERENCES

1. I. Mayer, and D.W. Thomas, Regulation of food intake and obesity, *Science*. 156:328 (1967).
2. P.G.M. Luiten, and P. Room, Interrelations between lateral, dorsomedial and ventromedial hypothalamic nuclei in the rat, a HRP study, *Brain Res*. 190:321 (1980).
3. T. Shimazu, Central nervous system regulation of liver and adipose tissue metabolism, *Diabetologia*. 20:343 (1981).
4. A. Misher, and F.P. Brooks, Electrical stimulation of hypothalamus and gastric secretion in the albino rat, *Am J Physiol*. 211:403 (1966).
5. T. Ishikawa, M. Nagata, and Y. Osumi, Dual effects of electrical stimulation of ventromedial hypothalamic neurons on gastric acid secretion in rats, *Am J Physiol*. 245:G265 (1983).
6. H.P. Weingarten, and T.L. Powley, Ventromedial hypothalamic lesions elevate basal and cephalic phase gastric acid output, *Am J Physiol*. 239:G221 (1980).
7. Y. Osumi, S. Aibara, K. Sakae, and M. Fujiwara, Central noradrenergic inhibition of gastric mucosal blood flow and acid secretion in rats, *Life Sci*. 20:1407 (1977).
8. Y. Okuma, Y. Osumi, T. Ishikawa, and M. Nagata, Central noradrenergic-cholinergic interaction in regulation of gastric acid secretion in rats, *Life Sci*. 32:1363 (1983).
9. Y. Taché, W. Vale, and M. Brown, Thyrotropin-releasing hormone: CNS action to stimulate gastric acid secretion, *Nature*. 287:149 (1980).
10. R.C. Rogers, and G.E. Hermann, Dorsal medullary oxytocin, vasopressin, oxytocin antagonist, and TRH effects on gastric acid secretion and heart rate, *Peptides*. 6:1143 (1985).
11. P. Juhani, and R. Heikki, Vagal-dependent stimulation of gastric acid secretion by intracerebroventricularly administered atrial natriuretic peptide in anaesthetized rats, *Eur J Pharmacol*. 141:493 (1987).
12. S.L. Allen, E.M. John, K. Julie, G. Martha, and E.S. Stephen, Muscimol induces gastric acid secretion after central administration, *Brain Res*. 229:270 (1981).
13. Y. Goto, Y. Taché, H. Debas, and D. Novin, Gastric acid and vagus nerve response to GABA agonist baclofen, *Life Sci*. 36:2471 (1985).

14. T. Ishikawa, Y. Osumi, and T. Nakagawa, Cholecystokinin intracerebroventricularly applied stimulates gastric acid secretion, *Brain Res.* 333:197 (1985).

15. H.J. Lenz, R. Klapdor, S.E. Hester, V.J. Webb, R.F. Galyean, J.E. Rivier, and M.R. Brown, Inhibition of gastric acid secrtion by brain peptides in the dog: Role of the autonomic nervous system and gastrin, *Gastroenterology.* 91:905 (1986).

16. T. Ishikawa, Y. Osumi, M. Fujiwara, and M. Nagata, Possible roles of central cholinergic nicotinic mechanisms in regulation of gastric functions, *Eur J Pharmacol.* 80:331 (1982).

17. Y. Okuma, and Y. Osumi, Central cholinergic descending pathway to the dorsal motor nucleus of the vagus in regulation of gastric functions, *Jpn J Pharmacol.* 41:373 (1986).

18. Y. Okuma, Y. Osumi, T. Ishikawa, and T. Mitsuma, Enhancement of gastric acid output and mucosal blood flow by tripeptide thyrotropin releasing hormone microinjected into the dorsal motor nucleus of the vagus in rats, *Jpn J Pharmacol.* 43:173 (1987).

19. R.L. Stephens, T. Ishikawa, H. Weiner, D. Novia, and Y. Taché, TRH analogue, RX 77368, injected into dorsal vagal complex stimulated gastric secretion in rats, *Am J Physiol.* 254:G639 (1988).

20. Y. Taché, W. Vale, J. Rivier, and M. Brown, Brain regulation of gastric secretion: Influence of neuropeptides, *Proc Natl Acad Sci USA.* 77:5515 (1980).

21. Y. Taché, Y. Goto, M.W. Gunion, W. Vale, J. Rivier, and M. Brown, Inhibition of gastric acid secretion in rats by intracerebral injection of corticotropin-releasing factor, *Science.* 222:935 (1983).

22. Y. Taché, M. Gunion, M. Lauffenburger, and Y. Goto, Inhibition of gastric acid secretion by intracerebral injection of calcitonin gene related peptide in rats, *Life Sci.* 35:871 (1984).

23. J.J. Hughes, A.S. Levine, J.E. Morley, B.A. Gosnell, and S.E. Silvis, Intraventricular calcitonin gene- related peptide inhibits gastric acid secretion, *Peptides.* 5:665 (1984).

24. C. Roze, M. Dubrasquet, J. Chariot, and C. Vaille, Central inhibition of basal pancreatic and gastric secretion by β-endorphin in rats, *Gastroenterology.* 79:659 (1980).

25. G.A. Humphreys, J.S. Davison, and W.L. Veal, Injection of neuropeptide Y into the paraventricular nucleus of the hypothalamus inhibits gastric acid secretion in the rat, *Brain Res.* 456:241 (1988).

26. H.J. Lenz, CNS regulation of gastric and autonomic functions in dogs by gastrin-releasing peptide, *Am J Physiol.* 255:G298 (1988).

27. W.M. Michael, and T.D. Haile, Central nervous system inhibition of pentagastrin-stimulated acid secretion by insulin-like growth factor II, *Life Sci.* 42:2091 (1988).

28. J. Puurunen, Central nervous system effects of arachidonic acid, PGE_2, $PGF_{2\alpha}$, PGD_2 and PGI_2 on gastric secretion in the rat, *Br J Pharmacol.* 80:255 (1983).

29. K. Yokotani, K. Yokotani, Y. Okuma, and Y. Osumi, Sympathoadrenomedullary system mediation of the prostaglandin E_2-induced central inhibition of gastric acid output in rats, *J Pharmacol Exp Ther.* 244:335 (1988).

30. E.S. Saperas, H. Yang, C. Rivier, and Y. Taché, Central action of recombinant interleukin-1 to inhibit acid secretion in rats, *Gastroenterology.* 99:1599 (1990).

31. Y. Osumi, Y. Nagasaka, W.L.H. Fu, and M. Fujiwara, Inhibition of gastric acid secretion and mucosal blood flow induced by intraventricularly applied neurotensin in rats, *Life Sci.* 23:2275 (1978).

32. D.E. Hernandez, G.A. Mason, J.W. Adcock, R.C. Orlando, and A.J. Prange, Effect of hypophysectomy, adrenalectomy, pituitary hormone secretion and gastric acid secretion on neurotensin induced protection against stress gastric lesions, *Life Sci.* 40:973 (1987).

33. L. Zhang, L. Xing, L. Demers, J. Washington and G. Kauffman, Central neurotensin inhibits gastric acid secretion: an adrenergic mechanism in rats, *Gastoenterology.* 97:1130 (1989).

34. K. Fuxe, Evidence for the existence of monoamine neurons in the central nervous system, *Acta Physiol Scand Suppl.* 247:39 (1965).

35. L.A. Loizou, Projections of the nucleus locus coeruleus in the albino rat, *Brain Res.* 15:563 (1969).

36. L. Olson, and K. Fuxe, Further mapping out of central noradrenaline systems. Projections of the "subcoeruleus" area, *Brain Res.* 43:289 (1972).

37. D.M. Jacobowitz, and M. Palkovits, Topographic atlas of catecholamine and acetylcholinesterase containing neurons in the rat brain, *J Comp Neurol.* 157:13 (1974).

38. P. Levitt, and R.Y. Moore, Origin and organization of brain stem catecholamine innervation in the rat, *J Comp Neurol.* 186:505 (1979).

39. Y. Osumi, T. Ishikawa, Y. Okuma, Y. Nagasaka, and M. Fujiwara, Inhibition of gastric functions by stimulation of the rat locus coeruleus, *Eur J Pharmacol.* 75:27 (1981).

40. Y. Takahashi, K. Satoh, T. Sakumoto, K. Yamamoto, Y. Kimoto, M. Tohyama, I. Kamei, and N. Shimizu, Noradrenaline innervation of the ala cinerea, *Acta Histochem Cytochem.* 11:120 (1978).

41. M. Nagata, Y. Okuma, and Y. Osumi, Effects of intracerebroventricularly applied nicotine on enhanced gastric acid secretion and mucosal blood flow in rats, *Eur J Pharmacol.* 101:185 (1984).

42. M. Nagata, Y. Okuma, and Y. Osumi, Nicotine applied into the dorsal motor nucleus of the vagus inhibits enhanced gastric acid output and mucosal blood flow in rats, *Eur J Pharmacol.* 121:313 (1986).

43. Y. Osumi, Central neurotransmitters and regulation of gastric acid secretion, *Folia Pharmacol Jpn.* 96:205 (1990).

44. M. Brown, Y. Taché, and D. Fisher, Central nervous system action of bombesin: mechanism to induce hyperglycemia, *Endocrinology.* 105:660 (1979).

45. Y. Okuma, K. Yokotani, and Y. Osumi, Sympatho-adrenomedullary system mediation of the bombesin- induced central inhibition of gastric acid secretion, *Eur J Pharmacol.* 139:73 (1987).

46. K. Yokotani, I. Muramatsu, M. Fujiwara, and Y. Osumi, Effects of the sympatho-adrenal system on vagally induced gastric acid secretion and mucosal blood flow in rats, *J Pharmacol Exp Ther.* 224:436 (1983).

47. K. Yokotani, I. Muramatsu, and M. Fujiwara, Alpha-1 and alpha-2 type adrenoceptors involved in the inhibitory effect of splanchnic nerves on parasympathetically stimulated gastric acid secretion in rats, *J Pharmacol Exp Ther.* 229:305 (1984).

48. M.W. Gunion, and Y. Taché, Bombesin microinfusion into the paraventricular nucleus suppresses gastric acid secretion in rats, *Brain Res.* 422:118 (1987).

49. Y. Okuma, K. Yokotani, and Y. Osumi, Central site of inhibitory action of bombesin on gastric acid secretion in rats, *Jpn J Pharmacol.* 45:129 (1987).

50. L.S. Wolfe, Eicosanoids: Prostaglandins, thromboxanes, leukotrienes, and other derivatives of carbon-20 unsaturated fatty acids, *J Neurochem.* 38:1 (1982).

51. Y. Okuma, K. Yokotani, and Y. Osumi, Chemical sympathectomy with 6-hydroxy-dopamine potentiates intracerebroventricularly applied bombesin-induced increase in plasma adrenaline, *Life Sci.* 22:1611 (1991).

ROLE OF CORTICOTROPIN-RELEASING FACTOR IN STRESS-INDUCED CHANGES IN GASTROINTESTINAL TRANSIT

Thomas F. Burks

Studies of pathological stress responses in patients and of experimental stress in normal human subjects indicate that stress can bring about profound changes in gastrointestinal motility and propulsion. Patients with irritable bowel syndrome (IBS) or normal subjects exposed to stress often display disordered patterns of contractions or transit in the small intestine and excessive contractions of the colon[1-3]. Mechanistic studies of stress require laboratory models that mimic, to the extent possible, the dysfunctions of motility associated with stress responses in humans.

STRESS MODEL

Systematic study of multiple animal models of stress in our laboratory revealed an array of gastrointestinal and other changes[4]. The variety of responses point up the importance of appropriate model selection in relation to the endpoint(s) of interest. At a minimum, it is necessary to show that the model actually results in a stress response, classically characterized by increases in circulating levels of adrenocorticotropin (ACTH), in the subjects. Our examination of nine rat models of stress allowed classification of the models into two general categories: those associated with alterations in small intestinal transit (wrap restraint, cold water swimming, cold restraint and exposure to vapors of diethyl ether) and those without discernable changes in transit (electroconvulsive shock, footshock, warm water swimming, exposure to cold ambient temperature and exposure to a cat)[4-6]. Several of the models produced analgesia, another hallmark of stress, but only three models were associated with both analgesia and changes in transit: wrap restraint, cold water swimming, and exposure to vapors of diethyl ether. Cold water swimming was rejected as a model because it produced profound hypothermia, which could in itself influence gastrointestinal motility. Exposure to vapors of diethyl ether was rejected because the initial analgesia was not blocked by naloxone, an opioid antagonist, and was probably the result of direct pharmacological actions of ether itself[4]. For these reasons, we selected the wrap restraint model of stress for detailed mechanistic investigations. We subsequently demonstrated that, in contrast to some models, the rat wrap restraint model does not result in formation of gastric or duodenal ulcers,

Neuroendocrinology of Gastrointestinal Ulceration
Edited by S. Szabo and Y. Taché, Plenum Press, New York, 1995

a distinct advantage in motility studies because ulcers could in themselves contribute to stress[7].

The wrap restraint model of stress in rats resulted in prompt increases in plasma levels of ACTH and β-endorphin, indicative of pituitary activation as part of the stress response[7]. Naloxone-sensitive analgesia, as assessed by the tailflick latency test, also developed.

Two interesting features of wrap restraint stress were noted. First, female rats were more susceptible to the effects of the stressor than male rats[4]. Second, a pronounced circadian rhythm of sensitivity to stress-induced changes in intestinal transit and release of ACTH was evident[7]. Rats were most susceptible to both stress-induced release of ACTH and changes in transit in the late afternoon hours (5:00-7:00 P.M.). As rats are nocturnal animals, the peak sensitivity to stress effects in rats would correspond to early morning hours in humans, a time when symptoms of functional bowel disease often are most pronounced[8].

CHANGES IN GASTROINTESTINAL TRANSIT

Wrap restraint stress produced consistent changes in gastrointestinal transit as assessed by the geometric center of distribution of radiochromium marker[9]. Small intestinal transit was inhibited by up to 50%, while colonic transit was increased and fecal excretion was increased[7]. The rate of gastric emptying was not affected. However, emptying of the cecum was accelerated by wrap restraint stress.

CRF MEDIATION OF THE STRESS RESPONSE

The inhibition of small intestinal transit induced by wrap restraint stress was not affected by hypophysectomy or adrenalectomy[7]. Also, intravenous (i.v.) administration of ACTH or β-endorphin did not mimic the effects of the stressor on small intestinal or colonic transit. We therefore concluded that the pituitary-adrenal axis was not directly implicated in mediation of the effects of stress[7].

However, administration of corticotropin-releasing factor (CRF) resulted in dose-related inhibition of gastric emptying, inhibition of small intestinal transit, stimulation of colonic transit, and stimulation of fecal excretion[10]. When administered in the dose of 0.3 μg i.c.v. or i.v., CRF produced effects that were quantitatively similar to the stress-related pattern of gastrointestinal effects: no change in gastric emptying, inhibition of small intestinal transit, increased colonic transit, and increased fecal excretion (Table 1). CRF (0.3 μg i.v.) also increased plasma levels of ACTH to values similar to those obtained with stress, and the time course of elevation of ACTH was similar [10].

Exogenously administered CRF and stress both produced similar changes in the parameters measured. Also, CRF is thought to initiate both endocrine and autonomic responses to stress. We therefore administered a CRF antagonist, α-helical CRF-(9-41)[12], to determine whether it would modify responses to exogenously administered CRF and to stress.

α-Helical CRF-(9-41) had no effect on gastrointestinal transit when administered alone either i.c.v. or i.v., but it effectively antagonized the effects of exogenously administered (i.c.v. or i.v.) CRF on small intestinal and colonic transit, even when CRF was given in doses as high as 10 μg[10]. Either i.c.v. or i.v. administration of α-helical CRF-(9-41) immediately before application of wrap restraint stress completely blocked the stress-induced increase in colonic transit and significantly inhibited the stress-induced increase in fecal excretion. However, α-helical CRF-(9-41) did not antagonize the decrease in small intestinal transit associated with stress.

Table 1. Effects of Stress and CRF on Gastrointestinal Functions in Rats

Treatment	Gastric emptying rate	Small intestinal transit	Colonic transit	Fecal excretion
CRF 0.3 μg i.c.v.	0	↓	↑	↑
CRF 0.3 μg i.v.	0	↓	↑	↑
Stress	0	↓	↑	↑

0 = little or no change
↓ = significant decrease
↑ = significant increase

These data indicate that extrapituitary actions of CRF are responsible for the changes in colonic transit associated with the wrap restraint model of stress in rats. Others also have since concluded that CRF release is responsible for stress-induced changes in gastrointestinal functions. In mice exposed to acoustic or cold stress, peripheral (i.p.) administration of CRF antiserum blocked the effects of the stressors on gastric emptying[13]. In a study carried out with wrap restraint stress in rats, i.c.v. but not i.v. administration of α-helical CRF-(9-41) blocked stress-induced decreases in gastric secretion, gastric emptying and small intestinal transit, as well as stress-induced increase in colonic transit[14]. While the site at which it acts is unclear, these investigations from multiple laboratories conclude that CRF is a critical mediator of stress-related changes in gastrointestinal motor effects. CRF produced similar functional changes whether administered i.c.v. or systemically and peripherally administered peptide CRF antagonist and CRF antiserum both blocked effects of stress[10,13]. It is therefore not possible at this time to define precisely the site of action of stress-released CRF as being in the brain or in the periphery. However, peripherally administered α-helical CRF-(9-41) did not block effects of stress in one study, whereas stress effects on gastrointestinal function were reduced by the CRF antagonist administered into the cerebral ventricles[14]. Obviously, this series of experiments suggests a predominantly central site of CRF action in mediation of stress effects.

A more recent study, using a non-physical form of stress in rats (conditioned fear), indicated that CRF released in the brain, not in the periphery, is responsible for stress-induced changes in motility of the colon[15]. The bulk of the evidence would thus support a primary central site of CRF action in gastrointestinal motility responses to stress.

SUMMARY

CRF appears to serve as a major link in initiation of stress effects on gastrointestinal functions. Both stress and exogenously administered CRF inhibit small intestinal transit and increase colonic transit in rats. The effects of both stress and CRF are antagonized by a CRF antagonist and by CRF antiserum, findings that strongly support a mediator role of CRF in gastrointestinal responses to stress. Most experiments indicate that a central site of CRF action is responsible for the motility changes associated with stress.

ACKNOWLEDGEMENTS

Work from the author's laboratory was supported by USPHS grant DA 02163. I am grateful to Dr. Cynthia L. Williams and Julia M. Peterson for their collaboration in these experiments.

REFERENCES

1. D. Kumar, and D.L. Wingate, Irritable bowel syndrome, *in:* "An Illustrated Guide to Gastrointestinal Motility," D. Kumar, S. Gustavsson, eds., John Wiley & Sons, Chichester (1988).
2. P. Latimer, S. Sarna, D. Campbell *et al.*, Colonic motor and myoelectrical activity: a comparative study of normal subjects, psychoneurotic patients, and patients with irritable bowel syndrome, *Gastroenterology.* 80:893 (1981).
3. F. Narducci, W.J. Snape, W.M. Battle *et al.*, Increased colonic motility during exposure to a stressful situation, *Dig Dis Sci.* 30:40 (1985).
4. C.L. Williams, and T.F. Burks, Stress, opioids, and gastrointestinal transit, *in:* "Neuropeptides and Stress," Y. Taché, J.E. Morley, M.R. Brown, eds., Springer-Verlag, New York (1989).
5. J.J. Galligan, F. Porreca, and T.F. Burks, Dissociation of analgesic and gastrointestinal effects of electroconvulsive shock-released opioids, *Brain Res.* 271:354 (1983).
6. J.J. Galligan, F. Porreca, and T.F. Burks, Footshock produces analgesia but no gastrointestinal motility effect in the rat, *Life Sci.* (Suppl. 1) 33:473 (1983).
7. C.L. Williams, R.G. Villar, J.M. Peterson, and T.F. Burks, Stress-induced changes in intestinal transit in the rat: a model for irritable bowel syndrome, *Gastroenterology.* 94:611 (1988).
8. D. Kumar, and D.L. Wingate, IBS: a paroxysmal disorder, *Lancet.* ii:973 (1985).
9. M.S. Miller, J.J. Galligan, and T.F. Burks, Accurate measurement of intestinal transit, *J Pharmacol Methods.* 6:211 (1981).
10. C.L. Williams, J.M. Peterson, R.G. Villar, and T.F. Burks, Corticotropin-releasing factor directly mediates colonic responses to stress, *Am J Physiol.* 253:G582 (1987).
11. L.A. Fisher, J. Rivier, C. Rivier *et al.*, Corticotropin-releasing factor (CRF): central effects on mean arterial pressure and heart rate in rats, *Endocrinology.* 10:2222 (1982).
12. J. Rivier, C. Rivier, and W. Vale, Synthetic competitive antagonists of corticotropin-releasing factor-induced ACTH secretion in the rat, *Science.* 224:889 (1985).
13. L. Búeno, and M. Gúe, Evidence for the involvement of corticotropin-releasing factor in the gastrointestinal disturbances induced by acoustic and cold stress in mice, *Brain Res.* 441:1 (1988).
14. H.J. Lenz, A. Raedler, H. Greten *et al.*, Stress-induced gastrointestinal secretory and motor responses in rats are mediated by endogenous corticotropin-releasing factor, *Gastroenterology.* 94:1510 (1988).
15. M. Gúe, J.L. Junien, and L. Búeno, Conditioned emotional response in rats enhances colonic motility through the central release of corticotropin-releasing factor, *Gastroenterology.* 100:964 (1991).

STRESS, CORTICOTROPHIN-RELEASING FACTOR (CRF) AND GASTRIC FUNCTION

Herbert Weiner

STRESSORS AND GASTRIC EROSIONS

Selye[1] demonstrated that many different kinds of "stressors" were associated with the development of gastric erosions (GE's) in the rat. Since that time the number of methods for producing GE's in the rat has grown exponentially[2-4]. Initially, the burden of the pathogenesis of GE was placed on the corticosteroids and on increases in gastric acid secretion (GAS). Yet, Cannon in 1929 first showed that GAS was reduced in the threatened cat. Since that time many different (stressful) perturbations of a physical and psychobiological nature have been shown to reduce GAS (reviewed in 5): They include restraining or rotating rats, exposing them to unpredictable electric shock or making them avoid it in free-operant experiments, confronting them with an aggressive opponent, or housing them in a hot environment.

Per contra, cold restraint sometimes increases GAS[4,6-13], an effect that is vagally mediated[10,13]. The increase in GAS may be specific to cold exposure, or restraint in a cold environment[7,13,14] although it does not always occur[6,15]. Increases in GAS neither proves or disproves the pathogenetic role of hydrochloric acid in GE formation: for e.g., GE's may occur during "stressful" procedures without increases in GAS. The evidence suggests that the presence of acid in the stomach (but not increases thereof) plays a permissive role in GE formation. Complete suppression of GAS (for e.g. by cimetidine) prevents GE's even though no significant increase in GAS has previously occurred[16].

Corticotropin Releasing Factor (CRF)

CRF is a 41-aminoacid peptide, first isolated from ovine hypothalamus and characterized in 1981[17]. The chemical structure of rat, pig and human CRF are identical[18-21]. CRF is widely distributed throughout the brain and body. In both places it is chemically homologous[22]. By means of radio-immunoassay (RIA) and immunohistochemistry, the largest concentration of CRF has been found in the perikarya of neurons of the dorsal aspect of the medial parvo-cellular portion of the paraventricular nucleus (PVN) of the hypothalamus, whose axons project to the median eminence from where the peptide is released into the portal circulation of the

Neuroendocrinology of Gastrointestinal Ulceration
Edited by S. Szabo and Y. Taché, Plenum Press, New York, 1995

hypophysis[23-25]. In addition, CRF-containing neurons are widely distributed throughout the brain[25] -- in the forebrain[26], amygdala, central gray matter of the brain stem, 3 pontine nuclei (the locus ceruleus, dorsal tegmental and parabrachial) and the medulla (the DVC, and A1 and A2 preganglionic neurons at the levels of T_{2-7}, and L_{2-3}[29]. Immunoreactivity for CRF is also present in the adrenal, liver, lung, pancreas and gastrointestinal tract (stomach and duodenum only)[22,30,31]. In these gut regions, CRF is contained in neuronal cell bodies and fibers and in typical endocrine cells[22,31,32].

CRF Receptors. These are found in the anterior hypophysis[18]; neocortex; limbic system; NTS; substantia gelatinosa of N.V.; locus ceruleus, spinal cord, cervical, thoracic, and celiac sympathetic nervous system, and in the adrenal medulla of several mammalian species[26,33-36].

Release of CRF During "Stress". Immunoreactive CRF is not usually detectable in the blood serum of rats and man under normal or stressful conditions[37,38]. However, it increases from baseline levels during periods of hemorrhage or stress[39] in the hypophyseal portal circulation of rats, in the adrenal venous blood of dogs[24,37] and in the rat brain [40,41].

Physiological Actions of CRF. CRF regulates the secretion by the anterior pituitary of the peptide hormones derived from pro-opiomelanocortin[17,42,43]. When injected ICV it has additional wide-ranging activities on behavior and physiology -- increasing locomotor activity (especially in a familiar setting), grooming behavior, and fright[44-47], while decreasing food and water intake[46,48], sexual receptivity[49] and gonadotropin secretion[50]. Also (when administered ICV), it activates sympathetic outflow[51], raises heart rate and blood pressure[52], oxygen consumption[51] and blood levels of epinephrine (E)[53], norepinephrine, glucose, glucagon[53] but it depresses growth hormone levels[54]. These effects are counteracted by the CRF antagonist, α-helical CRF[55]: In fact, stressed (e.g. restrained) animals pretreated by ICV α-helical CRF failed to have increased ACTH secretion, plasma E levels, or reduced food intake[41,43,48,50,53-57]. Such data support the hypothesis that CRF plays a key role in the psychobiological responses to stress of several kinds, including restraint.

Action of CRF on the Gastrointestinal Tract. In 1983, Taché[58] reported that ICV instillation of CRF inhibits GAS in rats and in dogs[59,60] and prevents GE during cold restraint[5]. Later, ICV, IC, or PVN[61] injections in rats and dogs were shown to delay gastric emptying[62], to inhibit gastric blood flow[64] and bicarbonate secretion[65]; but not to alter mesenteric cyclic migratory motor complexes (MMC)[63], to stimulate gastric blood flow[64] and bicarbonate secretion[65]; but not to affect mesenteric blood flow[64] or the integrity of the gastric mucosa in contrast to TRH[66]. ICV CRF also enhances colonic transit[39,67].

Intravenous injection of CRF in the rat and dog inhibits GAS [58,60,68] gastric emptying[62,68] and MMC's but has no effect on pancreatic bicarbonate or protein secretion[68,69]. In dogs and man CRF selectively stimulates pancreatic polypeptide secretion but not that of a large variety of other gut peptides[63,68,70,71].

Because IC and IV CRF have somewhat similar actions on gastric function, it is crucial to ascertain that the peptide is acting within the brain when injected IC or ICV and not by leaking into the circulation. Pretreatment with IV infusion of CRF

124

antibody does not prevent the inhibition of gastric emptying and GAS in rats by IC CRF but it does so if CRF is administered IV[72].

The sites of action of CRF in the brain to produce the above effects on gastric function remain unknown with the exception of the PVN[12,58,61,65,73].

Effect of CRF on Gastric Contractions. Based on the foregoing review, we have studied the effects of IC and IV CRF on gastric contractions stimulated by injecting the dorsal vagal complex with the TRH analog, RX77368 (p-Glu-His-(3,3'dimethyl-Pro-NH$_2$), with IC 2-deoxy-D-glucose and IV infusions of carbachol in urethane-anesthetized rats.

We studied gastric contractions because: a) They are apparently necessary if not sufficient (as is the presence of gastric acid) in GEF. In stressful conditions usually associated with a high incidence of GEF, complete suppression of contractions (for e.g. by papaverine) averts mucosal erosions.

Intracisternal injection of the TRH analog, RX77368, was followed 15 min later by IC CRF (30-1000 ng) to determine the dose-response dynamics of CRF on gastric contractility. RX77368 stimulated gastric contractility within 10 min in anesthetized animals as previously demonstrated in conscious animals (Figure 1)[74]. CRF dose-dependently inhibited the high-amplitude contractions stimulated by 100 ng RX77368 (Figure 2). There was no change in frequency or duration of contractions with IC CRF injections which remained constant at 4.5 ± 0.2 contractions/min and 13.2 ± 0.7 s/contraction, respectively. The suppression began within the first 10 min following IC injection. Injection of 0.9% NaCl IC alone (n = 3) had no effect on basal contractile pattern and when injected 15 min following RX77368 (100 ng) it did not modify the increased contractility produced by the TRH analog (Figure 3).

Injection of 2-deoxy-D-glucose at a six mg dose IC reliably stimulated gastric motility index 8.2 ± 1.5-fold over basal levels. The maximal response was obtained

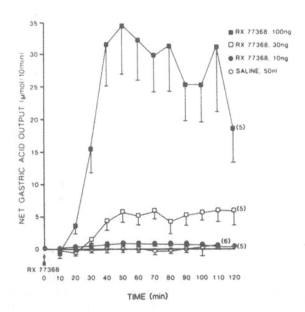

Figure 1. RX77368 microinjected unilaterally into the dorsal vagal complex (DVC). Time- and dose-related net stimulatory effect on gastric acid secretion in urethane-anesthetized rats. Each point represents means $\pm$ SEM of 5-6 animals. (Am J Physiol. 254:G640, 1988).

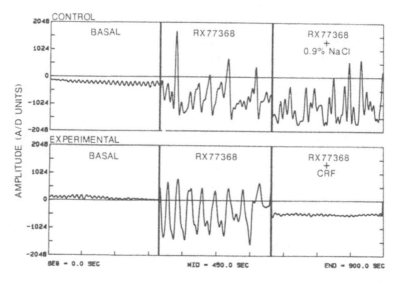

Figure 2. Typical tracings illustrating the effects of IC CRF (210 pmol) or vehicle control on gastric contractility stimulated by IC RX77368. Tracings were digitally stored during an experiment and sections reconstituted for illustrative purposes. Each panel consists of three, five min segments of recording. The leftmost tracing in each panel is a 24 h fasted basal recording. The middle panel is taken five min after the animal received RX77368 (260 pmol IC). This pattern typically persists for over one h. Right tracing, upper panel, is the control recording taken five min after IC injection of vehicle (0.9% NaCl) only (20 min after the injection of RX77368). The right tracing, lower panel, is the experimental recording taken five min after IC injection of CRF (210 pmol, IC; 20 min after IC injection of RX77368). The ordinate is labelled in analogue-to-digital units (A/D units). Full scale is $\pm$ 1.25 V. (Regul. Pept. 21:173, 1988).

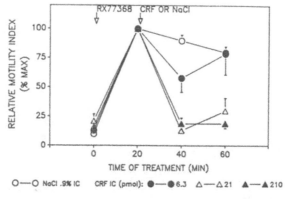

Figure 3. Dose-dependent inhibitory effect of IC CRF on gastric contractility stimulated by RX77368 in urethane-anesthetized rats. Basal contractility was recorded for 15 min. Rats were then injected IC with RX77368 (260 pmol) and 15 min later with 0.9% NaCl or various doses of CRF. (Regul. Pept. 21:173, 1988).

126

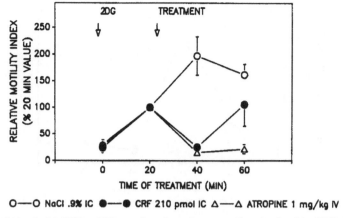

Figure 4. Inhibition by IC CRF and IV atropine of gastric contractility stimulated by 2DG. Following a 15 min basal recording the rats were injected IC with 2DG (6 mg) and 30 min later with IC 0.9% NaCl (10 μl), IC CRF (210 pmol) or IV atropine (1 mg/kg). (Regul. Pept. 21:173, 1988).

40 min after the injection (based on maximal stimulation for the whole group, n = 11; CRF (1 μg IC) injected 30 min after 2-deoxy-D-glucose reliably suppressed the stimulated contractility (p < 0.05) (Figure 4). There was no consistent significant change in frequency or duration of contractions.

Intravenous infusion of carbachol (15-300 μg/kg/h) dose-dependently stimulated gastric contractions (Figure 5). Motility index plateaued at 4.8 ± 0.9-fold over basal values for a 200 μg/kg/h dose (n = 22; p < 0.05). The stimulatory effect of carbachol was not altered by IC injection of CRF at 1 μg (Figure 6).

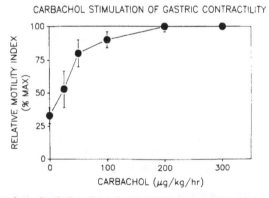

Figure 5. Dose-dependent stimulation of gastric contractility by IV carbachol in urethane-anesthetized rats. (Regul. Pept. 21:173, 1988).

RX77368 was injected IC and 15 min later CRF (21-210 pmol) was injected as an IV bolus. Doses of CRF, which potently suppresses the RX77368-stimulated gastric contractility when injected IC were ineffective when given intravenously. A significant inhibition of the gastric motility index (p < 0.05) was only observed at 40

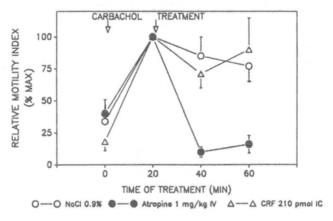

Figure 6. Effect of IC CRF or IV atropine on carbachol stimulated gastric contractility. Rats were first infused with carbachol (200 μg/kg/h). Following 15 min of recording they were injected either IC with CRF (210 pmol) or IV with atropine (1 mg/kg). Atropine but not IC CRF inhibited carbachol-stimulated gastric contractility. (Regul. Pept. 21:173, 1988).

min following the 210 pmol dose of CRF (Figure 7). Frequency and duration of inhibited contractions were not different from stimulated levels. Atropine (1 mg/kg) inhibited gastric contractility stimulated by 2 deoxy-D-glucose (n=7) and carbachol (n = 6) to basal levels whereas IV CRF (210 pmol; n = 6) had no effect on gastric contractility stimulated by either substance.

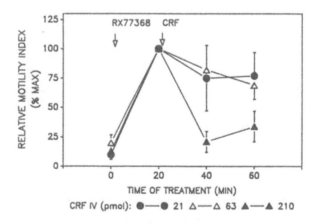

Figure 7. Effect of IV CRF on RX77368-stimulated gastric contractility. Following stimulation of gastric contractility by IC RX77368 (260 pmol), rats received an IV bolus injection of CRF at various doses (21-210 pmol). Only the highest dose of IV CRF significantly inhibited gastric contractility. (Regul. Pept. 21:173, 1988).

SUMMARY

In these experiments, IC administration of CRF inhibited the increase in gastric contractions brought about by the IC injection of TRH analog, RX77368, and 2-

deoxy-D-glucose but <u>not</u> those stimulated by IV carbachol. The inhibitory action of CRF was dose-dependent, rapid in onset and lasted at least 60 min. Intravenous CRF also inhibited the action of the TRH analog but it required 10 times the IC CRF dose to do so. Therefore, the effects of IC CRF were not likely to be due to its leakage into the blood stream. Intravenous CRF had no effect on the increased gastric contractions brought about by IC 2-deoxy-D-glucose or IV carbachol. Previous studies have shown that TRH also stimulates and CRF suppresses gastric acid secretion [74-77]. But the doses of CRF required to suppress contractions were less (by a factor of 10) than those needed to suppress acid secretion.

In fasted dogs, gastrointestinal interdigestive activity (migrating motor complexes) was immediately abolished by injection of CRF into the lateral brain ventricles. This effect, lasting several hours, is believed to be centrally mediated based on diminished activity of intravenous injection; it appears to be mediated, in part, by the blockade of the cyclic release of motilin[63].

The most likely brain site that TRH and CRF interact is in the dorsal motor nucleus and nucleus ambiguus of the vagus nerve[67], rather than by some mediating mechanism: for example the delay in gastric emptying brought about by CRF was not abolished by adrenalectomy or naloxone injection. These studies are relevant to our beginning understanding of the brain mechanisms by which certain stressors may produce gastric erosions.

Indeed, the co-injection of RX77368 (26 pmol) into the DVC of rats (which, as previously mentioned, produced a marked increase in the contractile force index within 30 m) with 63-210 pmol of CRF reduced the stimulated contractility in a dose-dependent manner. Neither CRF alone nor vehicle injected at the same site had any effect on the contractile force[78]. The medullary cell-bodies that synthesize TRH have been identified; they are located on the raphé nuclei, and innervate the DVC[79,80]. When these cell-bodies were chemically or electrically stimulated, gastric and pyloric motility were increased[81-83], presumably by the release of endogenous TRH which is transported to the DVC[84]. When 141 pmol of kainic acid was injected into the nucleus raphé pallidus, the contractile force index increased 5-fold, an effect that was suppressed by CRF microinjection at 105 pmol/site into the DVC[78].

Therefore, the conclusion that CRF counter-regulates TRH- (or its analog) stimulated gastric contractions at the level of the DVC, seems warranted.

REFERENCES

1. H. Selye, A syndrome produced by diverse nocuous agents, *Nature (London)*. 148:84 (1936).
2. T. Garrick, A. Veiseh, A. Sierra *et al.*, Corticotropin-releasing factor acts centrally to suppress stimulated gastric contractility in the rat, *Regul Pept.* 21:173 (1988).
3. W.P. Paré, Psychological studies of stress ulcer in the rat, *Brain Res Bull.* 5(1):73 (1980).
4. K. Takeuchi, O. Furukawa, and S. Okabe, Induction of duodenal ulcers in rats under water-immersion stress conditions. Influence of stress on gastric acid and duodenal alkaline secretion, *Gastroenterol.* 91:554 (1986).
5. Y. Taché, R.L. Stephens, and T. Ishikawa, Stress-induced alterations of gastrointestinal function: involvement of brain CRF and TRH, *in:* "Neuronal Control of Bodily Function: Basic and Clinical Aspects. Frontiers of Stress Research Vol III," H. Weiner, I. Florin, R.C. Murison, D. Hellhammer, eds., Hans Huber, Toronto (1989).
6. S. H. Ackerman, Early life events and peptic ulcer susceptibility: an experimental model, *Brain Res Bull.* 5(1):43 (1980).
7. I. Arai, M. Muramatsu, and H. Aihara, Body temperature dependency of gastric regional blood flow, acid secretion and ulcer formation in restraint and water-immersion stressed rats, *Jpn J Pharmacol.* 40:501 (1986).

8. I. Arai, H. Hirose, M. Muramatsu et al., Effects of restraint and water immersion stress and insulin on gastric acid secretion in rats, *Physiol Behav.* 40:357 (1987).

9. H. Kitagawa, M. Fujiwara, and Y. Osumi, Effects of water-immersion stress on gastric secretion and mucosal blood flow in rats, *Gastroenterol.* 77:298 (1979).

10. M. Muramatsu, I. Arai, J. Tamaki et al., Central regulation of gastric acetylcholine metabolism and acid output. Analysis using stress and 2-deoxy-D-glucose administration in rats, *Neurochem Int.* 8:553 (1986).

11. A. Robert, C. Lancaster, K.P. Kolbasa et al., Cold sensitizes to ulcer formation by aspirin, but not to gastric injury by ethanol or taurocholate, *Gastroenterol.* (Abstr) 90:1605 (1986).

12. Y. Taché, T. Ishikawa, R.L. Stephens et al., Stressor specific alterations of gastric function in rats: Role of brain TRH and CRF, *Gastroenterol.* (Abstr.) 94:A452 (1988).

13. R.T. Witty, and J.F. Long, Effect of ambient temperature on gastric secretion and food intake in the rat, *Am J Physiol.* 219:1359 (1970).

14. Y. Taché, D. Lesiege, and W. Vale, Gastric hypersecretion by intra-cisternal TRH: dissociation from hypophysiotropic activity and role of central catecholamines, *Eur J Pharmacol.* 107:149 (1985).

15. T. Garrick, S. Buack, and P. Bass, Gastric motility is a major factor in cold restraint-induced lesion formation in rats, *Am J Physiol.* 250(13):G191 (1986).

16. T. Garrick, Y. Goto, S. Buack et al., Cimetidine and ranitidine protect against cold restraint-induced gastric ulceration by suppressing gastric acid secretion, *Dig Dis Sci.* 32:1261 (1987).

17. W. Vale, J. Spiess, C. Rivier et al., Characterization of a 41-residue ovine hypothalamic peptide that stimulates secretion of corticotropin and beta-endorphin, *Science.* 213:1394 (1981).

18. F. Esch, N. Ling, P. Bohlen et al., Isolation and characterization of the bovine hypothalamic corticotropin releasing factor, *Biochem Biophys Res Commun.* 122:899 (1984).

19. M. Patthy, J. Horvath, M. Mason-Garcia et al., Isolation and amino acid sequence of corticotropin releasing factor from pig hypothalami, *Proc Natl Acad Sci USA.* 82:8762 (1985).

20. J. Rivier, J. Spiess, and W. Vale, Characterization of rat hypothalamic corticotropin-releasing factor, *Proc Natl Acad Sci USA.* 80:4851 (1983).

21. S. Shibahara, Y. Morimoto, Y. Furutani et al., Isolation and sequence-analysis of the human corticotropin releasing factor precursor gene, *EMBO J.* 2:775 (1983).

22. A.C.N. Kurseman, E.A. Linton, L.H. Rees et al., Corticotropin-releasing factor immunoreactivity in human gastrointestinal tract, *Lancet.* II:153 (1983).

23. S. Cummings, R. Elde, J. Ells et al., Corticotropin-releasing factor immunoreactivity is widely distributed within the central nervous system of the rat: An immunohistochemical study, *J Neurosci.* 3:1355 (1983).

24. P.M. Plotsky, and W. Vale, Hemorrhage-induced secretion of corticotropin-releasing factor-like immunoreactivity into the rat hypophyseal portal circulation and its inhibition by glucocorticoids, *Endocrinol.* 114:164 (1984).

25. L.W. Swanson, P.E. Sawchenko, J. Rivier et al., Organization of ovine corticotropin-releasing factor immunoreactive cells and fibers in the rat brain: An immunohistochemical study, *Neuroendocrinol.* 36:165 (1983).

26. E.B. De Souza, M.H. Perrin, J. Rivier et al., Corticotropin-releasing factor receptors in rat pituitary gland: auto-radiographic localization, *Brain Res.* 296:202-207 (1984).

27. I. Merchenthaler, Corticotropin releasing factor (CRF)-like immunoreactivity in the rat central nervous system. Extra-hypothalamic distribution, *Peptides.* 5(1):53 (1984).

28. F. Skofitsch, G.S. Hamill, and D.M. Jacobowitz, Capsaicin depletes corticotropin-releasing factor-like immunoreactive neurons in the rat spinal cord and medulla oblongata, *Neuroendocrinol.* 38:514 (1984).

29. T.L. Krukof, Segmental distribution of corticotropin-releasing factor-like and vasoactive intestinal peptide-like immunoreactivity in presumptive sympathetic preganglionic neurons of the cat, *Brain Res.* 382:153 (1986).

30. K. Hashimoto, K. Murakami, T. Hattori et al., Corticotropin-releasing factor (CRF)-like immunoreactivity in the adrenal medulla, *Peptides.* 5:707 (1984).

31. P. Petrusz, I. Merchenthaler, P. Ordronneau et al., Corticotropin-releasing factor (CRF)-like immunoreactivity in the gastro-, entero-pancreatic endocrine system, *Peptides.* 5:71 (1984).

32. H.J. Wolter, Corticotropin-releasing factor is contained within perikarya and nerve fibres of rat duodenum, *Biochem Biophys Res Commun.* 122:381 (1984).

33. J.R. Dave, L.E. Eiden, and R.L. Eskay, Corticotropin-releasing factor binding to peripheral tissue and activation of the adenylate cyclase-adenosine 3',5'-monophosphate system, *Endocrinol.* 116:2152 (1985).

34. E.B. De Souza, M.H. Perrin, T.R. Insel *et al.*, Corticotropin-releasing factor receptors in the rat forebrain: auto-radiographic identification, *Science.* 224:1449 (1984).
35. G. Skofitsch, T.R. Insel, and D.M. Jacobowitz, Binding sites for corticotropin releasing factor in sensory areas of the rat hindbrain and spinal cord, *Brain Res Bull.* 15:519 (1985).
36. R. Udelsman, J.P. Harwood, M.A. Millan *et al.*, Corticotropin releasing factor receptors in the primate peripheral sympathetic nervous system, *Nature (London).* 319:147 (1986).
37. D.M. Gibbs, and W. Vale, Presence of corticotropin releasing factor-like immunoreactivity in hypophysial portal blood, *Endocrinol.* 11:1418 (1982).
38. A. Sasaki, A.S. Liotta, M.M. Luckey *et al.*, Immunoreactive corticotropin-releasing factor is present in human maternal plasma during the third trimester of pregnancy, *J Clin Endocrinol Metab.* 59:812 (1984).
39. C.L. Williams, J.M. Peterson, R.G. Villar *et al.*, Corticotropin-releasing factor directly mediates colonic responses to stress, *Am J Physiol.* 253:G582 (1987).
40. P.B. Chappell, M.A. Smith, C.D. Kilts *et al.*, Alterations in corticotropin-releasing factor-like immunoreactivity in discrete rat brain regions after acute and chronic stress, *J Neurosci.* 6:2908 (1986).
41. Y. Nakane, T. Audhya, N. Kanie *et al.* Evidence for a role of endogenous corticotropin-releasing factor in cold, ether, immobilization and traumatic stress, *Proc Natl Acad Sci USA.* 82:1247 (1985).
42. C.L. Rivier, and P.M. Plotsky, Mediation by corticotropin releasing factor (CRF) of adenohypophysial hormone secretion, *Ann Rev Physiol.* 48:475 (1986).
43. D.L. Rivier, J. Rivier, and W. Vale, Inhibition of adrenocorticotropic hormone secretion in the rat by immunoneutralization of corticotropin-releasing factor, *Science.* 218:377 (1982).
44. K.T. Britton, G. Lee, W. Vale *et al.*, Corticotropin releasing factor (CRF) receptor antagonist blocks activating and "anxiogenic" actions of CRF in the rat, *Brain Res.* 369:303 (1986).
45. M. Eaves, K. Thatcher-Britton, J. Rivier *et al.*, Effects of corticotropin releasing factor on locomotor activity in hypophysectomized rats, *Peptides.* 6:923 (1985).
46. J.E. Morley, and A.S. Levine, Corticotrophin releasing factor, grooming and ingestive behavior, *Life Sci.* 31:1459 (1982).
47. R.E. Sutton, G.F. Koob, M. LeMoal *et al.*, Corticotropin releasing factor produces behavioral activation in rats, *Nature (London).* 297:331 (1982).
48. D.D. Krahn, B.A. Gosnell, M. Grace *et al.*, CRF antagonist partially reverses CRF- and stress-induced effects on feeding, *Brain Res Bull.* 17:285 (1986).
49. D.J.S. Sirinathsinghji, L.H. Rees, J. Rivier *et al.*, Corticotropin-releasing factor is a potent inhibitor of sexual receptivity in the famale rat, *Nature (London).* 305:232 (1983).
50. C. Rivier, J. Rivier, and W. Vale, Stress-induced inhibition of reproductive functions: Role of endogenous corticotropin-releasing factor, *Science.* 231:607 (1986).
51. M.R. Brown, L.A. Fisher, J. Rivier *et al.*, Corticotropin-releasing factor: effects on the sympathetic nervous system and oxygen consumption, *Life Sci.* 30:207 (1982).
52. L.A. Fisher, and M.R. Brown, Corticotropin-releasing factor and angiotensin II: Comparison of CNS actions to influence neuroendocrine and cardiovascular function, *Brain Res.* 296:41 (1984).
53. M.R. Brown, L.A. Fisher, V. Webb *et al.*, Corticotropin-releasing factor: A physiologic regulator of adrenal epinephrine secretion, *Brain Res.* 328:355 (1985).
54. C. Rivier, and W. Vale, Involvement of corticotropin-releasing factor and somatostatin in stress induced inhibition of growth hormone secretion in the rat, *Endocrinol.* 117:2478 (1985).
55. J. Rivier, C. Rivier, and W. Vale, Synthetic competitive antagonists of corticotropin-releasing factor: Effect of ACTH secretion in the rat, *Science.* 224:889 (1984).
56. M.R. Brown, T.S. Gray, and L.A. Fisher, Corticotropin-releasing factor receptor antagonist: Effects on the autonomic nervous system and cardiovascular function, *Regul Pept.* 16:321 (1986).
57. E.A. Linton, F.J.H. Tilders, S. Hodgkinson *et al.*, Stress-induced secretion of adrenocorticotropin in rats is inhibited by administration of antisera to ovine corticotropin-releasing factor and vasopressin, *Endocrinol.* 116:966, (1985).
58. Y. Taché, Y. Goto, M.W. Gunion *et al.*, Inhibition of gastric acid secretion in rats by intracerebral injection of corticotropin-releasing factor, *Science.* 222:935 (1983).
59. H.J. Lenz, S.E. Hester, and M.R. Brown, Corticotropin-releasing factor. Mechanisms to inhibit gastric acid secretion in conscious dogs, *J Clin Invest.* 75:889 (1985).
60. Y. Taché, Y. Goto, M. Gunion *et al.*, Inhibition of gastric acid secretion in rat and in dogs by corticotropin-releasing factor, *Gastroenterol.* 86:281 (1984).

61. M.W. Gunion, and Y. Taché, Intrahypothalamic microinfusion of corticotropin-releasing factor inhibits gastric acid secretion but increases secretion volume in rats, *Brain Res.* 411:156 (1987).

62. T. Pappas, H. Debas, and Y. Taché, Corticotropin-releasing factor inhibits gastric emptying in dogs. *Regul Pept.* 11:1 (1985).

63. L. Bueno, M.J. Fargeas, M. Gue *et al.*, Effects of corticotropin-releasing factor on plasma motilin and somatostatin levels and gastrointestinal motility in dogs, *Gastroenterol.* 91:884 (1986).

64. H.J. Lenz, L.A. Fisher, W. Vale *et al.*, Corticotropin-releasing factor, sauvagine, and urotensin I: Effects on blood flow, *Am J Physiol.* 249:R85 (1985).

65. M.W. Gunion, Y. Taché, and G.L. Kauffman, Intrahypothalamic corticotropin-releasing factor (CRF) increases gastric bicarbonate content, *Gastroenterol.* 88:1407 (1985).

66. Y. Goto, and Y. Taché, Gastric erosions induced by intracisternal thyrotropin-releasing hormone (TRH) in rats, *Peptides.* 6:153 (1985).

67. Y Taché, H. Yang, and M. Yoneda, Vagal regulation of gastric function involves thyrotropin-releasing hormone in the medullary raphe nuclei and dorsal vagal complex. *Digestion.* 54:65 (1993).

68. S.J. Konturek, J. Bilski, W. Pawlik *et al.*, Gastrointestinal secretory, motor and circulatory effects of corticotropin-releasing factor, *Life Sci.*, 37:1231 (1985).

69. T. Pappas, H. Debas, and Y. Taché, Corticotropin-releasing factor inhibits gastric emptying in dogs: studies on its mechanism of action, *Peptides.* 10(3):193 (1989).

70. N. Lytras, A. Grossman, L.H. Rees *et al.*, Corticotropin releasing factor: Effects on circulating gut and pancreatic peptides in man, *Clin Endocrinol.* 20:725 (1984).

71. T.R. Solomon, Control of exocrine pancreatic secretion, *in:* "Physiology of the Gastrointestinal Tract," L.R. Johnson, ed., Raven Press, New York (1987).

72. Y. Taché, M. Maeda-Hagiwara, and C.M. Turkelson, Central nervous system action of corticotropin releasing factor to inhibit gastric emptying in rats, *Am J Physiol.* 253:G241 (1987).

73. Y. Taché, H. Monnikes, B. Bonaz, and J. Rivier, Role if CRF in stress related alterations of gastric and motor function, *Ann NY Acad Sci.* 697:233 (1993).

74. R. Stephens, T. Ishikawa, H. Weiner *et al.*, TRH analogue, RX77368 injected into dorsal vagal complex stimulates gastric secretion in rats, *Am J Physiol.* 254(17):G639 (1988).

75. Y. Taché, Y. Goto, D. LeSiege *et al.*, Central nervous system action of thyrotropin-releasing hormone (TRH) to stimulate gastric acid and pepsin secretion in rats, *Endocrinol.* 112:149 (1983).

76. Y. Taché, D. Hamel, and M. Gunion, Inhibition of gastric acid secretion in rats by intracisternal or intrathecal injection of rat corticotropin releasing factor (rCRF), *Dig Dis Sci.* 29(Suppl):62A (1984).

77. T. Garrick, S. Buack, A. Veiseh *et al.*, Thyrotropin-releasing hormone (TRH) acts centrally to stimulate gastric contractility in rats, *Life Sci.* 40:649 (1987).

78. I. Heymann-Mönnikes, Y. Taché, M. Trauner, H. Weiner, and T. Garrick, CRF microinjected into the dorsal vagal complex inhibits TRH analog- and kainic acid-stimulated gastric contractility in rats, *Brain Res.* 554:139 (1991).

79. M. Palkovits, E. Mezey, R.L. Eskay, and M.J. Brownstein, Innervation of the nucleus of the solitary tract and the dorsal vagal nucleus by thyrotropin-releasing hormone-containing raphé neurons, *Brain Res.* 373:246 (1986).

80. R.C. Rogers, H. Kita, L.L. Butcher, and D. Novin, Afferent projections to the dorsal motor nucleus of the vagus, *Brain Res Bull.* 5:365 (1980).

81. P.J. Hornby, C.D. Rossiter, R.L. White, W.P. Norman, D.H. Kuhn, and R.A. Gillis, Medullary raphé: a new site for vagally mediated stimulation of gastric motility in cats, *Am J Physiol.* 258:G637 (1990).

82. M. McCann, G.E. Hermann, and R.C. Rogers, Nucleus raphé obscurus (nRO) influences vagal control of gastric motility in rats, *Brain Res.* 486:181 (1989).

83. H. Yang, M. Trauner, E. Livingstone, Y. Taché, and T. Garrick, TRH analogue, RX77368, microinjected into the raphé pallidus and obscurus stimulates gastric motility in rats, *Gastroenterol.* 98:A533 (1990).

84. R.L. White, C.D. Rossiter, Jr., P.J. Hornby, J.W. Harmon, D.K. Kasbekar, and R.A. Gillis, Excitation of neurons in the medullary raphé increases gastric acid and pepsin production in cats, *Am J Physiol.* 260:G91 (1991).

CORTICOTROPIN RELEASING FACTOR AND GASTRIC EROSIONS

Hans Kristian Bakke and Robert Murison

Several neuropeptides have been shown to either promote or inhibit the development of gastric erosions in rats; recent reviews have been presented by Taché[1] and Hernandez[2].

One of these peptides, corticotropin releasing factor (CRF) inhibits pentagastrin stimulated gastric acid secretion after either central[3] or peripheral[4] administration. However, the results reported concerning the effects of CRF on gastric erosions have not been consistent.

When administered peripherally, CRF (10 µg, subcutaneously) was shown to potentiate ulcer formation during cold restraint in one study[5]. We have previously reported a protective effect of 8-10 µg of CRF (intraperitoneal injection) during water restraint (1-2 h)[6,7]. The latter study revealed that while this protective effect was evident in rats aged 100 days, it was not in 200-day old rats. In fact, our data indicate that CRF exerts ulcerogenic properties in older (220 days) rats[7], as opposed to the effect seen in younger rats. Interestingly, this study also revealed a trend for ulcerogenic properties in unrestrained, 20 h fasted younger (100 days) animals. It may be that CRF exerts different gastric effects in rats depending on the condition (stress/non-stress) under which the rats are tested.

As for the effects seen after central administration, Nakane et al.[8] reported an ulcerogenic effect in 24 h food-deprived rats but not in restrained rats after intracisternal administration of 5 µg CRF. No significant potentiation of cold supine restraint-induced gastric erosions was observed. Conversely, others[9] have reported no increase in gastric ulcers in 24 h fasted, non-restrained rats 4 h after ICV administration of 10 µg CRF. Moreover, Krahn et al.[10] showed an inhibitory effect of intracerebroventricular (ICV) administration of 5 µg CRF during cold restraint. This effect was reversed by the CRF antagonist α-helical CRF (9-41) (ICV). The antagonist itself was inactive. Intraventromedial hypothalamic administration of 2 µg CRF (per side) inhibited cold restraint-induced mucosal damage in another study[11].

While the mechanisms behind the effects of CRF on gastric acid secretion have been carefully investigated, especially by Taché and coworkers[3,4,12], this seems not to be the case regarding the effect on restraint induced gastric erosions.

The inhibitory effect of ICV-administered CRF on gastric acid secretion is blocked by vagotomy and adrenalectomy but not by hypophysectomy or naloxone treatment. This effect therefore, seems to be a CNS phenomenon independent of hypophysiotropic actions but mediated through vagal and adrenal mechanisms[3]. However, the importance of gastric acid secretion for the development of gastric erosions have been called into question. Though the presence of some acid seems to be a prerequisite for erosions to develop, experimental studies have revealed restraint stress in rats to induce a <u>decreased</u> gastric acid secretion in rats[13,14]. The mechanisms by which centrally administered CRF affects stress-induced gastric erosions may therefore be different from those that cause the inhibition of gastric acid secretion. Previous experiments from our laboratory employing peripherally administered CRF have, on the other hand, given strong indications that the ulceroprotective effect of CRF is also independent of its hypophysiotropic actions[6,7].

In the experiment to be presented here, we undertook to elucidate the central mechanisms behind the effect of CRF on restraint stress-induced gastric erosions. Specifically, we wanted to investigate a possible interaction with the locus coeruleus (LC) noradrenergic system. Both the CNS noradrenergic system in general and the LC system in particular have been implicated in the development of restraint induced gastric erosions[15]. Anatomical studies have revealed coexistence of CRF immunoreactivity with noradrenaline (NA) within the LC[16] and alterations in CRF-like immunoreactivity in the LC after exposure to stress[17]. Moreover, electrophysiological studies have shown administration of CRF ICV or directly to LC neurons to increase discharge rates of these neurons and have given evidence for a neurotransmitter role for CRF in the LC during hemodynamic stress[18,19,20].

Experimental rats were in this study pretreated with N-(2-chloroethyl)-N-ethyl-2-bromobenzylamine hydrochloride (DSP-4), a highly selective and long-lasting noradrenergic neurotoxin. This drug affects mainly nerve terminal projections originating in the LC, while noradrenergic pericarya in this nucleus appear unaffected[21]. There is also a transient toxic effect on sympathetic neurons outside the brain with an almost normal appearance within one week after drug administration[15,21,22]. The transient nature of this peripheral effect has been shown in several organs; iris, stomach, salivary glands and the heart.

ICV administered neuropeptides can "leak" into the peripheral vascular system[23]. To control for such an occurrence, CRF was also administered intraperitoneally (IP) to DSP-4 treated animals.

PROCEDURES

35 male Sprague-Dawley rats (Möllegaard, Denmark) were treated IP with a single dose (0.4 ml) of DSP-4 (50 mg/kg in 0.9% NaCl) (Astra Läkemedel, Sweden). Control animals (N=12) were injected with an equivalent quantity of physiological saline. The average weight of the rats was 300 grams at the start of the experiment. Animals were single housed and maintained in light-(12:12 h) and temperature-regulated animal quarters. They had free access to food and water. Twenty experimental and all control animals were later implanted with chronic cannulas in the left lateral ventricle of the brain. Four DSP-4 animals died or were sacrificed due to physical illness with serious weight loss (three from the implanted group).

All groups were during the following weeks, tested for plasma catecholamine and corticosterone response to either ICV or IP administration of CRF. An open field test was also performed. The results from these experiments (to be presented elsewhere) were in accordance with previous reports. However, no interaction between the effects of CRF and pretreatment with DSP-4 was observed on any parameter.

Two weeks after the open field test, all animals were deprived of food but not water for 24 h prior to immobilization stress procedures. The immobilization procedure was water restraint for 120 min at $19.3 \pm 0.2°C$. Human CRF (Bissendorf, West Germany) 8 µg (1.68 nmol) dissolved in 8 µl 0.9% NaCl or only 8 µl saline was administered ICV 15 min before onset of the restraint period. Intraperitoneally treated rats were injected with equivalent doses in a volume of 0.4 ml at the same time point. The animals were sacrificed immediately after the restraint session. The stomachs were removed, cut along the lesser curvature, and cumulative ulcer length was scored. At the same time, brains from six DSP-4 treated and six control animals were rapidly taken out of the skull, the neocortex removed and frozen (liquid nitrogen). They were later analyzed for NA and dopamine content employing high-performance liquid chromatography (HPLC) methods.

RESULTS

HPLC analyses of cerebral cortex NA content revealed a 60% reduction after DSP-4 treatment (Mann-Whitney U Test: $Z=2.88$, $p<0.004$). Means, ng/g wet weight tissue (SEM); DSP-4: 132.53 (2.17), Control: 310.16 (13.96). No difference between groups was observed on dopamine content ($F<1.0$).

The cumulative ulcer length scores were statistically treated on square root transformed data due to heterogeneity of variance. In the ICV treated groups, a significant effect of CRF was shown ($F(1,25)=12.9$, $p<0.002$) and of DSP-4 ($F(1.25)=6.9$, $p<0.02$) (Two-way ANOVA). Subsequent Newman-Keuls tests revealed a significant decrease in water restraint-induced gastric erosions after CRF administration ($p<0.05$) in Control animals, while not in DSP-4 treated animals. In fact, the DSP-4 treated rats exhibited a significantly higher ulcer score after ICV injection of CRF than Control animals (Newman-Keuls test, $p<0.05$) (Fig. 1)

Intraperitoneal administration of CRF to DSP-4 animals, however, did cause a significant decrease in ulcer score (Newman-Keuls test, $p<0.05$) (Fig. 1).

DISCUSSION

The gastric ulcer scores from the present experiment give support to those previous reports that have shown a protective effect of centrally administered CRF under restraint[10,11,24]. This protective effect of ICV administered CRF was not evident in DSP-4 treated rats. Although not completely abolished after DSP-4 treatment, the ulceroprotective effect of ICV CRF administration seems dependent on an intact LC noradrenergic system. The still-present reduction of gastric pathology after peripheral administration of CRF to DSP-4 animals indicates that this interaction occurs centrally.

Our data support the proposal that CRF may have neurotransmitter/neuromodulator functions in the central nervous system [16,20]. Furthermore, the functional relationship between CRF and the LC noradrenergic system during acute and chronic stress as proposed by Chappell et al.[17] seems to be of major importance regarding the effect of CRF on stress-induced gastric pathology.

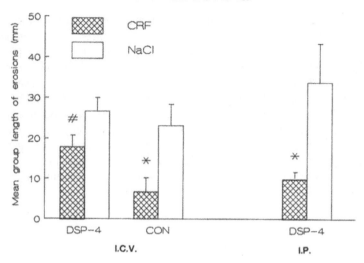

Figure 1. Mean group post restraint cumulative length of gastric erosions (mm ± SEM) after ICV or IP administration of CRF or NaCl to DSP-4 treated (DSP-4) and Control (CON) rats. *, p<0.05 compared to Saline group. #, p<0.05 compared to DSP-4 group. Reprinted with permission from Physiology and Behavior, H.K. Bakke, A. Bogsnes and R. Murison, Studies on the interaction between ICV effects of CRF and CNS noradrenaline depletion, Vol. 47, No 6 1990, Pergamon Press PLC.

REFERENCES

1. Y. Taché, The peptidergic brain-gut axis: influence on gastric ulcer formaction, *Chronobiol Int.* 4:11 (1987).
2. D.E. Hernandez, Neurobiology of brain-gut interactions, implications for ulcer disease, *Dig Dis Sci.* 34:1809 (1989).
3. Y. Taché, Y. Goto, M.W. Gunion *et al.*, Inhibition of gastric acid secretion in rats by intracerebral injection of corticotropin-releasing factor, *Science.* 222:935 (1983).
4. Y. Taché, Y. Goto, M. Gunion *et al.*, Inhibition of gastric acid secretion in rats and in dogs by corticotropin-releasing factor, *Gastroenterology.* 86:281 (1984).
5. N. Basso, Y. Goto, E. Passaro Jr. *et al.*, Hypothalamic peptides aggravate cold-restraint stress ulcer formation in the rat, *Gastroenterology.* 84:1100 (1983).
6. R. Murison, J.B. Overmier, D.H. Hellhammer *et al.*, Hypothalamo pituitary-adrenal manipulations and stress ulcerations in rats, *Psychoneuroendocrinology.* 14:331 (1989).
7. H.K. Bakke, and R. Murison, Plasma corticosterone and restraint induced gastric pathology: Age related differences after administration of corticotropin releasing factor, *Life Sci.* 45:907 (1989).
8. T. Nakane, N. Kanie, T. Audhya *et al.*, The effects of centrally administered neuropeptides on the development of gastric lesions in the rat, *Life Sci.* 36:1197 (1985).
9. Y. Goto, and Y. Taché, Gastric erosions induced by intracisternal injection of thyrotropin-releasing hormone (TRH) in rats, *Gastroenterology.* 86:1095 (1984).
10. D.D. Krahn, B. Wright, C.J. Billington *et al.*, Exogenous corticotropin-releasing factor inhibits stress induced gastric ulceration, *Soc Neurosci Abstr.* 12:1063 (1986).
11. M.W. Gunion, G.L. Kauffman Jr., and Y. Taché, Intrahypothalamic corticotropin-releasing factor elevates gastric bicarbonate and inhibits stress ulcers in rats, *Am J Physiol.* 258:G152 (1990).
12. R.L. Stephens Jr., H. Yang, J. Rivier *et al.*, Intracisternal injection of CRF antagonist blocks surgical stress-induced inhibition of gastric secretion in the rat, *Peptides.* 9:1067 (1988).
13. R. Menguy, Effects of restraint stress on gastric secretion in the rat, *Am J Dig Dis.* 5:911 (1960).
14. M. Hayase, and K. Takeuchi, Gastric acid secretion and lesion formation in rats under water-immersion stress, *Dig Dis Sci.* 31:166 (1986).

15. H.K. Bakke, R. Murison, and B. Walther, Effect of central noradrenaline depletion on corticosterone levels and gastric ulcerations in rats, *Brain Res.* 368:256 (1986).

16. S. Cummings, R. Elde, J.Ells *et al.*, Corticotropin-releasing factor immunoreactivity is widely distributed within the central nervous system of the rat: an immunohistochemical study, *J Neurosci.* 3:1355 (1983).

17. P.B. Chappell, M.A. Smith, C.D. Kilts *et al.*, Alterations in corticotropin-releasing factor-like immunoreactivity in discrete rat brain regions after acute and chronic stress, *J Neurosci.* 6:2908 (1986).

18. R.J. Valentino, Corticotropin-releasing factor (CRF) and physiological stressors alter activity of rat noradrenergic locus coeruleus neurons (LC) in a similar manner, *J Cell Biochem Suppl.* 12D:317 (1988).

19. R.J. Valentino, S.L. Foote, and G. Aston-Jones, Corticotropin-releasing factor activates noradrenergic neurons of the locus coeruleus, *Brain Res.* 270:363 (1983).

20. R.J. Valentino, R.G. Wehby, Corticotropin-releasing factor: evidence for a neurotransmitter role in the locus coeruleus during hemodynamic stress, *Neuroendocrinology.* 48:674 (1988).

21. G. Jonson, H. Hallman, F. Ponzio *et al.*, DSP4 (N-(2-chloroethyl)-N-ethyl-2-bromobenzylamine) a useful denervation tool for central and peripheral noradrenaline neurons, *Eur J Pharmacol.* 72:173 (1981).

22. G. Jaim-Etcheverry, and M. Zieher, DSP-4: a novel compound with neurotoxic effects on noradrenergic neurons of adult and developing rats, *Brain Res.* 188:513 (1980).

23. E. Passaro Jr., H. Debas, W. Oldendorf *et al.*, Rapid appearance of intraventricularly administered neuropeptides in the peripheral circulation, *Brain Res.* 241:335 (1982).

24. T. Shibasaki, N. Yamauchi, M. Hotta *et al.*, Brain corticotropin-releasing factor acts as inhibitor of stress induced gastric erosions in rats, *Life Sci.* 47:925 (1990).

THE LIMBIC SYSTEM AND STRESS ULCERS

Peter G. Henke

Our working hypothesis, over the past decade, has been that the limbic system modulates stress ulcer susceptibility[1,2]. It was largely based on early reports in the literature that linked (in a vague kind of way) this system of interconnected brain structures to the development of so-called "psychosomatic" disorders. In 1949 MacLean[3], elaborating on the Papez[4] theory of emotions, proposed that the limbic system might represent a primitive "visceral brain", responsible for the development of psychosomatic diseases. Much of the evidence was based on case reports from patients with temporal lobe epilepsies. Patients with so-called uncinate fits were described, whose feelings of terror were also sometimes associated with visceral auras, i.e., various gastric and intestinal symptoms, ranging from epigastric uneasiness, chewing movements and grinding of the teeth, to feelings of hunger and frequent defecation and urination. Correspondingly, the data from early animal studies indicated that parts of the limbic system might be involved in emotional behavior. Klüver and Bucy[5] had reported that monkeys with bilateral temporal lobectomies became tame and docile, showed a collection of symptoms which were labelled "psychic blindness", displayed various "oral tendencies", and engaged in bizarre sexual activities. It appeared that the monkeys could no longer discriminate between potentially dangerous objects and objects that might be useful to them. One interpretation of these studies is that the brain lesions had disconnected sensory inputs from the appropriate, adaptive behaviors. Later experiments showed that many of the effects of these radical lobectomies could be traced to damage in the amygdala[6]. These findings lead to the proposal that the amygdala connects sensory inputs with emotional and visceral responses, based on the recognition that a particular situation may be threatening to the organism[7].

Other limbic structures, originally included in the so-called Papez circuit included the hippocampal formation and the cingulate cortex[3,4]. The results of earlier stimulation studies suggested to some that these areas might be looked upon as the highest representations of visceral functions in the brain[3,8]. Subsequently, they were also implicated in fear and anxiety states[9,10]. In fact, lesions in the anterior cingulate gyrus had been used to alleviate ulcers and colitis in human patients[11,12]. In recent years, however, the emphasis has shifted to presumed functions in learning, motivation and memory, particularly, in the case of the hippocampal formation[13]. There are a number of studies that have suggested that damage to these areas might mimic the

Neuroendocrinology of Gastrointestinal Ulceration
Edited by S. Szabo and Y. Taché, Plenum Press, New York, 1995

effects of severe stress conditions on learning capacity and memory utilization[14]. In general agreement with these findings, the present proposal is that the cingulate cortex and the hippocampal formation may form part of a so-called coping system, where environmental experiences can influence the degree to which an organism becomes resistant to the effects of stressful conditions[2,15].

AMYGDALA

Anatomically, the amygdala is a heterogeneous structure, but it is also in a strategic position to influence autonomic and endocrine mechanisms. Traditionally, the amygdalar complex has been divided into a phylogenetically older corticomedial portion (medial, central and cortical nuclei), and the more recent basolateral group (basomedial, basolateral and lateral nuclei). On the basis of functional studies, however, the basomedial nucleus is frequently included in the corticomedial division[16,17]. This group of nuclei receives direct olfactory, gustatory, vagal-sensory, and pain-related inputs. However, all sensory systems gain access to widespread areas of the amygdalar complex[18-23].

Changes in gastrointestinal functions have been elicited by stimulation in a number of different nuclei. Increases in the volume and acidity of gastric secretions have been obtained after anteromedial, central and lateral stimulations[24-26]. Electrical stimulation of the medial and central nuclei also produced increased gastric motility[24]. But stimulation of the central nucleus, in addition, reduced the mucosal blood flow[27-29]. Inhibitory effects on secretions and motility have also been found following stimulation of more basal amygdalar regions[17,30]. Vagotomy abolished both excitatory and inhibiting effects[31,32].

Stimulations of the anteromedial[24], medial[34] and central[33,35] regions of the amygdala also produced stomach ulcers; an effect that was blocked by vagotomy[34], or sectioning the so-called amygdalofugal pathway[35]. This latter projection system connects the central amygdalar nucleus with autonomic control areas of the hypothalamus and lower brainstem[36-38]. For example, recent data have implicated the central nucleus as being important during stress ulcer development[33,39]. Fibers from this nucleus reach the lateral hypothalamus, perifornical area, ventral tegmental area, substantia nigra, central grey region, parabrachial nuclei, locus coeruleus, nucleus of the solitary tract, and dorsal vagal complex[40]. Radio-frequency lesions[41] or wire-knife cuts (unpublished observations) in this ventral pathway attenuated the effects of restraint-stress on stomach erosions in rats. A similar result occurred after bilateral lesions in the centromedial amygdala[41,42].

Recordings of multiple-unit activity in the amygdala indicated that stress-responsive neurons exist in the medial, central and lateral nuclei. However, low-level electrical stimulation of these units produced stomach erosions only from placements in the central nucleus[7]. Furthermore, prior "kindling" of the right centromedial amygdala, i.e., daily brief electrical stimulus pulses were presented until after-discharges could be recorded, also made the rat more susceptible to restraint-induced ulcers. This effect of prior kindling was observed regardless of whether or not motor seizures accompanied the after-discharges. Kindling by itself, however, produced very little stomach pathology[43]. These data show that a hyperfunctional state of these neurons, generated by kindling, produced an animal that was now abnormally susceptible to stressful experiences.

When the stress-related behaviors of neurons in the central amygdalar nucleus were more closely examined, several different multiple-unit profiles were found. Two of these profiles were related to the degree of stress ulceration found in these cases. Type 1 activity, represented by an initial increase in rate followed by suppression during the later stages of restraint, was seen in ulcer-susceptible rats. On the other hand, Type 2 responses, recorded as initial increases and subsequent baseline levels, were associated with "tough" animals. These data suggest that distinct neural "signatures" correlated with the animal's resistance to stress ulcer formation[44].

Follow-up experiments examined these neural signatures in rats which had been selectively bred for extreme differences in avoidance conditioning performance, the so-called Roman High - (RHA) and Low - (RLA) Avoidance strains[45]. In addition, Wistar rats which had been divided into high- and low-emotional categories, based on a criterion of "defecation" in an open-field arena before five rearing responses had occurred, were also included. The results showed that Type 1 profiles were mostly seen in the more emotional rats and in the RLA animals. On the other hand, Type 2 activity was usually seen in the less emotional and the RHA rats. The RLA and high-emotional Wistar rats also had more stomach erosions following physical restraint[44]. It is known from previous studies that RLA rats defecate more, "freeze" more, and ambulate less in an open-field than do RHA animals[45] indicating, perhaps, that open-field measures might be correlated with stress ulcer susceptibility, as has been suggested previously[46,47].

Low-level electrical stimulation of both types of units in the central nucleus, however, was found to produce stomach erosions in all cases. One interpretation of these results might be that although distinct neural signatures exist in the central nucleus, similar types of neurons generate these different response profiles. The assumption is that simply the behaviors of these neurons differ in stress-resistant as compared to stress-vulnerable animals[44].

The differences in the observed neuronal behaviors could be a function of changed dopamine (DA) transmission, possibly from fibers originating in the Ventral Tegmental Area projecting to the central amygdalar nucleus[48]. The central nucleus receives DA inputs from A8, A9 and A10, and is also particularly rich in various neuropeptide transmitters[49-54]. A series of studies, using agonists and antagonists, indicate that altered DA-transmission might be related to stress-ulcer susceptibility [55-58]. This may occur in interaction with a number of neuropeptides (Table 1). For example, applications of thyrotropin-releasing hormone into the central amygdala aggravated stress ulcer development. But this effect was reversed with concurrent DA-injections. Furthermore, the protective effects of both neurotensin and met-enkephalin were also reversed by DA-blocking agents[55,56,59,60].

In addition to this presumed protective function for DA in the central nucleus, recent data also showed that GABA (gamma-aminobutyric acid) receptors in this area might also be important during stressful conditions. It was found that peripheral injections of the benzodiazepine chlordiazepoxide (CDP) affected the multiple-unit activity in the amygdala during restraint-stress[7]. Recent studies also indicated that direct applications to the central amygdala of either GABA or CDP attenuated the restraint ulceration in rats. The specific benzodiazepine receptor antagonist RO 15-1788, however, reversed these protective effects. RO 15-1788 also showed an intrinsic aggravating effect when applied to the central nucleus[61]. Benzodiazepines applied to the central amygdala also attenuated the usual aggravation of stress ulcers seen after injections of either TRH or physostigmine into this area. These data suggested that benzodiazepines might block the presumed TRH-induced augmentation of acetylcholine effects during stressful conditions[62]. It is well-established that the amygdala contains benzodiazepine receptors and high concentrations of GABA[63].

Table 1. Interactions of Dopamine and Neuropeptides in the Central Amygdala During Stress Ulcer Development

Treatment	Pathology
DA	↓
TRH	↑
DA + TRH	↓
Cloz + DA + TRH	↑
DAMEA	↓
Cloz + DAMEA	↑
NT	↓
Hal + NT	-

TRH = thyrotropin-releasing hormone, DA = dopamine, Cloz = clozapine, DAMEA = [D -Ala^2] Met-enkephalinamide, NT = neurotensin, Hal = haloperidol; ↑ increase, ↓ decrease, - control level.

Benzodiazepines are thought to reduce anxiety states by interacting with GABA, as part of a supramolecular complex, facilitating the coupling of the GABA recognition site to the associated chloride ion channel[64]. An important location for such anti-anxiety activity of the benzodiazepines apparently is the central amygdala[65]. Increased glucose utilization rates, seen during restraint and water-immersion stress in the central nucleus, were significantly reduced by injections of the benzodiazepine bromazepam[66].

Numerous previous reports had implicated the amygdala in fear and anxiety states. For example, electrical stimulation of the human amygdala has been reported to elicit such feelings and the autonomic reactions associated with these emotions[67,68]. In animal models, autonomic changes (heart rate, blood pressure, respiration), interruptions of ongoing activity ("freezing"), or sensitizations to other stimuli (so-called fear-potentiation of reflexes) have been used to measure fear or anxiety. Electrical stimulation of the central amygdala has been reported to affect all of these indexes. It produces a cessation of ongoing activity[69], alters heart rate[69,70], increases blood pressure[71], changes respiration[72], and increases the fear-potentiation of the so-called startle-reflex.

Stimuli that had been paired with aversive conditions also altered the neural firing patterns in the central nucleus[7,74,75], and lesions in this area attenuated or eliminated the behavioral signs of fear and anxiety. Bilateral lesions reduced "freezing" in response to aversive conditions in rats[76], conflict behaviors[65], conditioned emotional responses[77,78], and fear-potentiated startle reflexes[79]. Taken together, these data make a compelling case in favor of an important role for the central amygdala in fear and anxiety states. Consequently, this area, through its direct projections in the ventral amygdalofugal pathway, seems to be in a crucial position to alter visceral functions when the organism recognizes that a threatening situation exists.

Large bilateral hippocampal lesions, which included damage to hippocampal subfields CA 1-4 and dentate gyrus, aggravated stress ulcers in rats. This effect could be blocked by vagotomy[80,81]. In other words, it appears that without the hippocampal formation the brain is handicapped in dealing with the impact of stress conditions. A number of behavioral studies also support this idea. Stressful experiences early in the life of the organism have been reported to affect the normal development of behaviors usually associated with hippocampal functions. Similar aversive conditions in adulthood also frequently were found to produce the types of behavioral deficits usually seen after hippocampal damage, i.e., impairments in avoidance learning, habituation, extinction performance, spontaneous alternation, and reversal learning[14]. Apparently, environmental stressors affected hippocampal functions but, in addition, hippocampal activity also seemed to assist the organism in coping with these situations.

An analysis of the channels through which stress inputs might reach the hippocampal formation involved a study of the major fiber systems that connect with this limbic structure. The two major input-output pathways are the fimbria-fornix system and the perforant path. The latter system connects the dentate gyrus with the entorhinal cortex in the temporal lobe. Selective lesions in these pathways showed that the temporal lobe projection seemed the important one. Isolating the hippocampus from the entorhinal cortex increased the restraint-induced stomach pathology. On the other hand, fimbria-fornix cuts produced no significant effects[82]. These findings implicated the monosynaptic entorhinal-dentate pathway in stress ulcer susceptibility. It is well-known that the fibers originating in the entorhinal cortex form excitatory synapses with the granule cells in the dentate gyrus, before the information is relayed to the subfields of Ammon's Horn[13]. Synaptic transmission in this pathway shows a phenomenon, i.e., long-term potentiation, which refers to a long-lasting increase in the efficacy of transmission. It is frequently assumed to represent the electrophysiological analog of some forms of learning and memory[83]. A study of the evoked population spikes of dentate granule cells, the synchronous discharge of a large number of cells following electrical stimulation of the perforant fibers, indicated that the experience of restraint-stress suppressed the population spikes in some rats, but also increased these potentials in others. Subsequent investigations of the stomachs pointed out that suppressed electrophysiological activity was correlated with severe stress ulcers. But, on the other hand, the increases in the efficacy of transmission were associated with reduced pathology. A similar suppression of hippocampal activity was also found following another form of uncontrollable aversive stimulation, i.e., inescapable electric shocks. These rats also displayed the frequently observed "learned-helplessness" effect, i.e., impaired coping performance when escape was now made possible[15]. Large hippocampal lesions also aggravated the learned-helplessness effect[84].

These results suggest that, in a more general sense, impaired coping ability seems to correlate with suppressed hippocampal functions. At the same time, however, differences in the neural response profiles were better predictors of the severity of stress ulceration than the imposition of the specific environmental stress-procedure. Similar differential signatures had been found in the central amygdala[44]. These data point to the fact that individual differences in stress vulnerability exist.

These differences, however, can be eliminated when transmission in the entorhinal-dentate pathway is uniformly enhanced by high-frequency electrical stimulation. This type of stimulation, which also induces long-term potentiation, increased the resistance of rats to subsequent restraint-stress. These animals had significantly less ulceration than the corresponding low-frequency controls. Therefore, increased synaptic efficacy in the dentate gyrus, measured by increased amplitudes of the population spikes, seemingly made the animals more resistant under stress conditions[85].

Long-term potentiation in this pathway is apparently mediated by excitatory amino acid transmitters involving the N-methyl-D-aspartate (NMDA) receptor. When these NMDA receptors were blocked by intra-ventricular infusions of the selective antagonist aminophosphonovaleric acid (AP5), it (1) prevented long-term potentiation in the perforant path, (2) increased the learned-helplessness effect, and (3) aggravated the restraint ulceration[86,87].

Clearly, transmission in the entorhinal-dentate pathway is important in stressful circumstances. Coping ability, in fact, is impaired when the efficacy of synaptic transmission is reduced. Recent data showed that enkephalins released at these synapses also facilitate transmission during stress conditions[88].

Electrical stimulation of the hippocampus, however, produces only weak and transient autonomic effects[89]. But it does inhibit the effectiveness of transmission in sensory pathways, including the arousal reactions normally elicited by painful stimulation[90]. In other words, hippocampal influences on autonomic functions may be indirect, possibly, via pathways through the amygdala. In agreement with this hypothesis, it has been reported that many hippocampal influences on hypothalamic regions, particularly, those originating in the ventral regions of the hippocampal formation, are relayed through the amygdala, and not as previously thought through the fimbria-fornix fibers[91].

The ventral hippocampus apparently also influences hypothalamic areas via amygdalar nuclei during emotional behaviors. Defensiveness-aggressiveness in cats seems to be mediated by this projection axis. Furthermore, defensive (fearful) animals showed greater inhibitory influences measured in dentate granule cells. Double-pulse stimulations showed a greater recurrent inhibition in these cases[92]. It is possible that this reduced excitability of dentate granule cells in fearful animals might be related to the stress-induced suppressions of evoked population potentials found in stress-vulnerable rats[15]. Such a reduction of hippocampal outputs to the central amygdala might, in fact, be responsible for an impaired coping ability in stressful situations[93].

LIMBIC MIDLINE CORTEX

MacLean[3] proposed that the cingulate gyrus might also be part of the so-called "visceral brain". The anterior cingulate cortex, labelled a "vagal sensory-motor field", is known to receive afferent vagal inputs via the anterior hypothalamus, and efferent fibers have been reported to modify various autonomic visceral effects, e.g., arterial pressure, pyloric peristalsis, salivation and bladder contractions[94,95]. Anterior cingulate lesions in dogs have also been found to reduce gastric acid secretions[96]. In humans, they have been reported to alleviate gastric ulcers and colitis[11,12]. Cingulumotomy in patients with severe pain has been reported to provide significant relief[97] suggesting, perhaps, that various aversive experiences might be attenuated following this operation.

Bilateral lesions in the anterior cingulate cortex of rats have also been found to attenuate restraint-induced gastric erosions. On the other hand, more posteriorly placed lesions aggravated the stress ulcers[98]. These posterior areas provide input fibers to the hippocampal formation[99], and may be part of the same coping system. Bipolar electrical stimulation of the anterior cingulate cortex, on the other hand, produced stomach erosions. This effect could be blocked by pretreatment with atropine, but not with cimetidine[100].

Multiple-unit recordings and also low-level electrical stimulation data implicated a restricted zone in the anterior cingulate region (area infraradiata b α, β[101]) in these effects. These units responded during restraint, and some of them could be "conditioned" or "sensitized" to respond to an auditory stimulus that had been present during the restraint treatment[102]. It is possible that these units represent memory cells, required for learning to tolerate stressful circumstances (see below).

Cingulate lesions have been performed, in the past, to alleviate extreme anxiety states in human patients[9,10], and neurons in this region respond to anti-anxiety drugs[102,103]. For example, chlordiazepoxide modified the behaviors of neurons in the anterior cingulate cortex of rats during restraint. The drug also attenuated the ulcerogenic effects of low-level electrical stimulation of these neurons[102]. But recently, chlordiazepoxide has been found to interfere with the habituation to repeated restraint experiences. Specifically, the results showed that this benzodiazepine attenuated the effects of a single (acute) restraint session on stress ulcer development in rats, but it also prevented the normal adaptation to repeated (chronic) stress sessions[104]. A similar effect was found with lesions in the anterior cingulate cortex of rats[105].

The presumption is that between-session habituation is a form of learning and requires memory. It implies that (1) the anterior cingulate cortex is necessary for learning to tolerate such chronic stress conditions, and (2) benzodiazepine therapy may be counterproductive under such circumstances. There are additional data from behavioral studies which also showed that benzodiazepines interfered with learning to tolerate other aversive conditions, i.e., these drugs prevented counterconditioning to take place[106], possibly by producing memory deficits[107,108].

SUMMARY

The hypothesis that limbic mechanisms modulate the demands placed on organisms during stressful conditions is generally supported by the data. It is important to point out that this idea is largely based on *behavioral* studies. These studies provide a compelling argument that the amygdala is important in recognizing threatening stimuli, the defensive behaviors generated by such experiences, and the visceral responses seen under such circumstances. The central nucleus of the amygdala is crucially involved in transmitting this information to visceral control areas in the diencephalon and lower brainstem. The present model assumes that threatening stimuli, either learned or unlearned, trigger neural responses in the amygdala, producing defensive maneuvers, and changes in visceral physiology. In an evolutionary context, this system is assumed to allow quick reactions to threatening situations, increasing the survival chances.

However, when these immediate adjustments by the organism prove ineffective, a learning and memory based coping system is activated. It is proposed that hippocampal and limbic midline cortical areas are essential parts of such a system. This system contributes behavioral strategies and programs. The effectiveness of these behavioral programs, it is proposed, provide feedback information to the amygdala, influencing the original threat perceptions and the corresponding visceral responses.

The present conception is similar to other models which have assigned a special role to the hippocampus in motivation, emotions, learning and memory. It has been suggested that the hippocampus performs a kind of fast-time calculation of possible actions[109], intended plans or programs[110], and of significant environmental events and the established response repertoire[111]. The emphasis on hippocampal involvement in memory[112] fits the present analysis, as well.

The evidence also points to the importance of individual differences in stress ulcers susceptibility. In the literature, these differences are largely ignored, but they are real. Distinct neural signatures, found in the central nucleus of the amygdala and in the dentate gyrus of the hippocampal formation, were significant markers of the severity of the stress ulcers. The present model, in fact, predicts that these differences exist.

ACKNOWLEDGEMENTS

The author's studies were funded by the Natural Sciences and Engineering Research Council of Canada.

REFERENCES

1. P.G. Henke, The hypothalamus-amygdala axis and experimental gastric ulcers, *Neurosci Biobehav Rev.* 3:75 (1979).
2. P.G. Henke, The telencephalic limbic system and experimental gastric pathology: A review, *Neurosci Biobehav Rev.* 6:381 (1982).
3. P.D. MacLean, Psychosomatic disease and the "visceral brain": Recent developments bearing on the Papez theory of emotion, *Psychosom Med.* 11:338 (1949).
4. J.W. Papez, A proposed mechanism of emotion, *Arch Neurol Psychiat.* 38:725 (1937).
5. H. Klüver and P.C. Bucy, Preliminary analysis of functions of the temporal lobes in monkeys, *Arch Neurol Psychiat.* 42:979 (1934).
6. G.V. Goddard, Functions of the amygdala, *Psychol Bull.* 62:87 (1964).
7. P.G. Henke, The amygdala and forced immobilization of rats, *Behav Brain Res.* 16:19 (1985).
8. P.I. Yakovlev, Motility, behavior, and the brain, *J Nerv Ment Dis.* 107:313 (1948).
9. J.A. Gray. "The neuropsychology of anxiety," Oxford University Press, New York (1982).
10. G. Powell. "Brain and personality," Praeger Publishers, New York (1979).
11. M. Bucaille, Electrocoagulations du cerveau préfrontal: Indications et résultats dans les ulcères gastriques et duodénaux rebelles, *Acta gastro-enter (belg.).* 20:506 (1957).
12. M. Bucaille, Fontal lobe operation for colitis, *Surgery.* 52:690 (1962).
13. R.L. Isaacson. "The limbic system," Plenum Press, New York (1982).
14. R.J. Douglas, The development of hippocampal function: Implications for theory and for therapy, *in*: "The Hippocampus (Vol. 2)," R.L. Isaacson, K.H. Pribram, eds., Plenum Press, New York (1975).
15. P.G. Henke, Granule cell potentials in the dentate gyrus of the hippocampus: Coping behavior and stress ulcers in rats, *Behav Brain Res.* 30:97 (1990).
16. B.R. Kaada, Stimulation and regional ablation of the amygdaloid complex with reference to functional representations, *in*: "The Neurobiology of the Amygdala," B.E. Eleftheriou, ed., Plenum Press, New York (1972).
17. H. Koikegami, Amygdala and other related limbic structures. II. Functional experiments, *Acta Med Biol.* 12:73 (1964).

18. P. Gloor, Inputs and outputs of the amygdala: What the amygdala is trying to tell the rest of the brain, *in*: "Limbic Mechanisms," K.E. Livingston, D. Hornykiewicz, eds., Plenum Press, New York (1978).

19. H.J. Lammers, The neural connections of the amygdaloid complex in mammals, *in*: "The Neurobiology of the Amygdala," B.E. Eleftheriou, ed., Plenum Press, New York (1972).

20. J.E. LeDoux, D.A. Ruggiero, and D.J. Reis, Projections to the subcortical forebrain from anatomically defined regions of the medial geniculate body in the rat, *J Comp Neurol*. 242:182 (1985).

21. R. Norgren, Taste pathways to the hypothalamus and amygdala, *J Comp Neurol*. 166:17 (1976).

22. O.P. Ottersen and Y. Ben-Ari, Afferent connections to the amygdaloid complex of the rat and cat. I. Projections from the thalamus, *J Comp Neurol*. 187:401 (1979).

23. R.J. Radna and P.D. MacLean, Effects of vagal volleys on unit activity of amygdala, hippocampus, and corpus striatum in squirrel monkeys (*Saimiri sciureus*), *Soc Neurosci Abstr*. 2:374 (1976).

24. R.N. Sen and B.K. Anand, Effect of electrical stimulation of the limbic system of brain ("visceral brain") on gastric secretory activity and ulceration, *Indian J Med Res*. 45:515 (1957).

25. C.N. Shealy and T.L. Peele, Studies on amygdaloid nucleus in the cat, *J Neurophysiol*. 20:125 (1957).

26. E.J. Zawoisky, Gastric secretory response of the unrestrained cat following electrical stimulation of the hypothalamus, amygdala and basal ganglia, *Exp Neurol*. 17:128 (1967).

27. G. Stock, U. Rupprecht, H. Stumpf, and K. Schlör, Cardiovascular changes during arousal elicited by stimulation of amygdala, hypothalamus and locus coeruleus, *J Autonom Nerv Syst*. 3:503 (1981).

28. G. Stock, K.H. Schlör, H. Heidt, and J. Buss, Psychomotor behaviour and cardiovascular patterns during stimulation of the amygdala, *Pflügers Arch*. 376:177 (1978).

29. R.J. Timms, A study of the amygdaloid defense reaction showing the value of Althesin anesthesia in studies of the functions of the fore-brain in cats, *Pflügers Arch*. 391:49 (1981).

30. H.A. Koikegami, A. Komoto, and C. Kido, Studies on the amygdaloid nuclei and periamygdaloid complex: Experiments on the influence of their stimulation upon motility of small intestine and blood pressure, *Folia Psychiat Neurol Jap*. 7:86 (1953).

31. S. Eliasson, Cerebral influence on gastric motility in the cat, *Acta Physiol Scand*. 26 (Suppl. 95):1 (1952).

32. F.M. Fennegan and M.J. Puiggari, Hypothalamic and amygdaloid influences on gastric motility in dogs, *J Neurosurg*. 24:497 (1966).

33. P.G. Henke, Recent studies of the central nucleus of the amygdala and stress ulcers, *Neurosci Biobehav Rev*. 12:143 (1988).

34. D.L. Innes and M.F. Tansy, Gastric mucosal ulceration associated with electrochemical stimulation of the limbic brain, *Brain Res Bull*. 5 (Suppl. 1):33 (1980).

35. P.G. Henke, The centromedial amygdala and gastric pathology in rats, *Physiol Behav*. 25:107 (1980).

36. D.A. Hopkins and G. Holstege, Amygdaloid projections to the mesencephalon, pons and medulla oblongata in the cat, *Exp Brain Res*. 32:529 (1978).

37. J.E. Krettek and J.L. Price, Amygdaloid projections to subcortical structures within the basal forebrain and brainstem in the rat and cat, *J Comp Neurol*. 178:225 (1978).

38. J.S. Schwaber, B.S. Kapp, G.A. Higgins, and P.R. Rapp, Amygdaloid and basal forebrain direct connections with the nucleus of the solitary tract and the dorsal motor nucleus, *J Neurosci*. 2:1424 (1982).

39. P.G. Henke, Limbic system modulation of stress ulcer development, *Ann NY Acad Sci*. 597:201 (1990).

40. J.L. Price, The efferent projections of the amygdaloid complex in the rat, cat and monkey, *in*: "The Amygdaloid Complex," Y. Ben-Ari, ed., Elsevier/North-Holland Biomedical Press, Amsterdam (1981).

41. P.G. Henke, The amygdala and restraint ulcers in rats, *J Comp Physiol Psychol*. 84:313 (1980).

42. P.G. Henke, Facilitation and inhibition of gastric pathology after lesions in the amygdala of rats, *Physiol Behav*. 25:575 (1980).

43. P.G. Henke and R.M. Sullivan, Kindling in the amygdala and susceptibility to stress ulcers, *Brain Res Bull*. 14:5 (1985).

44. P.G. Henke, Electrophysiological activity in the central nucleus of the amygdala: Emotionality and stress ulcers in rats, *Behav Neurosci*. 102:77 (1988).

45. P. Driscoll and K. Bättig, Behavioral, emotional and neurochemical profiles of rats selected for extreme differences in active, two-way avoidance performance, in: "Genetics of the Brain," I. Lieblich, ed., Elsevier Biomedical Press, Amsterdam (1982).

46. G. Glavin, Restraint ulcer: History, current research and future implications, *Brain Res Bull.* 5 (Suppl. 1):51 (1980).

47. G.B. Glavin and R.E. Ykema, Subject emotionality and subsequent pylorus ligation induced or restraint induced gastric ulcer, *Pavlovian J Biol Sci.* 15:102 (1980).

48. A. Ray, P.G. Henke, and R.M. Sullivan, Central dopamine systems and gastric stress pathology in rats, *Physiol Behav.* 42:359 (1988).

49. A.Y. Deutsch, M. Goldstein, F. Baldino, and R.H. Roth, Telencephalic projections of the A8 dopamine cell group, *Ann NY Acad Sci.* 537:27 (1988).

50. J.H. Fallon, Histochemical characterization of dopaminergic, noradrenergic and serotonergic projections to the amygdala, in: "The Amygdaloid Complex," Y. Ben-Ari, ed., Elsevier/North-Holland Biomedical Press, Amsterdam (1981).

51. G.W. Roberts, P.L. Woodham, J.M. Polak, and T.J. Crow, Distribution of neuropeptides in the limbic system of the rat: The amygdaloid complex, *Neuroscience.* 7:99 (1982).

52. G.W. Roberts, J.M. Polak, and T.J. Crow, Peptide circuitry of the limbic system, in: "Psychopharmacology of the Limbic System," M.R. Trimble, E. Zarifian, eds., Oxford University Press, Oxford (1985).

53. N. Sharif and D. Burt, Limbic, hypothalamic, cortical and spinal regions are enriched with receptors of thyrotropin-releasing hormone: Evidence from [^{3}H] ultrafilm autoradiography and correlation with cerebral effects of the tripeptide in rat brain, *Neurosci Lett.* 60:337 (1985).

54. T.H. Williams, M.D. Cassell, N.J. Mankovich, H.K. Huang, and T.S. Gray, The distribution of neuropeptides in the central nucleus of the amygdala: A quantitative approach using image analysis, in: "The Amygdaloid Complex," Y. Ben-Ari, ed., Elsevier/North-Holland Biomedical Press, Amsterdam (1981).

55. P.G. Henke, R.M. Sullivan, and A. Ray, Interactions of thyrotropin-releasing hormone with neurotensin and dopamine in the central nucleus of the amygdala during stress ulcer formation in rats, *Neurosci Lett.* 91:95 (1988).

56. A. Ray, P.G. Henke, and R.M. Sullivan, The central amygdala and immobilization stress induced gastric pathology in rats: Neurotensin and dopamine, *Brain Res.* 409:398 (1987).

57. A. Ray, P.G. Henke, and R.M. Sullivan, Effects of intra- amygdalar dopamine agonists and antagonists on gastric stress lesions in rats, *Neurosci Lett.* 84:302 (1988).

58. A. Ray and P.G. Henke, Role of dopaminergic mechanisms in the central amygdalar nucleus in the regulation of stress-induced gastric ulcer formation in rats, *Indian J Med Res.* 90:224 (1989).

59. A. Ray, P.G. Henke, and R.M. Sullivan, Opiate mechanisms in the central amygdala and gastric stress pathology in rats, *Brain Res.* 442:195 (1988).

60. A. Ray and P.G. Henke, Enkephalin-dopamine interactions in the central amygdalar nucleus during gastric stress ulcer formation in rats, *Behav Brain Res.* 36:179 (1990).

61. R.M. Sullivan, P.G. Henke, A. Ray, M.A. Hebert, and J.M. Trimper, The GABA/ benzodiazepine receptor complex in the central amygdalar nucleus and stress ulcers in rats, *Behav Neur Biol.* 51:262 (1989).

62. A. Ray, P.G. Henke, and R.M. Sullivan, Effects of intra-amygdalar thyrotropin releasing hormone (TRH) and its antagonism by atopine and benzodiazepines during stress ulcer formation in rats, *Pharmacol Biochem Behav.* 36:597 (1990).

63. D.G. Amaral, J.L. Price, A. Pitkanen, and T.S. Carmichael, Anatomical organization of the primate amygdaloid complex, in: "The Amygdala: Neurobiological Aspects of Emotion, Memory, and Mental Dysfunction," J.P. Aggleton, ed., Wiley-Liss, New York (1992).

64. E. Costa, M.G. Corda, B. Epstein, C. Forchetti, and A. Guidotti, GABA-benzodiazepine interactions, in: "The Benzodiazepines: From Molecular Biology to Clinical Practice," E. Costa, ed., Raven Press, New York (1983).

65. K. Shibata, Y. Kataoka, Y. Gomita, and S. Ueki, Localization of the site of the anticonflict action of benzodiazepines in the amygdaloid nucleus of rats, *Brain Res.* 234:442 (1982).

66. K. Nakamura, T. Hayashi, and K. Nakamura, Effects of bromazepam on cerebral neuronal activity in male Wistar rats with immobilized stress, *Folia Pharmacol Jap.* 83:401 (1984).

67. P. Gloor, A. Oliver, and L.F. Quesney, The role of the amygdala in the expression of psychic phenomena in temporal lobe seizures, in: "The Amygdaloid Complex," Y. Ben-Ari, ed., Elsevier/North-Holland Biomedical Press, Amsterdam (1981).

68. W.P. Chapman, H.R. Schroeder, G. Geyer, M.A.B. Brazier, C. Fager, J.L. Poppen, H.C. Solomon, and P.I. Yakovlev, Physiological evidence concerning the importance of the amygdaloid nuclear region in the integration of circulating function and emotion in man, *Science*. 129:949 (1954).

69. C.D. Applegate, B.S. Kapp, M.D. Underwood, and C.L. McNall, Autonomic and somatomotor effects of amygdala central n. stimulation in awake rabbits, *Physiol Behav*. 31:353 (1983).

70. B.S. Kapp, M. Gallagher, M.D. Underwood, C.L. McNall, and D. Whitehorn, Cardiovascular responses elicited by electrical stimulation of the amygdala central nucleus in the rabbit, *Brain Res*. 234:251 (1982).

71. G.J. Mogenson and F.R. Calaresu, Cardiovascular responses to electrical stimulation of the amygdala in the rat, *Exp Neurol*. 39:166 (1973).

72. R.M. Harper, R.C. Frysinger, R.B. Trelease, and J.D. Marks, State-dependent alteration of respiratory cycle timing by stimulation of the central nucleus of the amygdala, *Brain Res*. 306:1 (1984).

73. J.B. Rosen and M. Davis, Enhancement of acoustic startle by electrical stimulation of the amygdala, *Behav Neurosci* 102:195 (1988).

74. P.G. Henke, Unit-activity in the central amygdalar nucleus of rats in response to immobilization stress, *Brain Res Bull*. 10:833 (1983).

75. J.P. Pascoe and B.S. Kapp, Electrophysiological characteristics of amygdaloid central nucleus neurons during Pavlovian fear conditioning in the rabbit, *Behav Brain Res*. 16:117 (1985).

76. J. Iwata, J.E. LeDoux, M.P. Meeley, S. Arneric, and D.J. Reis, Intrinsic neurons in the amygdala field projected to by the medial geniculate body mediate emotional response conditioned to acoustic stimuli, *Brain Res*. 383:195 (1986).

77. M.H. Kellicut and J.S. Schwartzbaum, Formation of conditioned emotional response (CER) following lesions of the amygdaloid complex in rats, *Psychol Rep*. 12:351 (1963).

78. A.A. Sperack, C.T. Campbell, and L. Drake, Effect of amygdalectomy on habituation and CER in rats, *Physiol Behav*. 15:199 (1975).

79. J.M. Hitchcock and M. Davis, Lesions of the amygdala, but not of the cerebellum or red nucleus, block conditioned fear as measured by the potentiated startle paradigm, *Behav Neurosci*. 100:11 (1986).

80. C. Kim, H. Choi, J.K. Kim, M.S. Kim, H.J. Park, B.T. Ahn, and S.H. Kang, Influence of hippocampectomy on gastric ulcer in rats, *Brain Res*. 109:245 (1976).

81. H.M. Murphy, C.H. Wideman, and T.S. Brown, Plasma corticosterone levels and ulcer formation in rats with hippocampal lesions, *Neuroendocrinology*. 28:123 (1979).

82. P.G. Henke, R.J. Savoie, and B.M. Callahan, Hippocampal deafferentation and deefferentation and gastric pathology in rats, *Brain Res Bull*. 7:395 (1981).

83. R.A. Nicoll, J.A. Kauer, and R.C. Malenka, The current excitement in long-term potentiation, *Neuron*. 1:97 (1988).

84. D.G. Elmes, L.E. Jarrard, and P.D. Swart, Helplessness in hippocampectomized rats: Response perseveration?, *Physiol Psychol*. 3:51 (1975).

85. P.G. Henke, Synaptic efficacy in the entorhinal-dentate pathway and stress ulcers in rats, *Neurosci Lett*. 107:110 (1989).

86. P.G. Henke, N-methyl-D-aspartate receptors and stress ulcers in rats, in: "Stress and Gastrointestinal Motility," L. Bueno, ed., John Libbey, London (1989).

87. P.G. Henke, Potentiation of inputs from the posterolateral amygdala to the dentate gyrus and resistance to stress ulcer formation in rats, *Physiol Behav*. 48:659 (1990).

88. P.G. Henke, Effects of naloxone on handling-induced modifications of hippocampal synaptic potentiation and stress ulcers in rats, *Exp Clin Gastroenterol*. 3:190 (1993).

89. B.R. Kaada, R.S. Feldman, and T. Langfeldt, Failure to modulate autonomic reflex discharge by hippocampal stimulation in rabbits, *Physiol Behav*. 7:225 (1971).

90. P.D. MacLean, Chemical and electrical stimulation of the hippocampus in unrestrained animals, *Arch Neurol Psychiat*. 81:331 (1972).

91. C.E. Poletti, M. Kliot, and G. Boytin, Metabolic influences of the hippocampus on hypothalamus, preoptic and basal forebrain is exerted through amygdalofugal pathways, *Neurosci Lett*. 45:211 (1984).

92. R.E. Adamec and C. Stark-Adamec, Limbic hyperfunction, limbic epilepsy, and interictal behavior: Models and methods of detection, *in*: "The Limbic System," B.K. Doane, K.E. Livingston, eds., Raven Press, New York (1986).

93. P.G. Henke, Hippocampal pathway to the amygdala and stress ulcer development, *Brain Res Bull.* 25:691 (1990).

94. B.R. Kaada, Cingulate, posterior orbital, anterior insular and temporal pole cortex, *in*: "Handbook of Physiology, Section 1: Neurophysiology, Vol. 2," J. Field, H.W. Magoun, V.E. Hall, eds., American Physiological Society, Washington, D.C. (1960).

95. P.D. MacLean, Culminating developments in the evolution of the limbic system: The thalamocingulate division, *in*: "The Limbic System," B.K. Doane, K.E. Livingston, eds., Raven Press, New York (1986).

96. H.M. Richter, R.A. Davis, D. Ruge, and N.T. Walker, The effect of cortical and subcortical brain lesions upon gastric secretion in dogs with a vagus-preserved total gastric pouch, *in*: "Surgical Forum, Vol. 7," American College of Surgeons, Chicago (1956).

97. E.L. Foltz and L.E. White, Pain "relief" by frontal cingulumotomy, *J Neurosurg.* 19:89 (1962).

98. P.G. Henke and R.J. Savoie, The cingulate cortex and gastric pathology, *Brain Res Bull.* 8:489 (1982).

99. L.W. Swanson, The anatomical organization of septo-hippocampal projections, *in*: "Functions of the Septo-Hippocampal System. Ciba Foundation Symposium 58 (new series)," Elsevier, Amsterdam (1978).

100. P.G. Henke, Mucosal damage following electrical stimulation of the anterior cingulate cortex and pretreatment with atropine and cimetidine, *Pharmacol Biochem Behav.* 19:483 (1983).

101. V.B. Domesick, Projections from the cingulate cortex in the rat, *Brain Res.* 12:296 (1969).

102. P.G. Henke, The anterior cingulate cortex and stress: Effects of chlordiazepoxide on unit-activity and stimulation-induced gastric pathology in rats, *Int J Psychophysiol.* 2:23 (1984).

103. P.G. Henke, R.M. Sullivan, A. Ray, and M.E. MacDougall, Stress ulcer modulation by chlordiazepoxide and GABA in the anterior cingulate cortex of rats, *Exp Clin Gastroenterol.* 2:13 (1992).

104. P.G. Henke, Chlordiazepoxide and stress tolerance in rats, *Pharmacol Biochem Behav.* 26:561 (1987).

105. R.M. Sullivan and P.G. Henke, The anterior midline cortex and adaptation to stress ulcers in rats, *Brain Res Bull.* 17:493 (1986).

106. J.A. Gray, N. Davis, J. Feldon, S. Owen, and M. Boarder, Stress tolerance: Possible neural mechanisms, *in*: "Foundations of Psychosomatics," M.J. Christie, P.G. Mellet, eds., John Wiley & Sons, New York (1981).

107. A. Lenègre, I. Avril, and S. Fromage, Absence of amnesia induction in mice with hydroxyzine in comparison with three other minor tranquillizers, *Arzneim Forsch/Drug Res.* 38:558 (1988).

108. M.H. Thiébot, Some evidence for amnesia-like effects of benzodiazepines in animals, *Neurosci Biobehav Rev.* 9:95 (1985).

109. K.H. Pribram. "Languages of the Brain," Prentice-Hall, Englewood Cliffs, N.J. (1971).

110. D.P. Crowne and D.D. Radcliffe, Some characteristics and functional relations of the electrical activity of the primate hippocampus and a hypothesis of hippocampal function, *in*: "The Hippocampus, Vol. 2," R.L. Isaacson, K.H. Pribram, eds., Plenum Press, New York (1975).

111. J. Olds, The central nervous system and the reinforcement of behavior, *Amer Psychol.* 24:114 (1969).

112. B. Milner, S. Corkin, and H.L. Teuber, Further analysis of the hippocampal amnesic syndrome: 14-year follow-up study of H. M., *Neuropsychologia.* 6:215 (1968).

113. P.G. Henke, A. Ray, and R.M. Sullivan, The amygdala: Emotions and gut functions, *Digest Dis Sci.* 36:1633 (1991).

114. P.G. Henke, Stomach pathology and the amygdala, *in*: "The Amygdala: Neurobiological Aspects of Emotion, Memory, and Mental Dysfunction," J.P. Aggleton, ed., Wiley- Liss, New York (1992).

115. P.G. Henke and A. Ray, Review: The limbic brain, emotions and stress ulcers, *Exp Clin Gastroenterol.* 1:287 (1992).

116. P.G. Henke and A. Ray, Stress ulcer modulation by limbic system structures, *Acta Physiol Hung.* 80:117 (1992).

NEUROTENSIN ACTS IN THE MESOLIMBIC DOPAMINE SYSTEM TO AFFECT GASTRIC MUCOSAL FUNCTION

Gordon L. Kauffman, Jr., Lianping Xing
and Robert Bryan

Neurotensin (NT) is a tridecapeptide which has been identified in the central nervous system of all vertebrate classes, having been originally isolated from bovine hypothalamus[1,2,3]. Using immunohistochemical techniques, NT has been identified within the substance of the substantia nigra (SN), pars compacta, and the ventral tegmental area (VTA)[4,5,6,7,8]. Recently, NT binding sites have been identified in the VTA with a dissociation content (K_d) of 9.3 ± 0.4 nm and the number of binding sites (B_m) of 86 ± 10 fm/mg protein[9]. The VTA is found in the mesolimbic dopamine (DA) system within the central nervous system. This mesolimbic system originates in the A8, A9 (SN), and A10 cell bodies, passes along the medial forebrain bundle, and terminates in the limbic structures such as nucleus accumbens (N.ACB), olfactory tubercle, and the amygdala.

NT appears to be involved in the regulation of midbrain DA cells, since interactions between NT and DA have been established in a number of situations. Intraventricular NT antagonizes the locomotor effects produced by DA agonist administration[10]. Administration of NT into N.ACB antagonizes the arousal produced by thyrotropin releasing hormone[11]. Intraventricular NT has also been shown to increase DA metabolite concentrations in the limbic area[12]. Finally, the hypothermic response to intraventricular NT is antagonized by pretreatment with DA[13].

Restraint in a cold environment produces macroscopic and microscopic injury to the gastric mucosa. Intraventricular NT protects the gastric mucosa against this form of experimental injury[14]. It is not known whether this NT-induced protection occurs as a result of interactions in the mesolimbic dopamine system. The following studies, previously reported[15, 16], were designed to determine whether 1) administration of NT into N.ACB protects the gastric mucosa against injury produced by restraint in a cold environment, 2) administration of NT into N.ACB causes inhibition of gastric acid secretion, 3) restraint in a cold environment is associated with a change in glucose utilization and whether intracerebroventricular (icv) NT affects glucose utilization in the mesolimbic dopamine nuclei.

METHODS AND MATERIALS

Animals. Adult male Sprague-Dawley rats (200-250 g) were fasted for 24 h but allowed free access to water. The following studies were approved by the animal research committee of The Milton S. Hershey Medical Center. Immediately after intranuclear NT administration the animals were immobilized in stapled window screen and allowed to recover from ether anesthesia for five min. They were then immersed to the neck in 20°C water for two h at which time they were killed by cervical dislocation.

Macroscopic Evaluation. The stomachs were removed and injected intraluminally with 1% formalin for 30 min. The mucosal surface was exposed by cutting along the greater curvature. The surface area of mucosal injury was scored using x30 magnification by an examiner who had no knowledge of the treatment group to which each animal belonged. The method for scoring has been described by Guth et al.[17] the surface area of injury being expressed in mm.

Intranuclear Administration of NT

For surgery, rats were anesthetized with nembutal (40 mg/kg, ip) and mounted in a stereotaxic apparatus. They were then implanted stereotaxically with chronic bilateral guide cannulae (23G stainless steel needle) 1 mm over the N.ACB (AP:+3.6; M/L: ±1.5; D/V:-6.5mm) according to the Atlas of Pelligrino[18]. The cannulae were secured to the skull with stainless steel screws and dental cement. A silicon rubber plug was used to maintain patency of this cannula and prevent infection.

Injection Procedure

One week following surgery, the rats were deprived of food for 18 h prior to cold water restraint (CWR). The injection apparatus consisted of 30 gauge stainless steel needles bent to extend exactly 1 mm below the tip of the guide cannulae. The needles were connected to a 5 μl Hamilton syringe. Doses of NT were injected in a total of 1 μl/side delivered over a period of 2 min. The needle remained in situ for an additional one min period to allow diffusion of the injected solution into the brain.

Gastric Acid Secretory Studies

At least two days before testing, rats were anesthetized with pentobarbital sodium (40 mg/kg ip) a stainless steel gastric cannula was placed in the gastric lumen.

All gastric acid secretion studies were performed after a 24 h fast. After immobilization in Bollman cages, the gastric fistulas were opened and gastric juice was collected by gravity drainage. Vehicle (0.15 M NaCl) or NT was administered through the previously positioned cannula. Two collections of gastric juice were made - the first was between 60 and 30 min before NT administration and the second between 30 and 0 min before NT administration. Ten minutes after NT administration gastric juice was collected for 30 min at intervals of 30 min for 120 min. Volume was measured to the nearest 0.1 ml and H^+ concentration ($[H^+]$) by titration to pH 7.0. The acid output was calculated as volume/unit time x $[H^+]$ and expressed as uEq per unit time.

NT was given directly into N.ACB. Haloperidol, 0.5 μg per side, was administered into N.ACB in some animals before intranuclear NT.

Regional Cerebral Metabolic Rate for Glucose (rCMRglc)

Reagents: D-[6-14C] glucose, 55 mCI/mmol, was obtained from Amersham Corporation, Arlington Heights, IL. Dimilume-30 liquid scintillation cocktail was purchased from Packard Instrument, Downers Grove, IL. Glucose reagent kits were purchased from Beckman Instruments, Inc., Brea, CA and neurotensin from Bachem, Inc., Torrance, CA.

Animal Surgery: One week before the experiment, a cannula was implanted into the lateral ventricle. Each rat was anesthetized with Nembutal (50 mg/kg ip) and mounted in a stereotaxis apparatus. A guide cannulae (made from 23G stainless needle) was stereotaxically implanted into the lateral ventricle (AP - 10.; M ± 1.3; V 4.0 mm) according to the Atlas of Pelligrino et al.[18]. The cannula was secured to the skull with stainless steel screws and sealed with dental cement. Silicone rubber was placed over each cannula to maintain patency and prevent infection.

One day before the experiment, venous catheters were inserted into the right jugular vein of each rat. Each rat was anesthetized with an intraperitoneal injection of Nembutal (50 mg/kg). Two catheters were inserted in the jugular vein (one was silastic 0.025 in ID x 0.047 OD, another was PE 10) and advanced to the right atrium[19]. Both catheters were passed under the skin and brought out in the back of the neck and secured in place with a harness. After surgery, rats were housed individually.

Experimental Protocol: Rats were divided into four groups consisting of seven or eight rats per group (Table 1). Between 0800 and 0900 h neurotensin (60 μg in 10 μl saline) was administered icv using the lateral ventricular cannula of rats in two of the groups of rats. Saline was administered to the rats in the remaining two groups. One rat from each of the aforementioned groups was exposed to CWR. The four groups were non CWR, non CWR + NT, CWR, CWR + NT. For CWR, rats were briefly anesthetized with ether, restrained by stapling a wire screen securely around each rat and submersed to the neck in 20°C water. After two h, each rat was placed in a large cage which was open at the top, providing access to the catheters in the jugular vein. Regional CMRglc was measured using [6-14C] glucose. The stomachs were removed and perfused intraluminally with 4% formalin. Gastric mucosal injury was evaluated by an unbiased observer using 4 x magnification and ulcer index as previously described [17] was calculated. Rectal temperature at the completion of the experiment was measured by a commercially available thermocouple probe (Telethermometer, Yellow Springs Instrument Co., Yellow Springs, OH).

rCMRglc Measurement: Regional CMRglc was measured using [6-14C] glucose and quantitative autoradiography[20, 21]. A tracer amount of [6-14] glucose 35 μCi in unstressed and 50 μCi in stressed rats, in 0.3 ml saline was infused into the PE 10 jugular catheter over 3-4 sec. Seven blood samples were withdrawn from the silastic catheter 0.25, 0.75, 1.25, 2,3,5, and 7.5 min after 14C-glucose in order to determine the specific activity of the label in blood over time[21]. The blood samples were centrifuged immediately and the plasma was separate and stored at -70°C until further analysis.

Table 1. Effect of Microinjection NT into Nucleus Accumbens (N.ACB) or Substantia Nigra (SN) on CWR-Induced Gastric Mucosal Injury

Treatment	Calculated surface area of mucosal injury (mm^2) microinjections into:	
	N.ACB	SN
Vehicle (1 μl)	39.1 $\pm$ 5.1 (11)	30.9 $\pm$ 12.4
NT (0.3 μg)	20.5 $\pm$ 7.3* (10)	
NT (1.5 μg)	19.1 $\pm$ 3.5* (9)	
NT (3.0 μg)	16.3 $\pm$ 6.1* (10)	30.3 $\pm$ 8.9
NT (30.0 μg)		50.8 $\pm$ 10.0

*$P < 0.005$, compared with vehicle group (one way ANOVA).
N values in ( ). Means ± SEM are shown.

After the last blood sample was withdrawn, each rat was killed with two ml pentobarbital sodium (50 mg/ml). The brain was immediately removed, frozen in Freon (-29.8°C) and placed in a freezer (-70°C). The brains were sectioned into 20 μm thick sections using a cryostat (A O Reichert, Buffalo, NY). Representative sections were placed on glass slides and dried on a warming tray. Together with calibrated poly[14]C-methacrylate standards (Amersham, Arlington Heights, IL), the sections were exposed against Dupont Cronex Lo-Dose Mammography film for 22 days; the film was then developed using an automated developer. The plasma samples were assayed for ^{14}C-glucose concentration by liquid scintillation counting and for glucose using a Beckman glucose analyzer. Regional CMRglc was calculated from ^{14}C concentration in specific brain regions, determined by comparing the optical densities of the autoradiograph images to calibrated standards using a densitometer (Tobias Assoc., Ivyland, PA), and the specific activities of the seven plasma samples.

For the two neurotensin-treated groups the assumption that neurotensin had no effect on the brain-to-plasma glucose ratio was made. Since the brain-to-plasma glucose ratio does not change very much except in extreme conditions[19, 21], this is a reasonable assumption. A brain-to-plasma glucose ratio of 0.22 was used for the calculation of rCMRglc in all groups of rats. Although this ratio may decrease during hypothermia[22], it was determined that the change in the ratio altered the calculated rCMRglc minimally and did not alter the conclusions as reported.

Statistical Analysis: The means of systemic parameters and gastric mucosal injury were compared using the analysis of variance and the t-test for pooled samples with a Bonferroni correction for multiple comparisons. The rates of rCMRglc were analyzed by repeated measures analysis of variance using the SAS General Linear Models Analysis[21].

Drugs

Neurotensin was purchased from Bachem, Inc. (Torrance, CA). Haloperidol was purchased from Sigma Chemical Co. (St. Louis, MO). Haloperidol was dissolved in 0.1% tartaric acid. NT was dissolved in 0.15 M NaCl. Appropriate vehicle solutions were used in control studies.

RESULTS

The effect of intra-N.ACB NT on CWR-induced ulcer formation is shown in Table 1. Bilateral injections of NT (0.3-3 µg/side) produced a significant reduction in the surface area of CWR-induced gastric mucosal injury. The lowest dose studied (0.3 µg/side) intra-N.ACB produced a significant, (50%) reduction in surface areas of injury.

The effect of direct administration of NT, 1.5 or 3.0 µg per side, on basal acid secretion is shown in Table 2. Neurotensin (3.0 µg per side) significantly reduced basal acid output by 67% and 61% at one and two h respectively.

Table 2. Effect of Neurotensin (ug per side) Administered into Nucleus Accumbens on Basal Gastric Acid Output

	Gastric acid output (µEq/h)		
	Vehicle	NT (1.5 µg)	NT (3.0 µg)
Pre-NT	97 ± 15	88 ± 21	83 ± 26
1st hour	78 ± 15	43 ± 17	26 ± 10*
2nd hour	112 ± 13	67 ± 15	44 ± 18*

NT, neurotensin; NACB, nucleus accumbens.
*p ≤ 0.05 by unpaired t-test when compared with vehicle.

Plasma glucose concentrations, rectal temperature, and the ulcer index for the four groups of rats are shown in Table 3. Non-CWR rats had little or no gastric mucosal injury. Rats subjected to CWR, given saline icv had a significantly higher ulcer index than control rats. As previously demonstrated, icv NT in CWR rats was associated with a significantly lower ulcer index than untreated CWR rats. Gastric mucosal injury was reduced by approximately 85% with the administration of icv NT before CWR. In all rats exposed to CWR, rectal temperature significantly decreased. The administration of NT did not alter body temperature in the non-CWR or CWR

Table 3. Plasma Glucose, Rectal Temperature and Ulcer Index in the Four Groups of Rats at 2h CWR

	Non CWR	Non CWR + NT	CWR	CWR + NT
Plasma glucose (µmol/ml)	6.05 ± 0.07	6.9 ± 1.4	4.2 ± 3.1	6.7 ± 2.9
Rectal temperature (°C)	37.5 ± 0.5	37.1 ± 1.3	24.2 ± 1.8*	24.8 ± 1.5*
Ulcer index	0 ± 0	0 ± 0	31 ± 9**	5 ± 3

*P<0.05 compared to non CWR and CWR + NT compared to CWR rats
**P<0.01 CWR compared to non CWR and CWR + NT

155

Table 4. Regional Cerebral Metabolic Rate for Glucose (rCMRglc) in the Four Groups of Rats. The rCMRglc is expressed as μmol/100g/min. The rates of rCMRglc were analyzed by repeated measures. Measures analysis of variance using the SAS General Linear Models Analysis.

	Non CWR	Non CWR +NT	CWR	CWR + NT
Brain region	(n=8)	(n=8)	(n=8)	(n=7)
Telenephalon				
Motor cortex	110 ± 4	118 ± 4	37 ± 4	37 ± 5
Sensory cortex	124 ± 7	131 ± 7	42 ± 7	51 ± 7
Caudate putamen	98 ± 4	103 ± 4	25 ± 4	30 ± 4
Hippocampus	75 ± 3	80 ± 3	20 ± 3	24 ± 3
Amygdala	73 ± 3	83 ± 3[*]	18 ± 3	29 ± 3[a]
Globus pallidus	56 ± 2	58 ± 2	13 ± 2	17 ± 2
Septal area				
Corpus callosum	34 ± 2	40 ± 2[*]	7 ± 2	10 ± 2
Diencephalon				
Thalamus				
Anterior n.	121 ± 5	133 ± 5	37 ± 5	43 ± 5
Ventral n.	88 ± 5	87 ± 5	25 ± 5	32 ± 5
l-Hypothalamus	63 ± 2	65 ± 2	13 ± 2	20 ± 2[a]
Mesencephalon				
Red nucleus	101 ± 5	107 ± 5	28 ± 5	42 ± 5[a]
Interpeduncular n.	141 ± 8	157 ± 8	49 ± 7	63 ± 8
Reticular formation	71 ± 2	70 ± 2	16 ± 2	20 ± 3
Metencephalon				
Pons	92 ± 4	100 ± 4	26 ± 4	40 ± 5[a]
Cerebral grey	92 ± 4	105 ± 4[**]	25 ± 4	40 ± 4[b]
Cerebral white	25 ± 2	33 ± 2[**]	8 ± 2	7 ± 2
Myelencephalon				
Superior				
olivary n.	143 ± 4	136 ± 4	46 ± 4	61 ± 4[a]
Dopamine System				
Accumbens n.	96 ± 3	106 ± 3[*]	25 ± 3	37 ± 3[b]
Tub. olfactorium	103 ± 4	121 ± 4[**]	32 ± 4	39 ± 5
Ventral teg. area	62 ± 4	63 ± 4	16 ± 4	25 ± 4
Substantia nigra	84 ± 4	86 ± 4	17 ± 4	30 ± 4[a]

[*]$P<0.05$, [**]$P<0.0.1$ Non-CWR + NT compared to Non-CWR
[a]$P<0.05$, [b]$P<0.01$ CWR compared to CWR + NT

rats. These observations suggest that the protective effect of icv NT is not related to change in body temperature.

Mean plasma glucose concentrations were not significantly different in either of the non-CWR groups. Plasma glucose in both of the CWR groups was much more variable than that in the non-CWR groups. Rats without CWR + NT had a higher ulcer index than those with CWR + NT, however, no significant correlation between severity of mucosal injury and plasma glucose was identified. Glucose utilization was measured in specific brain regions in the four groups (Table 4). All brain regions in rats exposed to CWR showed a significant decrease in glucose utilization compared to the corresponding regions in unstressed rats. Cold water restraint resulted in reduced global glucose utilization of 72% and 65% in control and NT-treated rats respectively. Neurotensin, icv, significantly increased rCMRglc in certain regions in unstressed and stressed rats (Table 5). In non-CWR + NT rats, an increase in rCMRglc was observed in the amygdala, corpus callosum, cerebellar grey, cerebellar white matter, nucleus accubens, and tuberculum olfactorium compared to non-CWR rats. Individual regions showing a significant increase in rCMRglc were amygdala, hypothalamus, red nucleus, cerebellar grey matter, superior olivary nucleus, nucleus accumbens pons and substantia nigra in CWR rats treated with NT icv compared to CWR rats without NT icv. The increase in rCMRglc in cerebellar grey matter, nucleus accumbens and amygdala was observed in both non-CWR and CWR rats given NT, icv, compared to the corresponding controls.

DISCUSSION

The present study demonstrates that NT, administered locally into N.ACB, produced dose-dependent protection against CWR-induced gastric mucosal injury, whereas intra-SN NT administration is not protective. In addition, administration of NT into N.ACB significantly inhibits basal rates of gastric acid secretion.

It has been suggested that some of the effects of NT on gastric function are mediated by the brain's dopamine system. NT can increase the function of DA neurons by influencing DA metabolism, turnover, and release. In the studies reviewed here, rCMRglc was used to determine whether changes could be observed in specific regions of the brain after NT administration. An increase in rCMRglc occurs in amygdala, corpus callosum, cerebellar grey and white matter, tuberculum

Table 5. NT Significantly Increased rCMRglc in Several Brain Regions in Unstressed and Stressed Rats

Non-CWR ± NT	CWR ± NT
Corpus Callosum	Hypothalamus
Cerebellar Grey[*]	Red Nucleus
Cerebellar White	Cerebellar Grey[*]
Nucleus Accumbens[*]	Superior Olivary Nucleus
Tub. Olfactorium	Nucleus Accumbens[*]
Amygdala[*]	Substantia Nigra
	Amygdala[*]

[*] Common to both CWR and non-CWR rats.

olfactorium and nucleus accumbens after NT icv in non-CWR rats, and in amygdala, hypothalamus, red nucleus, cerebellar grey matter, substantia nigra, superior olivary nucleus, pons and nucleus accumbens after NT, icv, in CWR rats. Among these brain regions, nucleus accumbens, tuberculum olfactorium, substantia nigra and amygdala are rich DA-containing neurons.

These data support the hypothesis that the mesolimbic dopamine system is involved in the gastric effects of neurotensin, since 1) NT administered into N.ACB reduces CWR-induced mucosal injury, 2) NT administration into N.ACB significantly reduces basal gastric acid secretion and 3) NT administration into the cerebrospinal fluid is associated with increased glucose utilization in N.ACB during CWR.

Taken together, these observations suggest that the nuclei of the mesolimbic dopamine pathway, which contain NT and possess NT receptors, exert an effect on gastric mucosal function.

REFERENCES

1. R. Carraway, and S.E. Leeman, The isolation of a new hypotensive peptide, neurotensin from bovine hypothalami, *J Biol Chem.* 248:6854 (1973).
2. R. Carraway, and S.E. Leeman, Characterization of radioimmunoassayable neurotensin in the rat. Its distribution in the central nervous system, small intestine and stomach, *J Biol Chem.* 251:7045 (1976).
3. P.J. Manberg, W.W. Youngblood, C.B. Nemeroff *et al.*, Regional distribution of neurotensin in human brain, *J Neurochem.* 38:1777 (1982).
4. G.R. Uhl, M.J. Kuhar, and S.H. Snyder, Neurotensin immunohistochemical localization in rat central nervous system, *Proc Natl Acad Sci USA.* 74:4059 (1977).
5. L. Jennes, W.E. Stumpf, and P.W. Kalivas, Neurotensin: Topographical distribution in rat brain by immunohistochemistry, *J Comp Neurol.* 210:211 (1982).
6. T. Hokfelt, B.J. Everitt, E. Theodorssen-Norheim *et al.*, Occurrence of neuortensin-like immunoreactivity in subpopulations of hypothalamic, mesencephalic and medullary catecholamine neurons, *J Comp Neurol.* 222:543 (1984).
7. P.W. Kalivas, Neurotensin in the ventromedial mesencephalon of the rat: Anatomical and functional considerations, *J Comp Neurol.* 226:495 (1984).
8. P.C. Emson, M. Goedert, and P.W. Mantyh, Neurotensin-containing neurons. *in:* "Handbook of Chemical Neuroanatomy, Vol. 4, GABA and Neuropeptides in the CNS," A. Bjorklund, T. Hokfelt, eds., Elsevier, Amsterdam (1985).
9. C. Dana, M. Vial, K. Leonard *et al.*, Electron microscopic localization of neurotensin binding sites in the midbrain tegmentum of the rat, I. Ventral tegmental area and interfascicular nucleus. *J Neurosci.* 9:2247 (1989).
10. F.B. Jolicoeur, G. DeMichele, A. Barbeau *et al.*, Neurotensin affects hyperactivity but not stereotype induced by pre and post synaptic dopamingergic stimulation, *Neurosci Biobehav Rev.* 7: 385 (1983).
11. E.C. Griffiths, P. Slater, and V.A.D. Webster. Behavioral interactions between neurotensin and thyrotropin-releasing hormone in rat central nervous system, *J Physiol.* 320:90P (1981).
12. A. Reches, R.E. Burke, D.H. Jiang *et al.*, Neurotensin interacts with dopaminergic neurons in rat brain, *Peptides.* 4:43 (1983).
13. P.W. Kalivas, L. Jennes, C.B. Nemeroff *et al.*, Neurotensin: Topographical distribution of brain sites involved in hypothermia and antinociception, *J Comp Neurol.* 210:225 (1982).
14. Z. Li, P.C. Colony, J.H. Washington *et al.*, Central neurotensin affects rat gastric integrity, prostaglandin E_2 and blood flow, *Am J Physiol.* 256:G226 (1989).
15. L. Xing, C. Balaban, J. Washington *et al.*, Central neurotensin-induced mucosal protection is mediated by dopamine receptors in nucleus accumbens, *Gastroenterology.* A504 (abstract) (1988).
16. L. Xing, J.C. King, R.M. Bryan *et al.*, Effect of neurotensin on regional cerebral glucose utilization in cold water-restrained rats, *Am J Physiol.* 258:G591 (1990).
17. T. Gati, and P.H. Guth, Mucosal lesions due to gastric distention in the rat, *Am J Dig Dis.* 22: 1083 (1977).

18. L.K. Pelligrino, A.X. Pellegrino, and A.J. Xushmen, "A Stereotaxis Atlas of the Rat Brain," Plenum, New York, (1979).

19. R.M. Bryan, K.A. Keefer, and C. MacNeill, Regional cerebral glucose utilization during insulin-induced hypoglycemia in unanesthetized rats, *J Neurochem.* 46:1904 (1986).

20. R.A. Hawkins, A.M. Mans, D.W. Davis *et al.*, Cerebral glucose use measured with (^{14}C) glucose labeled in the 1, 2, or 6 position, *Am J Physiol.* 248:C170 (1985).

21. R.M. Bryan, R.A. Hawkins, A.M. Mans, *et al.*, Cerebral glucose utilization in awake unstressed rats, *Am J Physiol.* 244:C270 (1983).

22. M. Hagerdal, J. Harp, and B.K. Siesjo, Effect of hypothermia upon organic and associated amino acids in rat cerebral cortex, *J. Neurochem.* 24:745 (1975).

ROLE OF MEDULLARY TRH IN VAGALLY MEDIATED FORMATION OF GASTRIC LESIONS

Yvette Taché, Hong Yang, Masashi Yoneda
and Bruno Bonaz

The central nervous system has long been recognized as implicated in the development of gastric ulceration. Brown-Sequard in 1876[1] and Schiff in 1867[2] showed that cerebral lesions were associated with gastric ulcerations. Cushing's observations in 1931 and subsequent clinical reports established a high incidence of gastritis in patients with head injury, cerebral stroke or tumor[3,4]. The demonstration that electrical stimulation or lesion of specific brain nuclei[5-8] and psychoactive drugs[9-11] influence stress-induced gastric lesions clearly established the existence of an interaction between the brain and the development of gastric lesions.

During the late decade, compelling experimental evidence has accumulated which support that brain peptides and neurotransmitters play a role in the regulation of gastric mucosal integrity[12-14]. Bombesin[15-18], corticotropin-releasing factor (CRF)[19-21], calcitonin[22-24], calcitonin gene related peptide[25], neurotensin[17,26,27], opioid peptides[15,17,28,29] and interleukin-1[30,31-34] injected into the cerebrospinal fluid (CSF) or specific brain nuclei prevent stress-induced gastric lesions. By contrast, thyrotropin-releasing hormone (TRH) has been reliably shown to induce or promote gastric erosions upon injection into the CSF at maximal effective dose[35,36,37]. This chapter will review medullary sites and mechanisms through which TRH promotes gastric lesions in rats and the role of medullary TRH in cold restraint stress-induced gastric erosions.

TRH ACTS IN THE DORSAL MOTOR NUCLEUS OF THE VAGUS TO INDUCE GASTRIC EROSIONS

CSF Injection of TRH Induced Gastric Erosions

TRH administered as a bolus or constant infusion into the CSF either into the lateral ventricle or cisterna magna induced 100% incidence of gastric hemorrhagic lesions in rats fasted overnight and otherwise left in a normal environment[35,38-40]. Gastric erosions occurred within 3-4 h after TRH injection and were lessened 6-24 h later[35,39]. Injection into the CSF of TRH analogs such as RX 77368 and DN-1417 also results in gastric erosions[35,41], whereas other structurally unrelated peptides, growth hormone-releasing factor, CRF, bombesin, neurotensin, somatostatin and substance P did not modify the integrity of the gastric mucosa. These results show the specificity of TRH action[35,39]. The most stable TRH analog, RX 77368[42], and DN-1417 have enhanced potency compared with TRH to induce gastric lesions[35,41]. Macroscopically, gastric lesions produced by TRH or TRH analogs injected into the CSF are similar to the acute mucosal damage induced by cold restraint and appeared as punctuated or elongated hemorrhages in the glandular part of the stomach. They can be identified histologically as mucosal erosions not penetrating the muscularis mucosa[35,41]. The intensity of gastric lesions is dose dependently related to the amount of TRH or analogs injected into the CSF[35,40,41]. In addition, injection of TRH into the lateral ventricle was shown to aggravate indomethacin-, aspirin-, or 5-hydroxytryptamine (5-HT)-induced gastric lesions[38].

TRH Action is Located in the Dorsal Motor Nucleus of the Vagus

Peptides injected into the CSF, including TRH, leak into the peripheral circulation[43-45]. However gastric damage in response to TRH injected intracisternally originates in the central nervous system. This was demonstrated by the fact that intravenous injection of the peptide or its analogs at doses effective into the CSF did not induce gastric lesions in rats[35,39,41]. Moreover, gastric lesion formation can be reproduced by selective unilateral microinjection of TRH into the dorsal motor nucleus of the vagus (DMN) or the central amygdala[46-48]. The DMN appears more sensitive than other extramedullary responsive sites such as the central amygdala. Several other sites rostral or caudal to the DMN, dorsal to the central canal, or the superior colliculus, medial septum, substantia nigra, CA1, area postrema, and nucleus gracilis were inactive[47-50].

Mechanism of TRH Action in the DMN

A high density of TRH binding sites is present in neurons of the DMN. In particular, the medial column of the DMN (which contributes axons forming the gastric branch of the vagus) contains the highest density of TRH receptors[51-55]. These findings are consistent with the mediation of TRH action through interaction with TRH receptors located on DMN neurons. However, the TRH analogs, DN-1417 and RX 77368, did not bind (DN-1417) or bound with low affinity (RX 77368) to high affinity TRH receptors in the brain and spinal cord[56] while exerting potent CNS mediated TRH-like action[57,58] including gastric damage[35,41] upon injection into the CSF. These data suggest that TRH analogs may act on other TRH receptor subtypes or different mechanisms that remain to be elucidated.

Convergent electrophysiological evidence indicates that TRH exerts an excitatory effect on DMN neurons as demonstrated either in vivo or in brainstem slices[59,60]. Intracellular recording from guinea pig or rat brainstem slices under current clamp conditions indicate that identified DMN cells responded to bath application of TRH by a long lasting, and reversible depolarization accompanied by an increase in the spontaneous firing of cells[61,62]. The excitatory response to TRH persisted after blockade of synaptic transmission indicating that TRH action is not dependent upon synaptic mechanisms but is due to a direct postsynaptic effect on DMN neurons[60-62]. Previous studies have shown that the long lasting excitation of DMN cell bodies induced by microinjection of kainic acid into the DMN resulted in gastric lesions in rats[63]. Taken together, these data suggest that the sustained increase in the excitability of DMN cell bodies by TRH is part of the initiating mechanism which leads to gastric erosion.

Modulation of TRH Action by Other Centrally Acting Transmitters

The dorsal vagal complex (composed of the nucleus tractus solitarius that receives vagal afferents and the DMN) is richly innervated by a large number of classical transmitters and peptides which have been shown to influence gastric function[64,65]. Functional and electrophysiological studies indicate that the postsynaptic action of TRH on DMN neurons might be mediated in concert with 5-HT. 5-HT injected intracisternally enhanced gastric damage induced by the TRH analog RX 77368, while it had no effect on gastric mucosa when injected alone[66]. The potentiating interaction between TRH and 5-HT has previously been established electrophysiologically in motorneurons from the lumbar region of the spinal cord[67].

Several peptides also influence the action of TRH. Bombesin, calcitonin, calcitonin gene related peptide, neurotensin, opioid peptides, interleukin-1ß, and basic fibroblast growth factor when injected into the cisterna magna along with TRH or RX 77368 prevented gastric lesion formation induced by TRH or the analog[17,23,25,26,32,46,48,68] (Fig. 1). The mechanism of the interaction between TRH and these peptides is most likely to take place at the site of action of TRH since microinjection of these peptides along with TRH in the DMN prevented the stimulation of gastric function induced by TRH or RX 77368[69,70,71].

PERIPHERAL MECHANISMS MEDIATING INTRACISTERNAL TRH-INDUCED GASTRIC LESIONS

Neural Pathways

Several sets of evidence indicate that intracisternal TRH or TRH analog-induced gastric lesions are mediated through the autonomic nervous system, mainly the parasympathetic outflow to the stomach. First, it has been established that direct electrical stimulation of the vagus in rats produces a 100% incidence of gastric hemorrhagic lesions similar to those observed after central injection of TRH[72-75]. Second, the TRH and TRH analogs, injected into the CSF at doses that induce gastric erosions induce a sustained and dose related stimulation of single- and multi-unit activity recorded from the cervical or gastric branches of the vagus[76-78]. Moreover, the DMN, which contains 90% of the preganglionic motoneurons whose axons join to form the gastric branch of vagus[79], is the most responsive site to TRH[47,48]. Lastly, vagotomy or peripheral muscarinic blockade by atropine prevented intracisternal, or central amygdala injection of TRH-induced gastric lesions[35,38,39,41,50].

Central injection of TRH also stimulates sympathetic efferent discharges to the adrenals and catecholamine secretion and inhibits sympathetic outflow to the stomach[76,80]. However, alterations of the sympathetic activity do not appear to contribute significantly to TRH action. Gastric damage induced by the TRH analog, DN-1417, is unaltered by peripheral injection of either adrenergic blockers, phentolamine or yohimbine or dopamine antagonists, sulpiride and haloperidol[38,41]. Although intracisternal injection of TRH is associated with the activation of the pituitary-thyroid axis[45,77], the production of gastric lesions is not causally related to the

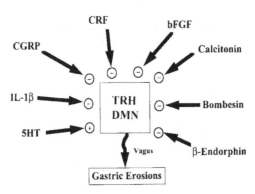

Figure 1. Peptides or transmitters preventing (-) or potentiating (+) TRH or TRH analog action in the dorsal motor nucleus of the vagus (DMN) leading to gastric erosions in fasted rats.

endocrine changes. The TRH analog, DN-1417 which has little TSH releasing activity[58] produces greater gastric damage than TRH[41]. Moreover, TRH or RX 77368 injected intravenously at doses stimulating the pituitary-thyroid axis did not induce gastric lesions[35,41]. Taken together these data clearly indicate that intracisternal injection of TRH induces gastric erosions primarily mediated through the stimulation of the parasympathetic outflow to the stomach. This occurs independently of the associated stimulation of the endocrine or sympathetic systems.

Peripheral Mechanisms

Pharmacologic studies indicated that gastric erosions in response to intracisternal injection of TRH or RX 77368 result from the interplay of aggressive and defensive factors stimulated by central vagal activation (Fig. 2).

Stimulation of Aggressive Factors. TRH or TRH analogs, RX 77368 and DN-1417, injected into the CSF or DMN at doses that induced gastric erosions elicited maximal stimulation of gastric acid and pepsin secretion through vagal muscarinic mechanisms in rats[38,41,47,77,81-84]. Sixty percent of the acid response to a maximal effective dose of RX 77368 is mediated by the increase in histamine release acting through H_2 receptors while the remaining acid response, resistant to H_2 blockade, represents a direct muscarinic activation of the parietal cells independently from gastrin[85-87]. The stimulation of gastric acid secretion may play a role since antisecretory doses of atropine, cimetidine, or omeprazole each inhibited gastric lesions induced by CSF injection of TRH or TRH analogs[35,38,39,41,46]. However, it is unlikely that such an increase in gastric acid secretion represents the only underlying mechanism leading to gastric erosions since bethanechol, given at doses that produced similar stimulation of acid secretion, neither causes gastric lesions nor aggravates indomethacin-induced ulcers in conscious rats[38]. Central injection of TRH analog also induced a marked release of gastric serotonin (5-HT)[88-90]. The influence of peripheral 5-HT in the pathogenesis of gastric erosions is still to be defined although gastric lesions have been induced in rats by compound 48/80 which increases the release of both histamine and 5-HT[91].

A synergism between enhanced gastric secretion and contractility has been involved in the genesis of other experimental ulcers induced by drugs or stress[41,92-94].

Injection of TRH or TRH analogs into the CSF or the DMN induced a sustained vagally mediated stimulation of high amplitude gastric contractions[95-98]. The sustained vagal cholinergic stimulation of gastric secretory and motor function induced by intracisternal or DMN injection of TRH may be part of the underlying mechanisms leading to the development of gastric erosions. Likewise, the cholinergic mimetic, carbachol, infused for two h in anesthetized rats induced stimulation of gastric acid secretion and motility and resulted in gastric erosions[99].

Stimulation of Defensive Factors. Simultaneously with the vagal dependent release of aggressive mechanisms (increased motility, acid pepsin, histamine, 5-HT, acetylcholine), there is a vagal dependent release of protective neurotransmitters that modulate the gastric erosive response[100]. TRH injected intracisternally or into the DMN induces a vagal muscarinic mediated release of gastric prostaglandin E_2 (PGE_2)[101-103]. Gastric PGE_2 released under these conditions is biologically active for decreasing the intensity of gastric lesions induced by TRH or TRH analog[66]. This was shown by the fact that the inhibition of prostaglandin synthesis with indomethacin injected at a dose that did not alter the integrity of the gastric mucosa under basal conditions, potentiates gastric erosion formation in response to intracisternal injection of RX 77368[66]. Intracisternal or DMN injection of TRH also stimulates gastric mucosal blood flow independently from the increase in acid secretion[104-106]. Increased gastric mucosal blood flow is known to increase the resistance of the gastric mucosa to injury[107].

PHYSIOLOGICAL ROLE OF MEDULLARY TRH AND 5-HT IN THE FORMATION OF GASTRIC LESIONS

The demonstration that exogenous TRH acts in the DMN to induce gastric lesions through vagal dependent pathways in rats has led to speculate that medullary TRH may play a role in the formation of gastric erosions. Exposure to cold in restrained rats facilitates the formation of gastric hemorrhagic lesions through vagal cholinergic dependent pathways[99,108-111] and increases hypothalamic TRH release and gene expression[112,113]. Convergent neuroanatomic and functional evidence indicates that endogenous medullary TRH is involved in the lesion formation induced by cold restraint stress.

Cold Restraint-Induced Gastric Lesions Involve Medullary TRH

Anatomical Evidence. The dorsal vagal complex contains the highest levels of TRH immunoreactivity outside of the hypothalamus where TRH immunoreactive fibers form a dense network[114-117]. TRH is located exclusively in the varicose and non varicose unmyelinated fibers and presynaptic terminals and never observed in perikarya or dendrites even in colchicine-treated rats[114-117]. Ultrastructural studies have further established that TRH terminals form asymmetric synaptic (excitatory) contacts with the dendrites of gastric preganglionic motoneurons[118,119]. In addition, TRH immunoreactivity and receptors are localized within synaptic terminals surrounding preganglionic vagal neurons[55]. In situ hybridization studies have further established that DMN neurons express mRNA TRH-receptor in rats[120].

Studies in rats using combined Fluoro-Gold retrograde neural tracing and TRH immunohistochemistry and ventral transections of various axonal pathways to the dorsal vagal complex clearly demonstrated that the sole source of TRH in DMN fibers and terminals originates from TRH containing cell bodies located in the raphe pallidus, and obscurus as well as the parapyramidal region of the ventral medulla[116,121]. Cell bodies in the paraventricular nucleus of the hypothalamus do not contribute to the TRH innervation of the dorsal vagal complex[116,121,122].

TRH coexists with 5-HT in neurons in the caudal raphe nuclei and parapyramidal region[121,123-125]. Although double labelling and retrograde tracing studies have not yet established whether TRH and 5-HT are co-localized in DMN terminals arising from medullary raphe neurons, studies have shown that 5-HT containing neurons from the raphe pallidus and obscurus project to the nucleus tractus solitarius[126].

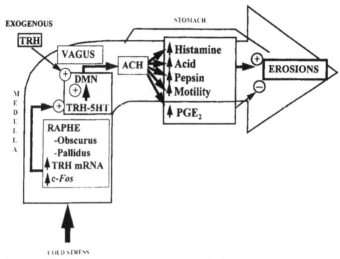

Figure 2. Schematic representation of activation of raphe-DMN pathways induced by cold restraint and peripheral mechanisms involved in the formation of gastric erosions in rats.

Functional Evidence. Convergent findings indicate that cold-restraint induced gastric erosions result from activation of the TRH/5-HT containing neurons in the raphe pallidus and raphe obscurus which leads to the release of TRH and 5-HT in the dorsal vagal complex and the activation of preganglionic vagal motorneurons and vagal dependent stimulation of gastric function (Fig. 2). First, activation of cell bodies in the raphe obscurus or pallidus by kainic acid or electrical stimulation induces TRH and 5-HT release in the dorsal vagal complex[127,128] as well as vagal muscarinic dependent stimulation of gastric secretory and motor function and gastric erosions in rats[128-132]. Second, there is evidence that exposure to cold under conditions inducing gastric erosions is associated with the activation of neurons in the raphe pallidus, raphe obscurus, parapyramidal region of the ventral medulla and DMN as shown by the high density of Fos positive cells, particularly in the raphe pallidus in rats. Control groups maintained for three h in semi-restraint at room temperature had no c-fos expression in the medulla and no gastric erosions[133]. Fos protein and Jun each regulate the transcription of genes containing AP-1 consensus element which is present on the TRH gene[134,135]. Further studies indeed indicate that cold-restraint exposure for three h increased TRH mRNA levels in the raphe pallidus, raphe obscurus and parapyramidal region[111]. Third, injection of TRH antiserum into the CSF prevented cold restraint stress induced gastric lesions in rats[40,99,136]. Fourth, the $5-HT_{2C}$ antagonist, ketanserin, injected intracisternally decreased cold-restraint induced gastric lesions while intracisternal injection of 5-HT potentiates gastric erosions induced by intracisternal TRH[66]. Lastly, although injection of TRH into the central amygdala was shown to induce gastric lesions[14,50], cold-restraint for three h did not increase c-fos expression in this nucleus[133]. These data, added to the lesser potency of TRH in this nucleus compared with the DMN[47,48] suggest that the raphe pallidus/obscurus-DMN pathways containing TRH/5-HT neurons[124] may be the primary anatomical and biochemical substrata which mediate vagal-dependent gastric erosions during cold-restraint stress.

SUMMARY

TRH and TRH analogs act in the DMN to induce gastric erosions in rats. TRH action is mediated through enhanced parasympathetic outflow to the stomach and expressed by vagal dependent stimulation of gastric secretory and motor function together with PGE_2 which reduces lesion formation. There is compelling neuroanatomical, electrophysiological and functional evidence to ascertain a physiological role of TRH/5-HT located in cell bodies in the raphe pallidus and raphe obscurus and their projections to the DMN in mediating cold restraint stress-induced vagal cholinergic dependent gastric erosions.

ACKNOWLEDGMENTS

Supported by the National Institute of Arthritis, Metabolism and Digestive Disease, Grants AM 30110 and AM 33061 and the Institute of Mental Health, Grant MH-0063. The author thanks Mr. Paul Kirshbaum for helpful assistance in the preparation of the manuscript.

REFERENCES

1. C.E. Brown-Sequard, Des alterations qui surviennent dans la muqueuse de l'estomac, consecutivement aux lesions cerebrales, *Prog Med.* 4:136 (1876).
2. M. Schiff, *in:* "Lecons sur la Physiologie de la Digestion," H. Loescher, Florence (1867).
3. J.J. Skillman, L.S. Bushnell, H. Goldman, and W. Silen, Respiratory failure, hypotension, sepsis, and jaundice. A clinical syndrome associated with lethal hemorrhage from acute stress ulceration of the stomach, *Am J Surg.* 117:523 (1969).
4. C.E. Lucas, C. Sugawa, J. Riddle, F. Rector, B. Rosenberg, and A.J. Walt, Natural history and surgical dilemma of "stress" gastric bleeding, *Arch Surg.* 102:266 (1971).
5. C.V. Grijalva, E. Lindholm, and D. Novin, Physiological and morphological changes in the gastrointestinal tract induced by hypothalamic intervention: an overview, *Brain Res Bull.* 5:19 (1980).

6. P.G. Henke, R.J. Savoie, and B.M. Callahan, Hippocampal deafferentation and deefferentation and gastric pathology in rats, *Brain Res Bull.* 7:395 (1981).

7. A.V. Ferguson, P. Marcus, J. Spencer, and J.L. Wallace, Paraventricular nucleus stimulation causes gastroduodenal mucosal necrosis in the rat, *Am J Physiol.* 255:R861 (1988).

8. A.S. Salim, The hypothalamus and gastric mucosal injuries: origin of stress-induced injury, *J Psychiatr Res.* 22:35 (1988).

9. S. Bonfils, and M. Dubrasquet, Psychotropic drugs in experimental peptic ulcer induced by psychological stress, *in*: "Psychotropic Drugs in Internal Medicine," A. Pletscher, A. Marino, eds., Excerpta Medica Foundation, Amsterdam (1969).

10. W.P. Paré, and G.B. Glavin, Restraint stress in biomedical research: a review, *Neurosci Biobehav Rev.* 10:339 (1986).

11. J. Kunchandy, and S.K. Kulkarni, Involvement of central type benzodiazepine and $GABA_A$ receptor in the protective effect of benzodiazepines in stress-induced gastric ulcers in rats, *Arch Int Pharmacodyn Ther.* 285:129 (1987).

12. Y. Taché, Central nervous system action of neuropeptides to influence or prevent experimental gastroduodenal ulceration, *in*: "Ulcer Disease: New Aspects of Pathogenesis and Pharmacology," S. Szabo, C.J. Pfeiffer, eds., CRC Press, Inc., Boca Raton (1989).

13. Y. Taché, and T. Ishikawa, Role of brain peptides in the ulcerogenic response to stress, *in*: "Hans Selye Symposia on Neuroendocrinology and Stress: Neuropeptides and Stress," Y. Taché, J.E. Morley, M.R. Brown, eds., Springer-Verlag, New York (1989).

14. G.B. Glavin, R. Murison, J.B. Overmier, W.P. Pare, H.K. Bakke, P.G. Henke, and D.E. Hernandez, The neurobiology of stress ulcers, *Brain Res Rev.* 16:301 (1991).

15. Y. Taché, P. Simard, and R. Collu, Prevention by bombesin of cold-restraint stress induced hemorrhagic lesions in rats, *Life Sci.* 24:1719 (1979).

16. M. Hagiwara, H. Watanabe, and R. Kanaoka, Effects of intracerebroventricular bombesin on gastric ulcers and gastric glycoproteins in rats, *J Pharmacobiodyn.* 8:864 (1985).

17. D.E. Hernandez, C.B. Nemeroff, R.C. Orlando, and A.J. Prange, The effect of centrally administered neuropeptides on the development of stress-induced gastric ulcers in rats, *J Neurosci Res.* 9:145 (1983).

18. D.E. Hernandez, Neuroendocrine mechanisms of stress ulceration: focus on thyrotropin-releasing hormone (TRH), *Life Sci.* 39:279 (1986).

19. M.W. Gunion, G.L. Kauffman, and Y. Taché, Intrahypothalamic microinfusion of corticotropin-releasing factor elevates gastric bicarbonate secretion and protects against cold-stress ulceration in rats, *Am J Physiol.* 258:G152 (1990).

20. R. Murison, and H.K. Bakke, The role of corticotropin-releasing factor in rat gastric ulcerogenesis, *Ann N Y Acad Sci.* 597:71 (1990).

21. T. Shibasaki, N. Yamauchi, M. Hotta, A. Masuda, H. Oono, I. Wakabayashi, N. Ling, and H. Demura, Brain corticotropin-releasing factor acts as inhibitor of stress-induced gastric erosion in rats, *Life Sci.* 47:925 (1990).

22. J.E. Morley, A.S. Levine, and S.E. Silvis, Intraventricular calcitonin inhibits gastric acid secretion, *Science.* 214:671 (1981).

23. Y. Taché, E. Kolve, M. Maeda-Hagiwara, and G.L. Kauffman, Jr., Central nervous system action of calcitonin to alter experimental gastric ulcers in rats, *Gastroenterology.* 94:145 (1988).

24. T. Ishikawa, and Y. Taché, Intrahypothalamic microinjection of calcitonin prevents stress-induced gastric lesions in rats, *Brain Res Bull.* 20:415 (1988).

25. Y. Taché, Intracisternal injection of calcitonin-gene related peptide inhibits experimental gastric ulcers, *in*: "Mechanisms of Injury, Protection and Repair of the Upper Gastrointestinal Tract," A. Garner, P.E. O'Brien, eds., John Wiley & Sons, Chichester (1991).

26. A. Ray, P.G. Henke, and R.M. Sullivan, The central amygdala and immobilization stress-induced gastric pathology in rats: neurotensin and dopamine, *Brain Res.* 409:398 (1987).

27. G.L. Kauffman, L. Zhang, L. Xing, J. Seaton, P. Colony, and L. Demers, Central neurotensin protects the mucosa by a prostaglandin-mediated mechanism and inhibits gastric acid secretion in the rat, *Ann NY Acad Sci.* 597:175 (1990).

28. S. Ferri, R. Arrigo-Reina, S. Candeletti, G. Costa, G. Murari, E. Speroni, and G. Scoto, Central and peripheral sites of action for the protective effect of opioids of the rat stomach, *Pharmacol Res Commun.* 15:409 (1983).

29. A. Ray, P.G. Henke, and R.M. Sullivan, Opiate mechanisms in the central amygdala and gastric stress pathology in rats, *Brain Res.* 442:195 (1988).

30. A. Uehara, T. Okumura, S. Kitamori, Y. Takasugi, and M. Namiki, Interleukin-1: a cytokine that has potent antisecretory and anti-ulcer actions via the central nervous system, *Biochem Biophys Res Commun.* 173:585 (1990).

31. A. Robert, E. Saperas, W. Zhang, A.S. Olafsson, C. Lancaster, D.E. Tracey, J.G. Chosay, and Y. Taché, Gastric cytoprotection by intracisternal interleukin-1β in the rat, *Biochem Biophys Res Commun.* 174:1117 (1991).

32. T. Okumura, A. Uehara, S. Kitamori, K. Okamura, Y. Takasugi and M. Namiki, Prevention by interleukin-1 of intracisternally injected thyrotropin-releasing hormone (TRH)-induced gastric mucosal lesions in rats, *Neurosci Lett.* 125:31 (1991).

33. T. Shibasaki, N. Yamauchi, M. Hotta, T. Imaki, T. Oda, N. Ling, and H. Demura, Interleukin-1 inhibits stress-induced gastric erosion in rats, *Life Sci.* 48:2267 (1991).

34. Y. Taché, and E. Saperas, Potent inhibition of gastric acid secretion and ulcer formation by centrally and peripherally administered interleukin-1, *Ann NY Acad Sci.* 659:353 (1992).

35. Y. Goto, and Y. Taché, Gastric erosions induced by intracisternal thyrotropin-releasing hormone (TRH) in rats, *Peptides.* 6:153 (1985).

36. D.E. Hernandez, J.D. Burke, R.C. Orlando, and A.J. Prange, Differential effects of intracisternal neurotensin and bombesin on stress- and ethanol-induced gastric ulcers, *Pharmacol Res Commun.* 18:617 (1986).

37. Y. Taché, R.L. Stephens, and T. Ishikawa, Central nervous system action of TRH to influence gastrointestinal function and ulceration, *Ann NY Acad Sci.* 553:269 (1989).

38. M. Maeda-Hagiwara, H. Watanabe, and K. Watanabe, Enhancement by intracerebroventricular thyrotropin-releasing hormone of indomethacin-induced gastric lesions in the rat, *Br J Pharmacol.* 80:735 (1983).

39. T. Nakane, N. Kanie, T. Audhya, and C.S. Hollander, The effects of centrally administered neuropeptides on the development of gastric lesions in the rat, *Life Sci.* 36:1197 (1985).

40. N. Basso, M. Bagarani, E. Pekary, A. Genco, and A. Materia, Role of thyrotropin-releasing hormone in stress ulcer formation in the rat, *Dig Dis Sci.* 33:819 (1988).

41. M. Maeda-Hagiwara, and H. Watanabe, Intracerebroventricular injection of a TRH analogue, gamma-butyrolactone-gamma-carbonyl-L-histidyl-prolinamide, induces gastric lesions and gastric acid stimulation in rats, *Naunyn-Schmiedeberg's Arch Pharmacol.* 330:142 (1985).

42. E.C. Griffiths, J.R. McDermott, and A.I. Smith, Mechanisms of brain inactivation of centrally-acting thyrotrophin-releasing hormone (TRH) analogues: a high performance liquid chromatography study, *Regul Pept.* 5:1 (1982).

43. E. Passaro, Jr., H. Debas, W. Oldendorf, and T. Yamada, Rapid appearance of intraventricularly administered neuropeptides in the peripheral circulation, *Brain Res.* 241:335 (1982).

44. G.S. Tannenbaum, and Y.C. Patel, On the fate of centrally administered somatostatin in the rat: massive hypersomatostatinemia resulting from leakage into the peripheral circulation has effects on growth hormone secretion and glucoregulation, *Endocrinology.* 118:2137 (1986).

45. A. Pekary, R. Stephens, M. Simard, X-P. Pang, V. Smith, J.J. Distephano,III, and J.M. Hershman, Release of thyrotropin and prolactin by a thyrotropin-releasing hormone (TRH) precursor, TRH-Gly: conversion to TRH is sufficient for in vivo effects, *Neuroendocrinology.* 52: 618 (1990).

46. P.G. Henke, R.M. Sullivan, and A. Ray, Interactions of thyrotropin-releasing hormone (TRH) with neurotensin and dopamine in the central nucleus of the amygdala during stress ulcer formation in rats, *Neurosci Lett.* 91:95 (1988).

47. D.E. Hernandez, and S.G. Emerick, Thyrotropin-releasing hormone: medullary site of action to induce gastric ulcers and stimulate acid secretion, *Brain Res.* 459:148 (1988).

48. A. Ray, P.G. Henke, and R.M. Sullivan, Effects of intra-amygdalar thyrotropin releasing hormone (TRH) and its antagonism by atropine and benzodiazepies during stress ulcer formation in rats, *Pharmacol Biochem Behav.* 36:597 (1990).

49. P.G. Henke, Recent studies of the central nucleus of the amygdala and stress ulcers, *Neurosci Biobehav Rev.* 12:143 (1988).

50. D.E. Hernandez, A.B. Salaiz, P. Morin, and M.A. Moreira, Administration of thyrotropin-releasing hormone into the central nucleus of the amygdala induces gastric lesions in rats, *Brain Res Bull.* 24:697 (1990).

51. S. Manaker, A. Winokur, W.H. Rostene, and T.C. Rainbow, Autoradiographic localization of thyrotropin-releasing hormone receptors in the rat central nervous system, *J Neurosci.* 5:167 (1985).

52. P.W. Mantyh, and S.P. Hunt, Thyrotropin-releasing hormone (TRH) receptors. Localization by light microscopic autoradiography in rat brain using [^{3}H][3-Me-His$_2$]TRH as the radioligand, *J Neurosci.* 5:551 (1985).

53. N.A. Sharif, and D.R. Burt, Visualization and identification of TRH receptors in rodent brain by autoradiography and radioreceptor assays: focus on amygdala, N. accumbens, septum and cortex, *Neurochem Int.* 7:525 (1985)

54. R. Norgren, and G.P. Smith, Central distribution of subdiaphragmatic vagal branches in the rat, *J Comp Neurol.* 273:207 (1988).

55. S. Manaker, and G. Rizio, Autoradiographic localization of thyrotropin-releasing hormone and substance P receptors in the rat dorsal vagal complex, *J Comp Neurol.* 290:516 (1989).

56. E.F. Hawkins, and W.K. Engel, Analog specificity of the thyrotropin-releasing hormone receptor in the central nervous system: possible clinical implications, *Life Sci.* 36:601 (1985).

57. G. Metcalf, P.W. Dettmar, D. Fortune, A.G. Lynn, and I.F. Tulloch, Neuropharmacological evaluation of RX 77368- a stabilised analogue of thyrotropin-releasing hormone (TRH), *Regul Pept.* 3:193 (1982).

58. M. Miyamoto, N. Fukuda, S. Narumi, Y. Nagai, Y. Saji, and Y. Nagawa, γ-butyrolactone-γ-carbonyl-histidyl-prolinamide citrate (DN-1417): a novel TRH analog with potent effects on the central nervous system, *Life Sci.* 28:861 (1981).

59. M.J. McCann, G.E. Hermann, and R.C. Rogers, Thyrotropin-releasing hormone:effects on identified neurons of the dorsal vagal complex, *J Auton Nerv Syst.* 26:107 (1989).

60. M. Raggenbass, C. Vozzi, E. Tribollet, M. Dubois-Dauphin, and J.J. Dreifuss, Thyrotropin-releasing hormone causes direct excitation of dorsal vagal and solitary tract neurones in rat brainstem slices, *Brain Res.* 530:85 (1990).

61. R.A. Travagli, R.A. Gillis, and S. Vicini, Effects of thyrotropin-releasing hormone on neurons in the rat dorsal motor nucleus of the vagus, in vitro, *Am J Physiol.* 263:G508 (1992).

62. C.A. Livingston, and A.J. Berger, Response of neurons in the dorsal motor nucleus of the vagus to thyrotropin-releasing hormone, *Brain Res.* 621:97 (1993).

63. T. Okumura, K. Okamura, S. Kitamori, H. Hara, Y. Shibata and M. Namiki, Gastric lesions induced by kainic acid injection into the dorsal motor nucleus of the vagus nerve in rats, *Scand. J. Gastroenterol.* 14(supp 162):15 (1989).

64. R.S. Leslie, Neuroactive substances in the dorsal vagal complex of the medulla oblongata: nucleus of the tractus solitarius, area postrema, and dorsal motor nucleus of the vagus, *Neurochem Int.* 7:191 (1985).

65. Y. Taché, and H. Yang, Brain regulation of gastric acid secretion by peptides: sites and mechanisms of action, *Ann NY Acad Sci.* 597:128 (1990).

66. M. Yoneda, and Y. Taché, Potentiating interactions between medullary serotonin and thyrotropin releasing hormone-induced gastric erosions in rats, *Neurosci Lett.* 161:199 (1993).

67. K.A. Clark, A.J. Parker, and G.C. Stirk, Potentiation of motoneurone excitability by combined administration of 5-HT agonist and TRH analog, *Neuropeptides.* 6:269 (1985).

68. T. Okumura, A. Uehara, S. Kitamori, Y. Takasugi, and M. Namiki, Central basic fibroblast growth factor inhibits gastric ulcer formation in rats, *Biochem Biophys Res Commun.* 177:809 (1991).

69. T. Ishikawa, and Y. Taché, Bombesin microinjected into the dorsal vagal complex inhibits vagally stimulated gastric acid secretion in the rat, *Regul Pept.* 24:187 (1989).

70. I. Heymann-Mönnikes, E.H. Livingston, Y. Taché, A. Sierra, H. Weiner, and T. Garrick, Bombesin microinjected into the dorsal vagal complex inhibits TRH-stimulated gastric contractility in rats, *Brain Res.* 533:309 (1990).

71. I. Heymann-Mönnikes, Y. Taché, M. Trauner, H. Weiner, and T. Garrick, CRF microinjected into the dorsal vagal complex inhibits TRH analog- and kainic acid-stimulated gastric contractility in rats, *Brain Res.* 554:139 (1991).

72. C.H. Cho, C.W. Ogle, and S. Dai, Acute gastric ulcer formation in response to electrical vagal stimulation in rats, *Eur J Pharmacol.* 35:215 (1976).

73. C.H. Cho, K.M. Hung, and C.W. Ogle, The aetiology of gastric ulceration induced by electrical vagal stimulation in rats, *Eur J Pharmacol.* 110:211 (1985).

74. T. Okumura, A. Uehara, K. Okamura, and M. Namiki, Site-specific formation of gastric ulcers by the electric stimulation of the left or right gastric branch of the vagus nerve in the rat, *Scand J Gastroenterol.* 25:834 (1990).

75. L. Hierlihy, J.L. Wallace, and A.V. Ferguson, Vagal stimulation-induced gastric damage in rats, *Am J Physiol.* 261:G104 (1991).

76. H. Somiya, and T. Tonoue, Neuropeptides as central integrators of autonomic nerve activity: effects of TRH, SRIF, VIP and bombesin on gastric and adrenal nerves, *Regul Pept.* 9:47 (1984).

77. Y. Taché, Y. Goto, D. Hamel, A. Pekary, and D. Novin, Mechanisms underlying intracisternal TRH-induced stimulation of gastric acid secretion in rats, *Regul Pept.* 13:21 (1985).

78. J.Y. Wei, and Y. Taché, Alterations of efferent discharges of the gastric branch of the vagus nerve by intracisternal injection of peptides influencing gastric function in rats, *Gastroenterology.* 98: A531 (Abstract) (1990).

79. R.E. Shapiro, and R.R. Miselis, The central organization of the vagus nerve innervating the stomach of the rat, *J Comp Neurol.* 238:473 (1985).

80. M.R. Brown, Thyrotropin releasing factor: a putative CNS regulator of the autonomic nervous system, *Life Sci.* 28:1789 (1981).

81. Y. Taché, W. Vale, and M. Brown, Thyrotropin-releasing hormone-CNS action to stimulate gastric acid secretion, *Nature* 287:149 (1980).

82. Y. Taché, Y. Goto, M. Lauffenburger, and D. Lesiege, Potent central nervous system action of p-Glu-His-(3,3'-dimethyl)-Pro NH_2, a stabilized analog of TRH, to stimulate gastric secretion in rats, *Regul Pept.* 8:71 (1984).

83. T. Ishikawa, H. Yang, and Y. Taché, Medullary sites of action of the TRH analogue, RX 77368, to stimulate gastric acid secretion in the rat, *Gastroenterology.* 95:1470 (1988).

84. R.L. Stephens, T. Ishikawa, H. Weiner, D. Novin, and Y. Taché, TRH analog, RX 77368, injected into the dorsal vagal complex stimulates gastric secretion in rats, *Am J Physiol.* 254:G639 (1988).

85. H. Yang, H. Wong, J.H. Walsh, and Y. Taché, Effect of gastrin monoclonal antibody 28.2 on acid response to chemical vagal stimulation in rats, *Life Sci.* 45:2413 (1989).

86. K. Yanagisawa, and Y. Taché, Intracisternal TRH analog RX 77368 stimulates gastric histamine release in rats, *Am J Physiol.* 259:G599 (1990).

87. K. Yanagisawa, H. Yang, J.H. Walsh and Y. Taché, Role of acetylcholine, histamine and gastrin in the acid response to intracisternal injection of TRH analog, RX 77368, in the rat, *Regul Pept.* 27:161 (1990).

88. R.L. Stephens, and Y. Taché, Intracisternal injection of a TRH analogue stimulates gastric luminal serotonin release in rats, *Am J Physiol.* 256:G377 (1989).

89. R.L. Stephens, T. Garrick, H. Weiner, and Y. Taché, Serotonin depletion potentiates gastric secretory and motor responses to vagal but not peripheral gastric stimulants, *J Pharmacol Exp Ther.* 251:524 (1989).

90. H. Yang, R.L. Stephens, and Y. Taché, TRH analogue microinjected into specific medullary nuclei stimulates gastric serotonin secretion in rats, *Am J Physiol.* 262:G216 (1992).

91. K. Takeuchi, H. Ohtsuki, and S. Okabe, Pathogenesis of compound 48/80-induced gastric lesions in rats, *Dig Dis Sci.* 31:392 (1986).

92. S. Yano, M. Akahane, and M. Harada, Role of gastric motility in development of stress-induced gastric lesions of rats, *Jpn J Pharmacol.* 28:607 (1978).

93. W.A. Mersereau, and E.J. Hinchey, Synergism between acid and gastric contractile activity in the genesis of ulceration and hemorrhage in the phenylbutazone-treated rat, *Surgery.* 901:516 (1981).

94. T. Garrick, S. Buack, and P. Bass, Gastric motility is a major factor in cold restraint-induced lesion formation in rats, *Am. J. Physiol.* 250:G191 (1986).

95. T. Garrick, S. Buack, A. Veiseh, and Y. Taché, Thyrotropin-releasing hormone (TRH) acts centrally to stimulate gastric contractility in rats, *Life Sci.* 40:649 (1987).

96. H.E. Raybould, L.J. Jacobsen, D. Novin, and Y. Taché, TRH stimulation and L-glutamic acid inhibition of proximal gastric motor activity in the rat dorsal vagal complex, *Brain Res.* 495:319 (1989).

97. R.C. Rogers, and G.E. Hermann, Oxytocin, oxytocin antagonist, TRH, and hypothalamic paraventricular nucleus stimulation effects on gastric motility, *Peptides.* 8:505 (1987).

98. T. Garrick, R. Stephens, T. Ishikawa, A. Sierra, A. Avidan, H. Weiner, and Y. Taché, Medullary sites for TRH analogue stimulation of gastric contractility in the rat, *Am J Physiol.* 256:G1011 (1989).

99. H. Niida, K. Takeuchi, K. Ueshima, and S. Okabe, Vagally mediated acid hypersecretion and lesion formation in anesthetized rat under hypothermic conditions, *Dig Dis Sci.* 36:441 (1991).

100. Y. Taché, H. Yang, and M. Yoneda, Vagal regulation of gastric function involves thyrotropin-releasing hormone in the medulary raphe nuclei and dorsal vagal complex, *Digestion.* 54:65 (1993).

101. M. Yoneda, and Y. Taché, Central thyrotropin-releasing factor analog prevents ethanol-induced gastric damage through prostaglandins in rats, *Gastroenterology.* 102:1568 (1992).

102. Y. Taché, and M. Yoneda, Central action of TRH to induce vagally mediated gastric cytoprotection and ulcer formation in rats, *J Clin Gastroenterol.* 17(Suppl. 1):S58 (1993).

103. M. Yoneda, and Y. Taché, Vagal regulation of gastric prostaglandin E2 release by central TRH in rats, *Am J Physiol.* 264:G231 (1993)

104. Y. Okuma, Y. Osumi, T. Ishikawa, and T. Mitsuma, Enhancement of gastric acid output and mucosal blood flow by tripeptide thyrotropin releasing hormone microinjected into the dorsal motor nucleus of the vagus in rats, *Jpn. J. Pharmacol.* 43, 173 (1987).

105. G. Thiefin, Y. Taché, F.W. Leung, and P.H. Guth, Central nervous system action of thyrotropin-releasing hormone to increase gastric mucosal blood flow in the rat, *Gastroenterology.* 97:405 (1989).

106. T. Tanaka, P. Guth, and Y. Taché, Role of nitric oxide in gastric hyperemia induced by central vagal stimulation, *Am J Physiol.* 264:G-280 (1993).

107. P. Holzer, Peptidergic sensory neurons in the control of vascular functions: mechanisms and significance in the cutaneous and splanchnic vascular beds, *Rev Physiol Biochem Pharmacol.* 121:50 (1992).

108. E.C. Senay, and R.J. Levine, Synergism between cold and restraint for rapid production of stress ulcers in rats. *Proc Soc Exp Biol.* 124:1221 (1967).

109. I. Arai, M. Muramatsu, and H. Aihara, Body temperature dependency of gastric regional blood flow, acid secretion and ulcer formation in restraint and water-immersion stressed rats, *Jpn J Pharmacol.* 40:501 (1986).

110. H. Niida, K. Takeuchi, and S. Okabe, Role of thyrotropin-releasing hormone in acid secretory response induced by lowering of body temperature in the rat, *Eur J Pharmacol.* 198:137 (1991).

111. H. Yang, S.V. Wu, T. Ishikawa, and Y. Taché, Cold exposure elevates thyrotropin-releasing hormone gene expression in medullary raphe nuclei: relationship with vagally mediated gastric erosions, *Neuroscience.* in press (1994).

112. R.T. Zoeller, N. Kabeer, and H.E. Albers, Cold exposure elevates cellular levels of messenger ribonucleic acid encoding thyrotropin-releasing hormone in paraventricular nucleus despite elevated levels of thyroid hormones. *Endocrinology.* 127:2955 (1990).

113. R.M. Uribe, J.L. Redondo, J.L. Charli, and P.J. Bravo, Suckling and cold stress rapidly and transiently increase TRH mRNA in the paraventricular nucleus, *Neuroendocrinology.* 58:140 (1993).

114. L. Calza, L. Giardino, S. Ceccatelli, M. Zanni, R. Elde and T. Hokelt, Distribution of thyrotropin-releasing hormone receptor messenger RNA in the rat brain: an in situ hybridization study, *Neuroscience.* 51:891 (1992).

115. W. Wu, R. Elde, M.W. Wessendorf, and T. Hokfelt, Identification of neurons expressing thyrotropin releasing-hormone receptor mRNA in spinal cord and lower brainstem of rat, *Neurosci Lett.* 142:143 (1992).

116. M. Palkovits, E. Mezey, R.L. Eskay, and M.J. Brownstein, Innervation of the nucleus of the solitary tract and the dorsal vagal nucleus by thyrotropin-releasing hormone-containing raphe neurons, *Brain Res.* 373:246 (1986).

117. M.J. Kubek, M.A. Rea, Z.I. Hodes, and M.H. Aprison, Quantitation and characterization of thyrotropin-releasing hormone in vagal nuclei and other regions of the medulla oblongata of the rat, *J Neurochem.* 40:1307 (1983).

118. L. Rinaman, R.R. Miselis, and M.S. Kreider, Ultrastructural localization of thyrotropin-releasing hormone immunoreactivity in the dorsal vagal complex in rat, *Neurosci Lett.* 104:7 (1989).

119. L. Rinaman, and R.R. Miselis, Thyrotropin-releasing hormone-immunoreactive nerve terminals synapse on the dentrites of gastric vagal motoneurons in the rat, *J Comp Neurol.* 294:235 (1990).

120. J. Zabavnik, G. Arbuthnott, and K.A. Eidne, Distribution of thyrotropin-releasing hormone receptor messenger RNA in rat pituitary and brain, *Neuroscience.* 53:877 (1993).

121. R.B. Lynn, M.S. Kreider, and R.R. Miselis, Thyrotropin-releasing hormone-immunoreactive projections to the dorsal motor nucleus and the nucleus of the solitary tract of the rat, *J Comp Neurol.* 311:271 (1991).

122. P. Siaud, L. Tapia-Arancibia, A. Szafarczyk, and G. Alanso, Increase of tyroptropin-releasing hormone immunoreactivitiy in the nucleus of the solitary tract following bilateral lesions of the hypothalamic paraventricular nuclei, *Neurosci Lett.* 79:47 (1987).

123. N. Liao, M. Bulant, P. Nicolas, H. Vaudry, and G. Pelletier, Immunoelectron microscopic localization of thyrotropin-releasing hormone (TRH) precursor in the rat raphe nuclei, *Peptides.* 11:397 (1990).

124. M.W. Wessendorf, N.M. Appel, T.W. Molitor and R.P. Elde, A method for immunofluorescent demonstration of three coexisting neurotransmitters in rat brain and spinal cord, using the fluorophores fluorescin, lissamine rhodamine, and 7-amino-4-methylcoumarin-3-acetic acid, *J Histochem Cytochem.* 38:1859 (1990).

125. P. Kachidian, P. Poulat, L. Marlier, and A. Privat, Immunohistochemical evidence for the coexistence of substance P, thyrotropin-releasing hormone, GABA, methionin-enkephalin, and leucin-enkephalin in the serotonergic neurons of the caudal raphe nuclei: A dual labeling in the rat, *J Neurosci Res.* 30:521 (1991).

126. K.B. Thorn, and C.J. Helke, Serotonin- and subtance P-containing projections to the nucleus tractus solitarii of the rat, *J Comp Neurol.* 265:275 (1987).

127. E. Brodin, B. Linderoth, M. Giony, Y. Yamamoto, B. Gazelius, D.E. Millhorn, T. Hokfelt, and U. Ungerstedt, In vivo release of serotonin in cat dorsal vagal complex and cervical ventral horn by electrical stimulation of the medullary raphe nuclei, *Brain Res.* 535:227 (1990).

128. H. Yang, G. Ohning, and Y. Taché, TRH in dorsal vagal complex mediates acid response to excitation of raphe pallidus neurons in rats, *Am J Physiol.* 265:G880 (1993).

129. M. McCann, G.E. Hermann, and R.C. Rogers, Nucleus raphe obscurus (nRO) influences vagal control of gastric motility in rats, *Brain Res.* 486:181 (1989).

130. H. Yang, T. Ishikawa, and Y. Taché, Microinjection of TRH analogs into the raphe pallidus stimulates gastric acid secretion in the rat, *Brain Res.* 531:280 (1990).

131. T. Garrick, M. Prince, H. Yang, G. Ohning, and Y. Taché, Raphe pallidus stimulation increases gastric contractility via TRH projections to the dorsal vagal complex, *Brain Res.* in press (1993).

132. T. Okumura, A. Uehara, Y. Taniguchi, Y. Watanabe, K. Tsuji, S. Kitamori, and M. Namiki, Kainic acid injection into medullary raphe produces gastric lesions through the vagal system in rats, *Am J Physiol.* 264:G655 (1993).

133. B. Bonaz, and Y. Taché, Induction of Fos immunoreactivity in the rat brain after cold-restraint induced gastric lesions and fecal excretion, *Brain Res.* in press (1994).

134. S.L. Lee, K. Stewart, and R.H. Goodman, Structure of the gene encoding rat thyrotropin releasing hormone, *J Biol Chem.* 263:16604 (1988).

135. M. Sheng, and M.E. Greenberg, The regulation and function of c-fos and other immediate early genes in the nervous system, *Neuron.* 4:477 (1990).

136. D.E. Hernandez, M.E. Arredondo, B.G. Xue, and L. Jennes, Evidence for a role of brain thyrotropin-releasing hormone (TRH) on stress gastric lesion formation in rats, *Brain Res Bull.* 24:693 (1990).

ROLE OF THE VAGAL NERVE IN THE DEVELOPMENT OF GASTRIC MUCOSAL INJURY AND ITS PREVENTION BY ATROPINE, CIMETIDINE, β-CAROTENE AND PROSTACYCLIN IN RATS

Gy.Mózsik, B.Bódis, M.Garamszegi, O.Karádi,
Á.Király, L.Nagy, G.Sütő, Gy.Tóth, and Á.Vincze

Peptic ulceration including acute erosions is a multifactorial disease involving well-known factors such as trauma, stress, sepsis, hemorrhagic shock, burns, pulmonary and liver diseases[1-9] and drugs such as reserpine[10], epinephrine[11-13], nonsteroidal antiinflammatory drugs[7], steroids and other chemicals[14].

A wide range of experimental models have been used to study the development of acute mucosal damage and its prevention. Most experimental models result in acute gastric mucosal damage.

The intragastric administration of 96 % ethanol, 0.6 M HCl, 0.2 M NaOH, 25 % NaCl or boiled water produces acute gastric mucosal injury[14]. These models are widely used to study the details of development of gastric mucosal damage and its prevention in animal experiments.

The study of the details of mucosal protection is the most important field from the point of view of ulcer research due to its relevance to human therapy[15-17]. Different symposia dealt with the details of mucosal protection during the last ten years.

A special type of gastric mucosal protection has been discovered by Chaudhury and Jacobson[18], called "gastric cytoprotection". The details of this type of gastric mucosal protection have been generally accepted after the publication of Robert et al.[14]. The essential point of "gastric cytoprotection" is that the different prostaglandins (PGs) are able to prevent the gastric mucosal damage (produced by 96 % ethanol, 0.6 M HCl, 0.2 M NaOH and thermal injury) without concomitant inhibition of gastric secretory responses, e.g. without decreasing aggressive factors.

The existence of "gastric cytoprotection" was suggested to result from PGs[14]. In addition, the existence of gastric cytoprotection was demonstrated with small doses of anticholinergics[19,20,21], antimuscarinic agents[22], H$_2$-blockers[21,23,24], epidermal growth factor[25], vitamin A and carotenoids[26-28] and sodium chromoglycate[29,30] in animal experiments. A lot of compounds [such as carotenoids[28,31-34], antacids[35], De-Nol[36], (tripotassium dicitrato bismuthate), sucralfate[34]] having only cytoprotective effects, are widely used in the treatment of patients with gastric and duodenal ulcer disease.

The exact mechanisms of the development of gastric cytoprotection are unknown, although many observations have been carried out. The results were summarized recently[37,38]. The existence of gastric cytoprotection has also been shown to occur with prostacyclin[39].

The details of PGI_2-induced gastric cytoprotection have been evaluated by different biochemical examinations[41]. It was proved that different drugs, such as atropine (pharmacological vagotomy), cimetidine, epinephrine, dinitrophenol, tetracycline, mannomustine (Degranol) and actinomycin had essential inhibitory effects on PGI_2-induced gastric cytoprotection[42-44]; a similar inhibition was produced by bilateral surgical vagotomy[40]. It has been concluded from these results that an intact vagus nerve is necessary for the development of PGI_2-induced gastric cytoprotection. The role of the intact vagus nerve has been proven in the development of "adaptive cytoprotection"[45].

The existence of gastric cytoprotective effects has been shown with PGI_2[39], atropine[20], cimetidine[21,23] and carotenoids[27] in models of ethanol, HCl, NaOH and concentrated NaCl. These experiments represent different types of compounds, all having gastric protective effects. PGI_2 is one of the endogenous defence compounds; carotenoids are nutritients, while atropine and cimetidine are pharmacological agents.

The aims of this study were:

1. To evaluate the effects of compounds (PGI_2, atropine, cimetidine and β-carotene) on gastric acid secretion in 4 h pylorus-ligated rats;

2. To evaluate the effect of surgical vagotomy on ethanol-induced gastric mucosal lesions;

3. To study the effects of cytoprotective (PGI_2, β-carotene) and cytoprotective plus antisecretory (atropine, cimetidine) compounds on ethanol-induced gastric mucosal lesions, in rats with intact vagal nerve and in vagotomized rats;

4. To clarify the possible roles of acute surgical vagotomy on the development of gastric mucosal protection produced by different agents, having significantly different mechanisms of action.

MATERIALS AND METHODS

The observations were carried out on both sexes of CFY-strain rats (LATI, Gödöllő, Hungary), weighing 180 to 210 g. This strain originated from the Sprague-Dawley strain, U.S.A. The animals were fasted for 24 h before the experiments, but they received water ad libitum. The observations were begun at 0800.

Gastric Secretory Studies

The gastric secretory studies were carried out in 4 h pylorus ligated rats. The animals were fasted for 24 h before the experiments. Pyloric ligation was carried out by the method of Shay et al[46]. The volume of gastric juice was quantitatively measured, and the H^+ output was measured by titration with 0.1 M NaOH to pH 7.0 (in presence of a titrimeter, Radekis, Budapest, Hungary). The gastric volume was expressed as ml/100g/4 h, while H^+ output as $\mu Eq/100g/4$ h.

Production of Gastric Mucosal Lesions

Gastric mucosal lesions (ulcers) were produced by intragastric administration of 96 % ethanol (1 ml). The animals were sacrificed at 1 h after ethanol administration.

The gastric mucosal lesions (ulcers) were calculated (number) and their severities were estimated by using a semiquantitative scale: 0, no lesion; 1, lesion is smaller than 1 mm size; 2, lesion is between 1 and 2 mm; 3, lesion is between 2 and 3 mm; 4, lesion is between 4 and 5 mm; 5, lesion is larger than 5 mm. The number and sum of their severity were calculated and expressed per one rat stomach (means±SEM).

Surgical Vagotomy

Bilateral surgical vagotomy was carried out on the lower part of the esophagus. Approximately 4-5 mm pieces were cut out from both sides of the vagus nerve. Sham-operated rats underwent laparotomy only. The animals were treated with ethanol at 30 min after surgical vagotomy or sham-operation.

Treatment with Atropine, Cimetidine, Prostacyclin and β-Carotene

Prostacyclin (PGI$_2$) (CHINOIN, Budapest, Hungary) and β-carotene (Hoffmann-LaRoche, Switzerland) were used at gastric cytoprotective doses and given intragastrically. PGI$_2$ was freshly dissolved from stock solution by the method of Wallus et al.[47]. β-carotene was dissolved in sunflower oil. Control animals were treated with the solvent of the above materials. Atropine (Atropinum sulfuricum, EGIS, Budapest, Hungary) and cimetidine (Tagamet, Smith & Kline French, England) were dissolved further with saline solution to obtain the necessary doses and given subcutaneously. The same treatment was carried out in vagotomized and intact (sham-operated) animals. The animals were treated with ethanol at 30 min later, and were sacrificed at 1 h after administration of ethanol. The number and severity of gastric mucosal lesions was noted (Fig. 1).

Statistical Analysis of the Results

The results were expressed as means±SEM. The unpaired Student's t test was used for statistical analysis of results, except for ulcer severity when the Mann-Whitney U test was applied. A p value of ≤ 0.05 was considered to be significant.

RESULTS

Gastric Secretory Studies

No gastric inhibitory effect was obtained with administration of PGI$_2$ (in doses of 5, 50 and 100 μg/kg) β-carotene (in doses of 1 and 10 mg/kg), smaller doses of atropine (0.025 mg/kg) or cimetidine (2.5 mg/kg). However, a significant inhibition was produced by larger doses of atropine (0.2 and 1.0 mg/kg), cimetidine (50 mg/kg) and by bilateral surgical vagotomy (Table 1).

Effect of Surgical Vagotomy on Ethanol-Induced Gastric Mucosal Injury

Bilateral surgical vagotomy aggravates both the number and the severity of ethanol-induced gastric injury (Fig. 2).

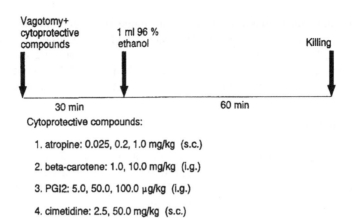

Cytoprotective compounds:

1. atropine: 0.025, 0.2, 1.0 mg/kg (s.c.)

2. beta-carotene: 1.0, 10.0 mg/kg (i.g.)

3. PGI2: 5.0, 50.0, 100.0 µg/kg (i.g.)

4. cimetidine: 2.5, 50.0 mg/kg (s.c.)

Figure 1. Experimental protocols for the study of drug actions on the development of ethanol-induced gastric mucosal damage in sham-operated (animals with intact vagal nerve) and surgically vagotomized rats (1 h experiments).

Gastric (Cyto-) Protection by Atropine, Cimetidine, β-carotene and Prostacyclin in Intact and Vagotomized Animals

Atropine Effects. Atropine inhibited (dose-dependently) both the number (Fig. 3) and severity (Fig. 4) of ethanol-induced gastric mucosal lesions in intact rats, but no effect was obtained in vagotomized rats.

β-Carotene Effects. β-carotene dose-dependently inhibited the number and severity of ethanol-induced gastric mucosal lesions in intact rats, while no protection was noticed in vagotomized rats (Fig. 5).

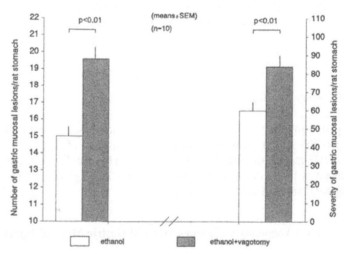

Figure 2. Aggravating effect of bilateral surgical vagotomy on the number (left side) and severity (right side) of ethanol-induced gastric mucosal lesions in rats.

178

Table 1. The Effects of PGI_2, Atropine, Cimetidine, β-Carotene and Bilateral Surgical Vagotomy on Gastric Secretory Responses (Volume, H^+ Output) in 4 h Pylorus-Ligated Rats

Groups of animals	n	Volume (ml/100g/4h)	H^+ output (μEq/100g/4h)
Control			
(saline treated)	10	4.1 ± 0.4	200 ± 18
PGI_2 5 μg/kg i.g.	10	4.0 ± 0.3^{NS}	180 ± 18^{NS}
50 μg/kg i.g.	10	4.1 ± 0.4^{NS}	175 ± 20^{NS}
100 μg/kg i.g.	10	4.0 ± 0.3^{NS}	180 ± 20^{NS}
Atropine			
0.025 mg/kg s.c.	10	4.2 ± 0.3^{NS}	190 ± 20^{NS}
0.2 mg/kg s.c.	10	$1.0 \pm 0.2^{***}$	$50 \pm 8^{***}$
1.0 mg/kg s.c.	10	$0.4 \pm 0.1^{***}$	$20 \pm 4^{***}$
Cimetidine			
2.5 mg/kg s.c.	10	4.1 ± 0.3^{NS}	180 ± 20^{NS}
50.0 mg/kg s.c.	10	$2.0 \pm 0.2^{***}$	$30 \pm 8^{***}$
β-Carotene			
1.0 mg/kg i.g.	10	3.8 ± 0.5^{NS}	197 ± 30^{NS}
10.0 mg/kg i.g.	10	4.1 ± 0.4^{NS}	207 ± 30^{NS}
Surgical vagotomy	10	$0.6 \pm 0.2^{***}$	$28 \pm 7^{***}$

p values are between results obtained on the control (saline treated) vs. treated (or vagotomized) groups. Abbreviations: NS: not significant; ***: $p < 0.001$.

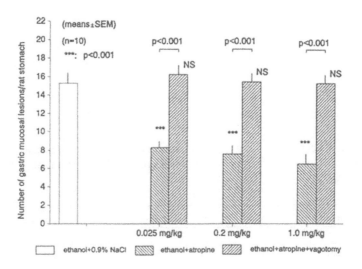

Figure 3. Effect of bilateral surgical vagotomy on atropine-induced gastric cytoprotection (0.025 mg/kg) and gastroprotection (0.2 and 1.0 mg/kg) appearing as the number of ethanol-induced gastric mucosal lesions in rats (1 h experiments).

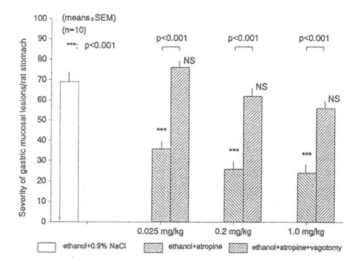

Figure 4. Effect of bilateral surgical vagotomy on atropine-induced gastric cytoprotection (0.025 mg/kg) and gastroprotection (0.2 and 1.0 mg/kg) appearing as the severity of ethanol-induced gastric mucosal damage in rats (1 h experiments).

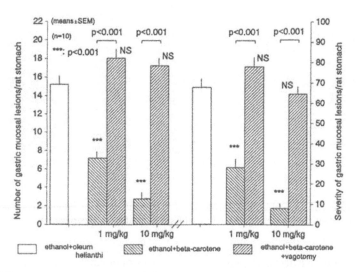

Figure 5. Effects of bilateral surgical vagotomy on β-carotene-induced gastric cytoprotection in rats treated with ethanol (1 h experiments).

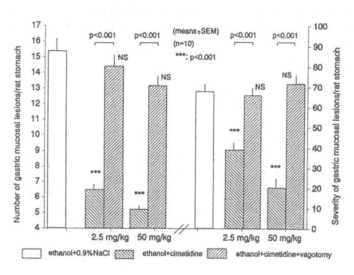

Figure 6. Effect of bilateral surgical vagotomy on cimetidine-induced gastric cytoprotection (2.5 mg/kg) and gastroprotection (50 mg/kg) in rats treated with ethanol (1 h experiments).

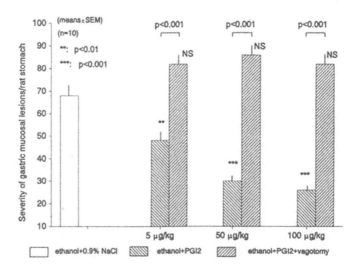

Figure 7. Effect of bilateral surgical vagotomy on PGI$_2$-induced gastric cytoprotection appearing as the number of ethanol-induced gastric mucosal lesions in rats (1 h experiments).

Cimetidine Effects. The number and severity of ethanol-induced gastric mucosal lesions could be decreased by cytoprotective (2.5 mg/kg) and antisecretory (50 mg/kg) doses of cimetidine in intact rats, but no prevention was noticed in surgically vagotomized rats (Fig. 6).

Prostacyclin Effects. PGI$_2$ was used only in gastric cytoprotective doses. The number (Fig. 7) and severity (Fig. 8) of ethanol-induced gastric mucosal lesions could be decreased by PGI$_2$ in intact rats, while no prevention was obtained by PGI$_2$ in vagotomized rats.

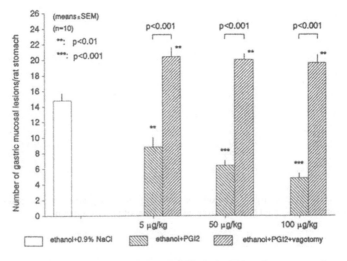

Figure 8. Effect of bilateral surgical vagotomy on PGI$_2$-induced gastric cytoprotection appearing as the severity of ethanol-induced gastric mucosal lesions in rats (1 h experiments).

DISCUSSION

The compounds used herein are widely different in their mechanisms of action: atropine is a typical anticholinergic compound, cimetidine is a representative agent of the H_2-receptor blockers, prostacyclin is an endogenous compound that protects the gastric mucosa, and β-carotene is a nutritional compound.

Previously, the existence of gastric cytoprotection was shown with smaller doses of atropine (0.025 mg/kg), cimetidine (2.5 mg/kg) and by PGI_2 and β-carotene. To evaluate the existence of gastric cytoprotection with smaller doses of atropine and cimetidine, four different experimental models (such as ethanol, HCl, NaOH and concentrated NaCl) were used. The basic requirement is the presence of gastric cytoprotection with small doses of atropine and cimetidine in four different experimental models, without concomitant gastric acid inhibitory effect[19,23], such as proposed by Guslandi[48].

The changes in membrane-bound ATP-dependent energy systems were analyzed during development of acute gastric mucosal lesions produced by 96 % ethanol, 0.6 M HCl, 0.2 M NaOH and 25 % NaCl (Table 2)[49]. Furthermore, these changes were studied at different times (0, 1, 5, 15, 30 and 60 min) after ethanol administration[43]. The results of these studies indicated that a very complex feedback mechanism system exists between the membrane-bound ATP-dependent energy systems, for example, ATP-membrane ATP-ase-ADP and ATP-adenylate cyclase-cAMP[19,26,41,42,43].

Table 2. Correlation Between the Membrane-Bound ATP-Dependent Energy Systems During Development of Gastric Mucosal Lesions Produced by 96% Ethanol, 0.6M HCl(HCl), 0.2M NaOH (NaOH) and Concentrated (25%) NaCl (NaCl) in Rats (1 h Experimental Period) (after Morón et al. Int.J.Tiss.Reac. 5:357-362,1983)

Membrane-bound ATP-dependent energy system	Gastric mucosal lesions produced by			
	Ethanol (96%)	HCl (0.6M))	NaOH (0.2M)	NaCl (25%)
ATP-ADP transformation	increased	increased	increased	increased
ATP-cAMP transformation	decreased	decreased	decreased	decreased
cAMP-AMP transformation	decreased	decreased	decreased	decreased

A considerable increase of ATP-ADP transformation was obtained during the whole experimental period, in association with a decreased extent of ATP-cAMP and cAMP-AMP (Table 2). The cellular mechanisms of the development of ethanol-induced gastric mucosal damage and of gastric hypersecretion in 4 h pylorus ligated rats was approached by using drugs having action at the level of cellular membrane-bound ATP-dependent energy systems (atropine, epinephrine, cimetidine, ouabain, PGI_2) oxidative phosphorylation (2,4-dinitrophenol), synthesis of DNA (mannomustine) RNA (actinomycin D) or translation of proteins (tetracycline)[40,42,43]. The intrinsic activity and affinity curves were identified for some drugs inhibiting the H^+ secretion in 4 h pylorus-ligated rats and on the development of ethanol-induced gastric mucosal lesions in rats with intact vagal nerves[43]. Surprisingly, the values of pD_2 and pA_2 were similar for the drugs in both experimental models (Table 3).

These results suggested that the same cellular mechanisms are involved both in the development of gastric hypersecretion in 4 h pylorus ligated rats and of ethanol-induced gastric mucosal damage in intact rats[43].

The gastric hypersecretion in pylorus-ligated rats is a vagus-dependent process; this process can be interrupted by bilateral surgical vagotomy and by atropine treatment[46,50,51]. Different examinations indicated that the ATP-ADP transformation is associated with the function of the intact vagal nerve, while the ATP-ADP transformation is inhibited by pharmacological (atropine treatment) and surgical vagotomy. Nevertheless, the two pathways significantly differ from the point of view of gastric biochemistry.

An active metabolic response of the gastric mucosa was obtained with alcohol, NaOH, 25 % NaCl and HCl in rats[40,41,43,53]. On the other hand, gastric mucosal cytoprotection and gastroprotection were obtained with a further increase in metabolic response (by PGI_2[14], β-carotene[54]) or by a decrease in metabolic response (cimetidine[19]) of the gastric mucosa (Table 4).

The increased or decreased metabolic adaptation to chemicals appears as a result of the equilibrium within the membrane-bound ATP-dependent energy system in the gastric mucosa, which can be modified by PGI_2[41], atropine[14], cimetidine[60] and β-carotene[61,62,63]. Many biochemical and pharmacological studies indicated a leading role for the ATP-membrane ATPase-ADP system in the regulation of both membrane-bound ATP-dependent energy systems[43], which are associated with the function of the intact vagal nerve[43].

The development of ethanol-induced gastric mucosal damage can be divided into an early (from 0 to 6 min) and a late (from 15 to 60 min) phase after administration of ethanol[43,55,56,62]. PGI_2 acts during the early phase[55-59,63], while β-carotene works in the late phase[55,56,63]. The actions of PGs appear at the level of the microvasculature[57,59], involving significant changes in gastric mucosal biochemistry[19,42,62]. β-carotene decreased gastric mucosal damage in the late phase (from 15 to 60 min) after ethanol administration[55,63].

The tested compounds produced gastric mucosal protection in ethanol-treated rats with intact vagal nerves (Table 5). These results emphasize the essential role of intact vagal nerves in the development of gastric mucosal protection produced by different compounds. On the other hand, β-carotene acts via changes in gastric mucosal biochemistry[54,56,61], and has free radical scavenging properties[56]. Because the mucosal protective effect of β-carotene is negated by surgical vagotomy, there is strong evidence for the dominant biochemical effects of β-carotene in the development of gastric mucosal protection.

The mechanisms of the role of vagus nerve in the development of gastric mucosal protection are unknown, however, some pathways can be suggested:

1. Decreased gastric emptying after surgical vagotomy, which however, was excluded in ethanol-treated animals[68];

2. Decreased synthesis of PGs (PGE_2, PGI_2) in vagotomized animals at 1 h[65], however, no decrease was found at 5 min in the ethanol model[72] (when mucosal damage had developed);

3. Increased vascular permeability to ethanol[66,69], HCl[75] and IND[77] in vagotomized rats. These observations were carried out in animals with acute surgical vagotomy, however, the increase in the vascular permeability was also found in ethanol-treated animals after chronic surgical vagotomy[74];

4. Changes in the gastric mucosal protective properties of SH-groups produced by vagotomy[64,75];

5. Changed biochemical mechanisms underlying gastric mucosal lesions in intact and in vagotomized rats[68,70];

6. Changed regulatory mechanisms between the membrane-bound ATP-dependent energy systems after surgical vagotomy[71,73,76];

7. Significantly altered regulatory pathways to protective agents in vagotomized rats relative to intact animals.

Table 3. Values of Affinities (pD_2) and Intrinsic Activities ($\alpha_{atropine}$ = 1.00) for Drugs Inhibiting Gastric Acid Secretion in 4 h Pylorus-Ligated Rats (A) and for Drugs Inhibiting the Development of Ethanol-Induced Gastric Mucosal Lesions (Number) in Rats with Intact Vagal Nerves (B)[+] (from Mózsik & Jávor, Dig.Dis.Sci. 33,92-105,1988, after modification).

Drugs	pD_2		$\alpha_{atropine}$		pA_2	
	A	B	A	B	A	B
Atropine	5.75	6.00	1.00	1.00	5.80	6.12
Actinomycin D	5.63	5.75	0.87	0.92	5.86	5.60
Cimetidine	3.00	3.87	0.64	0.96	3.20	3.75
Degranol	4.25	5.25	0.78	0.87	4.90	5.37
Dinitrophenol	3.50	4.87	0.56	0.48	3.75	4.72
Epinephrine	4.95	6.00	0.80	1.00	4.75	6.00
PGI_2	0.00	7.25	0.00	0.90	0.00	7.75
Ouabain	4.75	6.00	0.80	0.45	4.50	6.12
Tetracycline	3.63	4.75	0.78	0.65	3.75	4.75

[+] values are (-) molar

Table 4. Correlations Between Membrane-Bound ATP-Dependent Energy Systems During Development of Gastric Mucosal Protection Produced by PGI_2, Atropine, Cimetidine and β-Carotene in Ethanol-Induced Gastric Mucosal Lesions in Rats with Intact Vagal Nerves (see references 41-44, 60-63.)

Compounds	ATP-ADP transformation	ATP-cAMP transformation	cAMP-AMP transformation
Atropine	decreased	increased	increased
β-Carotene	no direct effect	increased	increased
Cimetidine	no direct effect	decreased	decreased
Prostacyclin	increased	increased	increased

The role of the vagus nerve is unknown in the development of gastric mucosal (cyto)protection, however it seems that the vagus nerve plays an essential role both in the development of gastric mucosal damage and as well as in gastric mucosal protection.

Table 5. Summary of the Presence of Gastric Cytoprotective and Gastroprotective Effects on the Number and Severity of Ethanol-Induced Gastric Mucosal Lesions in Intact and Surgically Vagotomized Rats

Tested doses	Number of gastric mucosal lesions		Severity of gastric mucosal lesions		
	intact rats	vagotomized rats	intact	rats	vagotomized rats
β-carotene					
1 mg/kg i.g.	Yes	No		Yes	No
10 mg/kg i.g.	Yes	No		Yes	No
PGI$_2$					
5 μg/kg i.g.	Yes	No		Yes	No
50 μg/kg i.g.	Yes	No		Yes	No
100 μg/kg i.g.	Yes	No		Yes	No
0.025 mg/kg s.c.	Yes	No	Atropine	Yes	No
0.2 mg/kg s.c.	Yes	No		Yes	No
1.0 mg/kg s.c.	Yes	No		Yes	No
Cimetidine					
2.5 mg/kg s.c.	Yes	No		Yes	No
50 mg/kg s.c.	Yes	No		Yes	No

doses underlined have antisecretory effects

SUMMARY

Observations were carried out to clarify the possible role of acute surgical vagotomy on the development of gastric cytoprotection (gastroprotection) produced by PGI$_2$, atropine, cimetidine and β-carotene on ethanol-induced gastric mucosal damage in rats.

The observations were carried out on rats of both sexes, weighing 180 to 210 g. Gastric mucosal damage was produced by intragastric administration of 96% ethanol (1 ml). Bilateral surgical vagotomy or laparotomy (in sham-operated or control animals) was carried out at 30 min before ethanol administration. At the time of surgical intervention, different compounds such as PGI$_2$ (in doses of 5, 50 and 100 μg/kg ig.), β-carotene (in doses of 1 and 10 mg/kg ig.), atropine (in doses of 0.025, 0.2 and 1.0 mg/kg sc.) and cimetidine (in doses of 2.5 and 50 mg/kg sc.) were given. The animals were sacrificed at 1 h after ethanol administration, when the number and severity of ethanol-induced gastric mucosal damage was noted.

It was found that: 1. all compounds (PGI$_2$, atropine, cimetidine, β-carotene) dose-dependently inhibited the development (both the number and severity) of ethanol-induced gastric mucosal lesions in rats with intact vagal nerves; 2. cyto- and gastroprotective effects of PGI$_2$, atropine, cimetidine and β-carotene were blocked by acute bilateral surgical vagotomy.

These results emphasize that the vagus nerve has an important role both in the development of gastric mucosal damage (by increasing aggressive factors) as well as in the development of gastric cytoprotection (organoprotection) produced by PGI$_2$, β-carotene, atropine and cimetidine. The details of this role for the vagus nerve in gastric protection are unknown.

ACKNOWLEDGEMENTS

The expert technical assistance of Mrs. Margaret Jermás and Mrs. Rosalia Nagy is especially acknowledged.

This study was supported by OTKA grant 2466 (Ofszágos Tudományos Kutatási Alap - Hungarian National Research Foundation).

REFERENCES

1. A.C. Ivy, M.I. Grossman, and W.H. Blachrach, "Peptic Ulcer", The Blakiston Company, Philadelphia (1950).
2. C.J. Pfeiffer, "Peptic Ulcer", Lippincott Company-Copenhagen, Munsgaard, Philadelphia (1971).
3. T.H. Gheorghiu, "Experimental Ulcer. Models, Methods and Clinical Validity", Gerhard Witzstrock Publishers, Baden-Baden, Brussels, Cologne (1975).
4. Gy. Mózsik, and T. Jávor, "Progress in Peptic Ulcer", Akadémiai Kiadó, Budapest (1976).
5. Gy. Mózsik, O. Hanninen, and T. Jávor, "Advances in Physiological Sciences, Vol.29, Gastrointestinal Defence Mechanisms", Pergamon Press-Akadémiai Kiadó, Oxford, Budapest (1981).
6. S. Umehara, and H. Ito, "Advances in Experimental Ulcer", International Conference on Experimental Ulcer, Tokyo (1972).
7. K.D. Rainsford, and G.P. Velo, "Side Effects of Antiinflammatory Analgetic Drugs", Raven Press, New York (1987).
8. S. Szabo, and Gy. Mózsik, "New Pharmacology of Ulcer Disease", Elsevier Sciences Publishers, New York (1987).
9. S. Szabo, and C.J. Pfeiffer, "Ulcer Disease: New Aspects of Pathogenesis and Pharmacology", CRC Press, Boca Raton (1989).
10. Gy. Mózsik, L. Nagy, I. Patty, and F. Tárnok, Cellular energy systems and reserpine ulcer, *Acta Physiol Hung.* 62:107 (1983).
11. S. Sethbhakdi, C.J. Pfeiffer, and J.L.A. Roth, Gastric mucosal ulceration following vasoactive agents. A new experimental model, *Amer J Dig Dis.* 15:261 (1970).
12. S. Sethbhakdi, J.L.A. Roth, and C.J. Pfeiffer, Gastric mucosal ulceration after epinephrine. A study of etiologic mechanisms, *Amer J Dig Dis.* 15:1055 (1970).
13. Gy. Mózsik, M. Fiegler, L. Nagy, I. Patty, and F. Tárnok, Gastric and small intestinal energy metabolism in mucosa, in: "Advances in Physiological Sciences, Vol.29, Gastrointestinal Defence Mechanisms", Gy. Mózsik, O. Hanninen, T.Jávor, eds., Pergamon Press-Akadémiai Kiadó, Oxford, Budapest (1981).
14. A. Robert, J.E. Nemazis, C. Lanchaster, and A.J. Hanchar, Cytoprotection by prostaglandins in rats. Prevention of gastric mucosal necrosis by alcohol, HCl, NaOH, hypertonic NaCl and thermal injury, *Gastroenterology.* 77:433 (1979).
15. A. Allen, G. Flemström, A. Garner, W. Silen, and L.A. Turnberg, "Mechanisms of Mucosal Protection in the Upper Gastrointestinal Tract", Raven Press, New York (1984).
16. Gy. Mózsik, A. Pár, and A. Bertelli, "Recent Advances in Gastrointestinal Cytoprotection", Akadémiai Kiadó, Budapest (1984).
17. Gy. Mózsik, T. Jávor, M. Kitajima, C.J. Pfeiffer, K.D. Rainsford, L. Simon, and S. Szabó, Advances in gastrointestinal cytoprotection: topics 1987, *Acta Physiol Hung.* 73:111 (1989).
18. K. Chaudhury, and E.D. Jacobson, Prostaglandin cytoprotection of gastric mucosa, *Gastroenterology.* 74:59 (1978).
19. Gy. Mózsik, F. Morón, M. Fiegler, T. Jávor, L. Nagy, I. Patty, and F. Tárnok, Membrane-bound ATP-dependent energy systems and gastric cytoprotection by prostacyclin, atropine and cimetidine, *Int J Tiss Reac.* 5:263 (1983).

20. Gy. Mózsik, L. Lovász, G. Kutor, L. Nagy, and F. Tárnok, Experimental evidence for cytoprotective effect of atropine on the rat gastric mucosa, *Acta Med Acad Sci Hung.* 37:401 (1980).

21. P.H. Guth, D. Aures, and G. Poulsen, Topical aspirin plus HCl gastric lesions in the rat. Cytoprotective effect of prostaglandin, cimetidine and Probanthine, *Gastroenterology.* 76:88 (1979).

22. E. Martinotti, C. Bernardini, M. Del Tacca, M. Pellegrini, G. Soldani, and A. Bertelli, Gastric cytoprotection by pirenzepine is not mediated by catecholamines, *Acta Physiol Hung.* 64:219 (1984).

23. F. Morón, E. Cuesta, M. Bata, and Gy. Mózsik, Cytoprotective effect of cimetidine: experimental evidence in rat gastric mucosal lesion induced by intragastric administration of necrotizing agents.

24. S.J. Konturek, T. Radecki, T. Brzozowski, I. Piastucki, A. Dembinska Kiec, and A. Zmuda, Gastric cytoprotection by prostaglandins, ranitidine and probantine, *Scand J Gastroent.* 16:7 (1981).

25. S.J. Konturek, T. Radecki, T. Brzozowski, I. Piastucki, A.Dembinska Kiec, and A. Zmuda, Experimental studies on gastric cytoprotection by prostaglandins, antisecretory agents and epidermal growth factor, *in*: "Advances in Ulcer Disease", K.H. Holtermüller, J.R. Malagelada, eds., Excepta Medica, Amsterdam, Oxford, Princeton (1980).

26. T. Jávor, M. Bata, G. Kutor, L. Lovász, Gy. Mózsik, and F. Tárnok, Gastric mucosal resistance to physical and chemical stress, *in*: "Advances in Physiol Sciences Vol.29. Gastrointestinal Defence Mechanisms", Gy. Mózsik, O. Hanninen, T. Jávor, eds., Pergamon Press - Budapest, Akadémiai Kiadó, Oxford (1989).

27. T. Jávor, M. Bata, L. Lovász, F. Morón, L. Nagy, I. Patty, J. Szabolcs, F. Tárnok, Gy. Tóth, and Gy. Mózsik, Gastric cytoprotective effects of vitamin A and others carotenoids, *Int J Tis Reac.* 5:289 (1983).

28. I. Patty, J. Szabolcs, G. Deák, T. Jávor, P. Kenéz, L. Nagy, L. Simon, F.Tárnok, and Gy. Mózsik, Controlled trial of vitamin A therapy in gastric ulcer, *Lancet* II:876 (1982).

29. F. Morón, N. Barreras, M. Achong, L. Nagy, T. Jávor, and Gy. Mózsik, Cytoprotective effect of disodium chromoglycate (INTAL) in gastric lesions induced by several necrotizing agents in rats, *Digestion.* 31:172 (1985).

30. F. Morón, N. Barreras, M. Achong, T. Jávor, L. Nagy, and Gy. Mózsik, Cytoprotective effect of disodium chromoglycate (INTAL) on the gastric mucosal lesions induced by necrotizing agents in rats, *in*: "New Pharmacology of Ulcer Disease. Experimental and Therapeutic Approaches", S. Szabó, Gy. Mózsik, eds., Elsevier Science Publishers, New York (1987).

31. I. Patty, Sz. Benedek, G. Deák, T. Jávor, P. Kenéz, F. Morón, L. Nagy, L. Simon, F. Tárnok, and Gy. Mózsik, Cytoprotective effect of vitamin A and its clinical importance in the treatment of patients with chronic gastric ulcer, *Int J Tiss Reac.* 5:301 (1983).

32. Gy. Mózsik, A. Bertelli, G. Deák, L. Nagy, I. Patty, L. Simon, F. Tárnok, and T. Jávor, Some aspects of the critical evaluation of peptic ulcer therapy in patients, *Int J Clin Pharmacol Res.* 5:447 (1985).

33. Gy. Mózsik, F. Morón, L. Nagy, Cs. Ruzsa, and F. Tárnok, Evidence of the gastric cytoprotective effects of vitamin A, atropine and cimetidine on the development of gastric mucosal damage produced by administration of indomethacin in healthy subjects, *Int J Reac.* 8:85 (1986).

34. I. Patty, F. Tárnok, L. Simon, G. Deák, Sz. Benedek, P. Kenéz, L. Nagy, and Gy. Mózsik, A comparative dynamic study of the effectiveness of gastric cytoprotection by vitamin A, De-Nol, sucralfate and ulcer healing by pirenzepine in patients with chronic gastric ulcer (a multiclinical and randomized study), *Acta Physiol Hung.* 64:379 (1984).

35. L. Nagy, Gy. Nagy, I. Rácz, L. Simon, I. Solt, I. Patty, F. Tárnok, Gy. Mózsik, and T.Jávor, A controlled multicentre therapeutic trial to determine the efficacy of a novel antacid (Al-Mg-Hydrocarbonate) in duodenal ulcer, *Int J Clin Pharmacol Res.* 9:85 (1989).

36. I. Patty, G. Deák, T. Jávor, Gy. Mózsik, L. Nagy, and F. Tárnok, A controlled trial with De-Nol (tripotassium dicitrato bismuthate) in patients with gastric ulcer, *Int J Tiss Reac.* 5:397 (1983).

37. Gy. Mózsik, T. Jávor, A. Bertelli, "Recent Advances in Gastrointestinal Cytoprotection", Akadémiai Kiadó, Budapest (1984).

38. Gy. Mózsik, T. Jávor, M. Kitajama, C.J. Pfeiffer,K.D. Rainsford, L. Simon, and S.Szabó, Advances in gastrointestinal cytoprotection: Topics 1987, *Acta Physiol Hung.* 73:111 (1989).

39. F. Morón, Gy. Mózsik, M. Bata, M. Lovász, G. Kutor, M. Fiegler, and T.Jávor, Ethanol-induced gastric mucosal damage in rats: prevention by prostacyclin, *Acta Med Acad Sci Hung.* 38:365 (1981).

40. Gy. Mózsik, M. Morón, and T. Jávor, Cellular Mechanisms of the development of gastric mucosal damage and of gastroprotection induced by prostacyclin in rats. A pharmacological study, *Prostaglandins, Leucotrienes Med.* 9:71 (1982).

41. Gy. Mózsik, F. Morón, M. Fiegler, T. Jávor, L. Nagy, I. Patty, and F. Tárnok, Interactions between membrane-bound ATP- dependent energy systems, gastric mucosal damage produced by NaOH, hypertonic NaCl, HCl and alcohol, and prostacyclin induced gastric cytoprotection in rats, *Prostaglandins, Leucotrienes and Med.* 12:423 (1983).

42. Gy. Mózsik, M. Fiegler, T. Jávor, F. Morón, L. Nagy, I. Patty, and F.Tárnok, Pharmacology and biochemistry of prostacycline-induced gastric cytoprotection on ethanol-induced gastric mucosal injury in rats, *in:*"Advances in Pharmacological Research and Practice, Vol.3. Prostanoids", V. Kecskeméti, K. Gyires, G. Kovács, eds., Pergamon Press - Akadémiai Kiadó, Oxford, Budapest (1986).

43. Gy. Mózsik, and T. Jávor, A biochemical and pharmacological approach to the genesis of ulcer disease. I. A model study of ethanol-induced injury to gastric mucosa in rats, *Dig Dis Sci.* 33:92 (1988).

44. Gy. Mózsik, M. Garamszegi, T. Jávor, L. Nagy, G. Sütő, and Á. Vincze, A pharmacological approach to cellular mechanisms of PGI_2-induced gastric cytoprotection on ethanol-induced gastric mucosal damage in rats, *Acta Physiol Hung.* 73:207 (1989).

45. T.A. Miller, Protective effects of prostaglandins against gastric acid mucosal damage: current knowledge of proposed mechanisms, *Am J Physiol.* 245:6601 (1983).

46. H. Shay, S.A. Komarov, S.S. Fels, D. Merance, M. Gruenstein, and H.A.Simplet, *Gastroenterology.* 5:43 (1945).

47. K.M. Wallus, W. Pawlik, and S.J. Konturek, Prostacyclin-induced gastric mucosal vasodilatation and inhibition of acid secretion in the dog, *Proc Soc Exp Med.* 163:228 (1980).

48. M. Guslandi, Cytoprotection: so what? *Dig Dis Sci.* 27:1144 (1982).

49. F. Morón, Gy. Mózsik, T. Jávor, M. Fiegler, L. Nagy, I. Patty, and F. Tárnok, Membrane-bound ATP-dependent energy systems and the development of the gastric mucosal damage in rats produced by 0.2M NaOH, 25% NaCl, 0.6N HCl and 96% ethanol, *Int J Tiss Reac.* 5:357 (1983).

50. Gy. Mózsik, B. Kiss, M. Krausz, and T. Jávor, A biochemical- cellular-morphological explanation for antiulcerogenic effects of parasympatholytics and of surgical vagotomy in Shay rats, *Scand J Gastroent.* 4:641 (1969).

51. Gy. Mózsik, and F. Vizi, Surgical vagotomy and stomach wall ATP and ADP in pylorus-ligated rats, *Amer J Dig Dis.* 22:1072 (1977).

52. Gy. Mózsik, M. Garamszegi, T. Jávor, G. Sütő, Á. Vincze, Gy. Tóth, and T. Zsoldos, Correlations between the oxygen free radicals, membrane-bound ATP-dependent energy systems irrelation to development of ethanol- and HCl-induced gastric mucosal damage and of β-carotene-induced gastric cytoprotection, *in:*"Free Radicals in Digestive Diseases", M. Tsuckiya et al., eds., Elsevier Science Publishers, Amsterdam (1988).

53. Gy. Mózsik, M. Garamszegi, M. Fiegler, L. Nagy, G. Sütő, Á.Vincze, T.Zsoldos, and T.Jávor, Mechanisms of Mucosal Injury in the Stomach. I. Timesequence analysis of gastric mucosal membrane-bound ATP-dependent energy system, oxygen free radicals and appearance of gastric mucosal damage, *in:*"Medical, Biochemical and Chemical Aspects of Free Radicals", O. Hayaishi, E. Niki, M. Kondo, T. Yoshikawa, eds., Elsevier Science Publishers, Amsterdam (1989).

54. Gy. Mózsik, M. Fiegler, M. Garamszegi, T. Jávor, L. Nagy, G. Sütő, Á.Vincze, Gy.Tóth, and T.Zsoldos, Mechanisms of gastric mucosal cytoprotection.I.Time-sequence analysis of gastric mucosal membrane-bound ATP-dependent energy systems, oxygen free radicals and macroscopically appearance of gastric cytoprotection by PGI_2 and β-carotene in HCl-model of rats, *in:*"Medical, Biochemical and Chemical Aspects of Free Radicals", O. Hayaishi, E. Niki, M. Kondo, T. Yoshikawa, eds., Elsevier Science Publishers, Amsterdam (1989).

55. G.Sütő, M.Garamszegi, T.Jávor, Á.Vincze, and Gy.Mózsik, Similarities and differences in the cytoprotection induced by PGI_2 and β-carotene in experimental ulcer, *Acta Physiol Hung.* 73:155 (1989).

56. Gy. Mózsik, and T. Jávor, Therapy of ulcers with sulfhydryl and non-sulfhydryl antioxidants, *in:* "Ulcer Disease: Investigation and Basis for Therapy", E.A. Sabb, S. Szabó, eds., Marcel Dekkel Inc., New York, Basel, Hong Kong (1991).

57. S. Szabo, Role of sulfhydryls and early vascular lesions in gastric mucosal injury, *Acta Physiol Hung.* 64:203 (1984).

58. S. Szabo, G. Pihan, and J.S. Trier, Alterations in blood vessels during gastric injury and protection, *Scand J Gastroent.* 21:Suppl.125:92 (1986).

59. S. Szabo, Critical and timely review of the concept of gastric cytoprotection, *Acta Physiol Hung.* 73:115 (1989).

60. F. Morón, Gy. Mózsik, and T. Jávor, Effects of cimetidine administered in cytoprotective and antisecretory doses on the membrane-bound ATP-dependent energy systems in the rat gastric mucosal lesions induced by HCl in rats, *Acta Physiol Hung.* 64:293 (1984).

61. Gy. Mózsik, M. Bata, M. Fiegler, T. Jávor, L. Nagy, I. Patty, Gy. Tóth, and F. Tárnok, Interrelationships between the membrane-bound ATP-dependent energy systems of the gastric mucosa and the gastric cytoprotective effect of β-carotene on the development of gastric mucosal damage, *Acta Physiol Hung.* 64:301 (1984).

62. Gy. Mózsik, M. Fiegler, M. Garamszegi, T. Jávor, L. Nagy, G. Sütő, and Á. Vincze, Biochemical background of the development of gastric mucosal damage produced by intragastric administration of ethanol or HCl in the rats, *Dig Dis Sci.* 33:906 (1988).

63. Gy. Mózsik, M. Fiegler, M. Garamszegi, T. Jávor, L. Nagy, G. Sütő, and Á. Vincze, PGI$_2$ prevents the development of gastric mucosal damage, β-carotene stimulates the repair mechanisms in ethanol-induced gastric mucosal damage in the rats, *Dig Dis Sci.* 33:906 (1988).

64. Gy. Mózsik, Á. Király, M. Garamszegi, T. Jávor, L. Nagy, A. Németh, G. Sütő, and Á. Vincze, Gastric cytoprotection mediating in SH groups is failured by surgical vagotomy, *Acta Physiol Hung.* 75 (Suppl.):219 (1990).

65. B. Bódis, M. Balaskó, Zs. Csontos, O. Karádi, Á. Király, T. Jávor, and Gy. Mózsik, Changes of gastric mucosal prostaglandin contents in intact and vagotomized rats, without and with β-carotene treatment, *Dig Dis Sci.* 35:1015 (1990).

66. Á. Király, M. Balaskó, B. Bódis, Zs. Csontos, O. Karádi, G. Sütő, Á. Vincze, T.Jávor, and Gy. Mózsik, Acute surgical vagotomy cause an increased vascular permeability to chemicals in the rat stomach, *Dig Dis Sci.* 35:1020 (1990).

67. M. Balaskó, B. Bódis, Zs. Csontos, O. Karádi, Á. Király, G. Sütő, Á. Vincze, T.Jávor, and Gy. Mózsik, Vagus and gastrointestinal defence in rats treated with indomethacin, *Dig Dis Sci.* 35:1015 (1990).

68. Gy. Mózsik, G. Sütő, and Á. Vincze, Correlations between the acute chemical and surgical vagotomy-induced gastric mucosal biochemistry in rats, *J Clin Gastroenterol.* 14(Suppl.1):S135 (1992).

69. Gy. Mózsik, Á. Király, M. Garamszegi, T. Jávor, L. Nagy, A. Németh, G. Sütő, and Á.Vincze, Mechanisms of vagal nerve in the gastric mucosal defence: unchanged gastric emptying and increased vascular permeability, *J Clin Gastroenterol.* 14(Suppl.1):S140 (1992).

70. Gy. Mózsik, Á. Király, G. Sütő, and Á. Vincze, ATP breakdown and resynthesis in the development of gastrointestinal mucosal damage and its prevention in animals and human, *Acta Physiol Hung.* 80:39 (1992).

71. Á. Vincze, Á. Király, G. Sütő, and Gy. Mózsik, Is acute surgical vagotomy an aggressor to gastric mucosa in pylorus ligated rats with and without indomethacin treatment?, *Acta Physiol Hung.* 80:195 (1992).

72. G. Sütő, Á. Király, Á. Vincze, and Gy. Mózsik, Effect of acute surgical vagotomy on the mucosal content of 6-keto-PGF$_{1\alpha}$, PGE$_2$ and glutathione after intragastric 96% ethanol treatment in rats, *Acta Physiol Hung.* 80:205 (1992).

73. Á. Király, G. Sütő, Á. Vincze, Gy. Tóth, Z. Matus, and Gy. Mózsik, Correlation between the cytoprotective effect of β-carotene and its gastric mucosal level in indomethacin (IND) treated rats with or without acute surgical vagotomy, *Acta Physiol Hung.* 80:213 (1992).

74. Á. Király, G. Sütő, Á. Vincze, and Gy. Mózsik, Acute and chronic surgical vagotomy (SV) and gastric mucosal vascular permeability in ethanol treated rats, *Acta Physiol Hung.* 80:219 (1992).

75. Gy. Mózsik, O. Karádi, Á. Király, Z. Matus, G. Sütő, Gy. Tóth, and Á. Vincze, Vagal nerve and the gastric mucosal defense, *J Physiol. (Paris)* 87:329 (1993).

76. Á. Vincze, Á. Király, G. Sütő, and Gy. Mózsik, Changes of gastric mucosal biochemistry in ethanol-treated rats with and without acute surgical vagotomy, *J Physiol. (Paris)* 87:339 (1993).

77. O. Karádi, Gy. Mózsik, Á. Király, G. Sütő, and Á. Vincze, Surgical vagotomy enhances the indomethacin-induced gastrointestinal mucosal damage in rats, *Inflammopharmacology.* in press (1993).

THE ROLE OF VAGAL INNERVATION IN ADAPTIVE CYTOPROTECTION

Thomas A. Miller, Gregory S. Smith,
Michael S. Tornwall, Rafael A. Lopez,
Julia M. Henagan, and Karmen L. Schmidt

In an earlier study[1] we observed that alcohol-induced gastric damage in the rat was exacerbated by truncal vagotomy. We further noted that the mild irritant, 30% ethanol, which significantly reduced the magnitude of injury when mucosa was subsequently exposed to 100% alcohol, was no longer capable of eliciting this protective effect under conditions of vagal denervation. These observations suggested that vagal integrity played an important role in mucosal defense and was necessary for adaptive cytoprotection to occur, at least under conditions of injury induced by alcohol. While the mechanism(s) responsible for these findings has remained elusive, the possibility that capsaicin-sensitive fibers may play a role in mediating these vagotomy effects must be considered. This contention is based on the fact that as many as 80% of fibers comprising the gastric vagus are afferent in nature, being composed of unmyelinated and thinly myelinated afferent neurons[2], and capsaicin, the pungent ingredient of red peppers, is able to impede the conducting activity of such primary afferent neurons when administered in high doses[3,4,5]. Further, functional ablation of capsaicin-sensitive fibers has been shown to enhance gastric injury and ulcer formation in a number of experimental models including those in which gastric damage was induced by pylorus ligation[6], acid distension of the stomach[6], and in response to various noxious agents such as indomethacin[7] and ethanol[7].

The studies described in this report were undertaken to explore further the role that vagal innervation plays in adaptive cytoprotection under conditions of injury induced by absolute alcohol, with particular attention being focused on those vagal fibers that are capsaicin-sensitive. Both stimulation of capsaicin-sensitive fibers as well as their functional ablation were studied. In addition, since glutathione, an antioxidant substance found in abundant quantities in the gastric mucosa[8,9,10], has been proposed as a potentially important mediator of mucosal defense[8,9,10], the role of this substance in enhancing gastric injury in the vagotomized state was also assessed. A portion of this work has been published[11].

MATERIALS AND METHODS

All studies were performed with Sprague-Dawley rats of either sex averaging 200 g in weight, and housed in wire mesh bottom cages to prevent coprophagia. Following a 24 h fast in which water, but not food, was allowed, animals were administered methoxyflurane anesthesia and underwent laparotomy through a vertical abdominal incision. They then were subjected to either a sham vagotomy and pyloroplasty (SVP) or a truncal vagotomy and pyloroplasty (TVP) as previously described[1]. In vagotomized animals, a subdiaphragmatic bilateral truncal vagotomy was performed, following which a Heineke-Mikulicz type pyloroplasty was fashioned to overcome potential gastric stasis which is known to occur following vagal denervation due to alteration in antral motility[2]. In sham operated animals, the gastroesophageal junction was manipulated and a similar pyloroplasty fashioned. All animals were allowed a one week recovery from operation at which time rats from both groups received either a 2 ml bolus of saline, the mild irritant 25% ethanol (vol/vol) in saline or 640 μM capsaicin in saline via a metal oroesophageal tube. This concentration of alcohol was chosen because it has been shown previously to induce adaptive cytoprotection in the rat stomach[12], and in other studies under the conditions of our experiments did not produce any macroscopic evidence of injury, either with or without concomitant vagotomy. The dose of capsaicin administered orally has been noted earlier to elicit protection against gastric injury induced by alcohol[13] and aspirin[14]. Thirty min following these treatment regimens, all animals received a 2 ml oral bolus of 100% ethanol. This volume of absolute alcohol was given because with the surgical pyloroplasty performed, whether or not vagal transection was also carried out, such a volume was necessary to elicit reproducible evidence of macroscopic injury.

Rats in all groups were killed 5 min after administration of absolute ethanol, their abdomens opened, stomachs clamped at the gastroesophageal and pyloric junctions and then removed. Stomachs were then opened along the greater curvature and laid flat, and the gastric mucosal surface photographed with a Nikon camera (Model N2000; 55 mm macrolens). The degree of gastric injury was then determined by digitized planimetry. This was accomplished by placing the resultant Ektachrome slides on a light board and visualizing the image through a video camera (Microimage Video Systems, Co., Model IHR, Boyertown, PA) attached to a conventional black and white monitor (World Video, Model HRM 14) and a computer. A mouse was then used to trace areas of clear macroscopic injury relative to totally uninjured areas. The damaged area was expressed as the percentage of surface epithelium showing gross damage on planimetric evaluation of the mucosa. For this and subsequent experiments described, damage was confined to the glandular portion of the stomach, usually involving primarily the acid-secreting portion, but occasionally the antrum, with sparing of the forestomach.

In addition to assessing the magnitude of injury macroscopically under the conditions just described, samples of mucosa were also obtained to determine the microscopic correlates of gastric injury and protection. Sections of stomach measuring 1x5 mm were removed from identical regions of the glandular mucosa of each stomach studied and immersed in half strength Karnovsky's fixative[15] for 24 h. Routine paraffin-embedded, hematoxylin and eosin-stained sections and toluidine blue-stained semithin plastic sections (0.25 μm) were used for light microscopic evaluation. Slides were reviewed in a Zeiss photomicroscope. Superficial injury represented injury to surface mucous cells facing the gastric lumen as well as cells lining the gastric pits extending down to the level of the gastric glands. Deep injury included not only damage to the superficial epithelium, but injury as well to most if not all of the gastric glandular region. Indices of cellular injury included cytoplasmic vacuolization, cytoplasmic swelling, nuclear pyknosis and nuclear swelling with chromatin margination.

In a second series of experiments, intact rats not subjected to SVP or TVP were pretreated with high dose capsaicin to functionally ablate afferent neurons[3,4,5]. This was done by subcutaneously administering capsaicin, dissolved in a 10:10:80 mixture of alcohol, Tween 80, and normal saline, in the back of the neck for two consecutive days in a dose of 50   mg/kg each day.   This regimen has previously been demonstrated to functionally ablate primary afferent neurons[16]. Capsaicin injections were administered under methoxyflurane anesthesia following pretreatment with terbutaline (0.1 mg/kg i.m.) and aminophylline (10 mg/kg i.m.) to counteract the respiratory impairment associated with capsaicin administration[16]. Control animals received a similar regimen of treatment with capsaicin vehicle. Two weeks following completion of capsaicin or vehicle treatment, experiments were performed.

In both capsaicin treated animals and their corresponding controls, either 1 ml of saline or the mild irritant 25% ethanol was administered orally via an oroesophageal tube following an overnight fast. Fifteen min later, they received 1 ml of 100% ethanol and then were sacrificed 5 min afterwards. As outlined above, stomachs were then retrieved to determine the magnitude of macroscopic injury using digitized planimetry.

In a third set of experiments, rats were again surgically prepared with SVP or TVP. One week later, they orally received a 2 ml bolus of either normal saline or 25% ethanol. Thirty min following such treatment, a portion of these animals was sacrificed for mucosal glutathione determination. The remaining animals were orally given 2 ml of 100% ethanol and then sacrificed 5 min later for glutathione determination.

Measurement of total glutathione content was performed using an adaptation of the 5,5'-dithiobis(2-nitrobenzoic acid) (DTNB)-glutathione disulfide (GSSG) recycling procedure[17], originally described by Owens and Belcher[18] and subsequently modified by Tietze[19] and Griffith[20]. Briefly, at the time of sacrifice, animals were anesthetized with methoxyflurane and stomachs rapidly excised through a midline laparotomy incision and immediately gently washed with ice-cold deionized water; they then were killed by cervical dislocation. The gastric mucosa, exclusive of the forestomach, was then stripped from the underlying muscularis and immediately transferred to vials of ice-cold 5-sulfasalicylic acid (SSA; 5% weight/volume) and weighed. After homogenization and centrifugation at 7000 rpm for 5 min, 10 μl of the resultant supernatant was added to cuvettes containing 0.248 mg nicotinamide adenine dinucleotide phosphate (NADPH) per milliliter of 0.143 M sodium phosphate buffer [with 6.3 mM edetic acid (EDTA) and adjusted to pH 7.5], 6 mM DTNB, and deionized water to a total volume of 1 ml. After addition of GSSG reductase (266 U/ml), the rate of DTNB reduction to 5-thio-2-nitrobenzoic acid (TNB) that is proportional to the total glutathione level was followed spectrophotometrically at 412 nm (Beckman DU-50 spectrophotometer; Beckman Instruments, Inc., Irvine, CA). The amount of glutathione in each sample was determined by comparison of the observed change in slope over time to a standard curve generated with known amounts of glutathione. Sample glutathione concentrations were expressed as μmoles/gm weight of wet tissue.

Since glutathione represents approximately 95% of the non-protein sulfhydryl concentration[8,9], the measurement of this latter substance was taken to represent the glutathione content. As previous studies have shown that there is a diurnal variation in the concentration of sulfhydryl compounds in tissue[8,21], all samples were obtained at approximately the same time of the day (between 9 a.m. and 12 noon).

For all experimental results, the means $\pm$ standard error of the means (SEM) were calculated. Comparison of changes among the various experimental groups for each set of experiments was statistically analyzed using the Student Newman Keuls Multiple Range Test. Differences were considered significant when a p value < 0.05 was obtained.

RESULTS

In those experimental groups in which macroscopic injury was evaluated, damage was noted to be confined to the glandular portion of the stomach with consistent sparing of the forestomach. This damage was generally limited to the oxyntic mucosa although, on occasion, hemorrhagic and necrotic lesions were also observed in the antrum. Lesions when present were usually longitudinal and paralleled the long axis of the stomach measuring 1 to 3 mm in width and 4 to 10 mm in length. Occasionally damage was typified by patches of hemorrhage rather than longitudinal lesions.

In the first group of experiments, in which SVP or TVP was performed seven days prior to study, animals receiving saline followed by 100% ethanol under nonvagotomized conditions consistently showed mucosal injury involving on the average about 18% of the glandular gastric epithelium. Damage in vagotomized animals was more pronounced and encompassed almost three times the amount of mucosal surface when compared with sham operated controls. In rats with intact vagi, prior exposure to 25% ethanol virtually abolished the injury noted under control conditions with 100% ethanol. However, in animals receiving truncal vagotomy prior to mild irritant administration, damage was again observed and the degree of injury was not significantly different from that noted with 100% ethanol alone whether or not vagotomy had been performed. A similar protective effect under nonvagotomized conditions was observed in animals receiving 640 μM capsaicin pretreatment. As in animals receiving 25% ethanol under vagotomized conditions prior to absolute ethanol exposure, the protective effect of capsaicin was not observed if previous vagotomy had been performed. These results are detailed in Figure 1.

Microscopic evaluation of gastric mucosa under nonvagotomized conditions demonstrated that 100% ethanol following saline pretreatment damaged virtually the entire mucosal surface of the glandular epithelium. A substantial portion of this injury (almost 50%) was of the deep variety involving most, if not all, of the gastric glands. Animals receiving 25% ethanol or low dose capsaicin prior to exposure to absolute ethanol also demonstrated damage involving virtually the entire mucosal surface. In contrast to saline controls, however, this damage was confined primarily to the superficial epithelium with sparing of the gastric glands.

In studies assessing the effects of high dose capsaicin, oral saline administration prior to absolute alcohol under vehicle control conditions demonstrated marked lesion formation of the glandular mucosa involving about 50% of the gastric epithelium (Figure 2). Pretreatment with 25% ethanol almost totally prevented this macroscopic expression of injury. In capsaicin-treated animals, neither the degree of damage encountered in animals receiving saline prior to 100% ethanol nor the

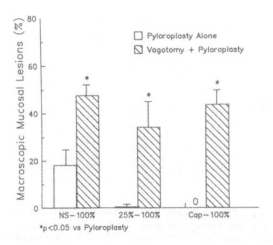

Figure 1. Percent injury of glandular gastric mucosa induced by 100% ethanol following saline, 25% ethanol, or 640 µM capsaicin pretreatment.

n = 6-8 animals per group; Pyloroplasty alone = sham vagotomy and pyloroplasty; NS - 100% = normal saline followed by 100% ethanol; 25% - 100% = 25% ethanol followed by 100% ethanol; CAP - 100% = capsaicin followed by 100% ethanol; 0 = no macroscopic injury.

protective effect of 25% ethanol against 100% ethanol injury was altered (see Figure 2).

In animals receiving saline 30 min prior to sacrifice, whether or not prior vagotomy had been performed, tissue glutathione levels were virtually identical and measured approximately 1.4 µmoles/g of tissue (Figure 3). These results compare favorably with previously reported concentrations of this substance in gastric tissue of both rats[8,9,22] and dogs[10]. Twenty-five percent ethanol alone, as well as 100% ethanol following saline pretreatment and 100% ethanol following 25% ethanol pretreatment, depressed glutathione levels to a comparable degree whether or not previous

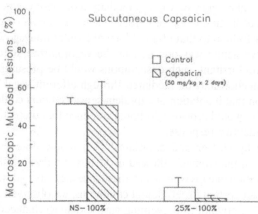

Figure 2. Percent injury of glandular gastric mucosa induced by 100% ethanol following saline or 25% ethanol pretreatment after subcutaneous administration of capsaicin.

n = 6-7 animals per group; NS - 100% = normal saline followed by 100% ethanol; 25% - 100% = 25% ethanol followed by 100% ethanol.

195

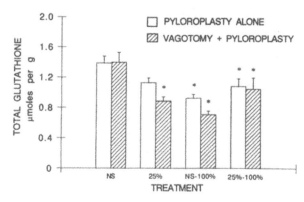

Figure 3. Effect of vagotomy on glutathione levels in glandular gastric mucosa.
n = 6-7 animals per group; Pyloroplasty alone = sham vagotomy and pyloroplasty; GSH =
glutathione; NS = normal saline; 25% = 25% ethanol; NS - 100% = normal saline followed by 100%
ethanol; 25% - 100% = 25% ethanol followed by 100% ethanol; *p < 0.05 - 0.01 vs pyloroplasty alone
in animals receiving NS. (Taken with permission from: Tornwall *et al.*, Dig Dis Sci. 38:2294, 1993).

vagotomy had been performed (see Figure 3). When these glutathione results were
compared with the degree of gastric mucosa protected or injured under the various
experimental conditions of SVP or TVP, no correlation between the amount of
glutathione in gastric tissue and the likelihood of mucosal injury could be ascertained
(see Figure 4).

DISCUSSION

In a previous study from our laboratory[1], we observed that alcohol induced injury
in the rat stomach could be enhanced with vagotomy, and the ability of a mild irritant
to prevent such injury was no longer demonstrable in the vagotomized state. The
studies summarized in this report were undertaken to determine what role
capsaicin-sensitive fibers may play in the mediation of these vagotomy effects. It was
our hypothesis that if stimulation of capsaicin-sensitive fibers could prevent alcohol
injury in a similar fashion to that of a mild irritant under nonvagotomized conditions,
and this protective action was obviated in the vagotomized state, again similar to
findings with a mild irritant, such observations would be presumptive evidence that
adaptive cytoprotection may be mediated through afferent vagal fibers. It was our
further contention that if ablation of capsaicin-sensitive fibers could prevent adaptive
cytoprotection, this would provide additional credence that these afferent neurons are
involved in the adaptive response.

The finding by Holzer and associates[13,14] that low dose capsaicin stimulates
afferent neurons in the stomach allowed us to test the first consideration. Using a
640 μM concentration of capsaicin, which these investigators had shown could prevent
gastric injury under certain experimental conditions, we observed that this agent was
as effective as 25% ethanol in preventing gastric damage induced by 100% ethanol
when assessed macroscopically. The protective effect with both agents was totally
prevented when prior vagotomy had been performed. As has been previously

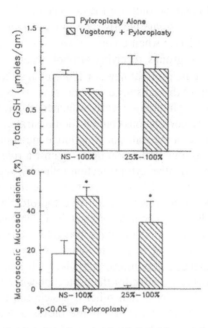

Figure 4. Lack of correlation between tissue levels of glutathione and percent injury of glandular gastric mucosa induced by 100% ethanol following saline or 25% ethanol pretreatment. Data taken from Figures 1 and 3.

n = 6-7 animals per group; Pyloroplasty alone = sham vagotomy and pyloroplasty; GSH = glutathione; NS - 100% = normal saline followed by 100% ethanol; 25% - 100% = 25% ethanol followed by 100% ethanol.

observed in studies evaluating the protective effects of prostaglandins against gastric injury[23,24,25], the protective effect of both 25% ethanol and low dose capsaicin was primarily limited to the deeper layers of the gastric epithelium since evidence of substantial injury to surface mucous cells lining the gastric lumen as well as the gastric pits was still evident microscopically even though macroscopically the mucosa appeared to be protected by these pretreatment strategies. Such findings are consistent with earlier studies evaluating the microscopic correlates of adaptive cytoprotection[12.]

To determine whether ablation of capsaicin-sensitive fibers could reverse adaptive cytoprotection, intact animals were used without prior vagal denervation. Capsaicin was given subcutaneously to rats, and then studied two weeks later; control animals received the capsaicin vehicle. With this approach, neither the degree of injury with 100% ethanol nor its prevention by the mild irritant 25% ethanol were in any way influenced by the capsaicin treatment. On the basis of these studies, it was concluded that no firm evidence had been established that adaptive cytoprotection was mediated through capsaicin-sensitive fibers and that some other mechanism was responsible for this process. These findings are in agreement with earlier observations of Evangelista and associates[26].

Since alterations in glutathione metabolism have been previously linked with the status of gastric mucosal integrity[8,9,10], we wondered whether perturbations in tissue levels of this antioxidant might account for the absence of adaptive cytoprotection in the vagotomized state. Our findings clearly showed that vagotomy by itself does not alter tissue levels of glutathione when compared with animals in which the vagi are intact. Interestingly, in all groups receiving alcohol, whether 25% ethanol alone, 100% ethanol following saline pretreatment, or 100% ethanol following 25% ethanol pretreatment, tissue levels of this substance were decreased despite the status of vagal innervation. When levels of glutathione were evaluated in terms of the presence or absence of gastric injury, no correlations could be demonstrated. Thus, if tissue levels of glutathione are indicative of the status of glutathione metabolism under a given set of experimental conditions, our findings mitigate against a role for this substance in explaining the absence of adaptive cytoprotection following vagotomy.

Since the studies summarized in this report indicate that functional ablation of capsaicin-sensitive fibers do not reverse adaptive cytoprotection, the ability of both 25% ethanol and low dose capsaicin to prevent injury induced by 100% ethanol suggests that the two agents mediate their protective effects through different mechanisms. Because low dose capsaicin failed to demonstrate its protective properties under vagotomized conditions, it is likely that stimulation of primary afferent vagal neurons was responsible for the protection when the vagus nerves were intact. An alternative possibility is that other capsaicin-sensitive afferent neurons, such as those which exist in the form of a dense plexus around submucosal blood vessels[27,28,29,30], afforded the protection by low dose capsaicin. Even with this circumstance, though, the vagus nerves would somehow have to be linked with these neurons to allow the protective response to remain operational when they are stimulated.

Equally uncertain is the means by which vagal interruption disrupts adaptive cytoprotection since capsaicin-sensitive fibers do not appear to be involved. At least three other possible mechanisms could be operational. First, mild irritants may stimulate other afferent fibers not sensitive to capsaicin that either turn on the vagal nucleus centrally or initiate a local neurogenic reflex not currently understood, but dependent upon intact vagal activity. A second possibility is that mild irritants may release a humoral substance that stimulates the vagal nucleus in the brain directly without any need for afferent fibers. Finally, mild irritants may release a humoral substance that stimulates another biochemical process that ultimately is responsible for adaptive cytoprotection and depends upon intact vagal fibers to express itself. Whether any or a combination of these factors is involved in the mechanism of adaptive cytoprotection must await further study.

ACKNOWLEDGEMENTS

These investigations were supported by grant DK 25838 from the National Institutes of Health awarded to Dr. Miller. The authors gratefully acknowledge the superb secretarial assistance of Mrs. Inci Akkaya. Dr. Schmidt is currently a member of the Anatomy Faculty at the University of Oregon Medical School in Portland, Oregon. Dr. Lopez currently practices orthopedic surgery in San Juan, Puerto Rico.

REFERENCES

1. J.M. Henagan, G.S. Smith, E.R. Seidel, and T.A. Miller, Influence of vagotomy on mucosal protection against alcohol-induced gastric damage in the rat, *Gastroenterology*. 87:903 (1984).
2. H.T. Debas, Physiology of gastric secretion and emptying, *in:* "Physiologic Basis of Modern Surgical Care," T.A. Miller, ed., C.V. Mosby, St. Louis (1988).
3. J.I. Nagy, Capsaicin: A chemical probe for sensory neuron mechanisms, *in:* "Handbook of Psychopharmacology," L.L. Iversen, S.D. Iversen, and S.H. Snyder, eds., Vol 15, Plenum Publ., New York (1982).
4. L.C. Russell, and K.J. Burchiel, Neurophysiological effects of capsaicin, *Brain Res Rev*. 8:165 (1984).
5. S.H. Buck, and T.F. Burks, The neuropharmacology of capsaicin: Review of some recent observations, *Pharmacological Rev*. 38:179 (1986).
6. J. Szolcsanyi, and L. Bartho, Impaired defense mechanism to peptic ulcer in the capsaicin-desensitized rat, *in:* "Advances in Physiological Sciences: Gastrointestinal Defense Mechanisms," G. Mozsik, O. Hanninen, and T. Javor, eds., Pergamon Press and Akademiai Kiado, Oxford and Budapest (1981).
7. P. Holzer, and W. Sametz, Gastric mucosal protection against ulcerogenic factors in the rat mediated by capsaicin-sensitive afferent neurons, *Gastroenterology*. 91:975 (1986).
8. S.C. Boyd, H.A. Sesame, and M.R. Boyd, Gastric glutathione depletion and acute ulcerogenesis by diethylmaleate given subcutaneously to rats, *Life Sci*. 28:2987 (1981).
9. S. Szabo, J.S. Trier, and P.W. Frankel, Sulfhydryl compounds may mediate gastric cytoprotection, *Science*. 214:200 (1981).
10. T.A. Miller, D. Li, Y-J Kuo, K.L. Schmidt, and L.L. Shanbour, Nonprotein sulfhydryl compounds in canine gastric mucosa: Effects of PGE_2 and ethanol, *Am J Physiol*. 249:G137 (1985).
11. M.S. Tornwall, G.S. Smith, J.C. Barreto, R.A. Lopez, J.M. Henagan, and T.A. Miller, Adverse effects of vagotomy on ethanol-induced gastric injury in the rat: Absence of a role for glutathione redox cycle, *Dig Dis Sci*. 38:2294 (1993).
12. K.L. Schmidt, G.S. Smith, and T.A. Miller, Microscopic correlates of adaptive cytoprotection in an ethanol injury model, *Histol Histopath*. 4:105 (1989).
13. P. Holzer, and I.T. Lippe, Stimulation of afferent nerve endings by intragastric capsaicin protects against ethanol-induced damage of gastric mucosa, *Neuroscience*. 27:981 (1988).
14. P. Holzer, M.A. Pabst, and I.T. Lippe, Intragastric capsaicin protects against aspirin-induced lesion formation and bleeding in the rat gastric mucosa, *Gastroenterology*. 96:1425 (1989).
15. M.J. Karnovsky, A formaldehyde-glutaraldehyde fixative of high osmolality for use in electron microscopy, *J Cell Biol*. 27:137A (1965).
16. C.-R. Martling, Sensory nerves containing tachykinins and CGRP in the lower airways. *Acta Physiol Scand Suppl*. 563:1 (1987).
17. M.E. Anderson, Determination of glutathione and glutathione disulfide in biological samples, *Meth Enzymol*. 113:548 (1985).
18. C.W.I. Owens, and R.V. Belcher, A colorimetric micro-method for the determination of glutathione, *Biochem J*. 94:705 (1965).
19. F. Tietze, Enzymatic method for quantitative determination of nanogram amounts of total and oxidized glutathione: Application to mammalian blood and other tissues, *Anal Biochem*. 27:502 (1967).
20. O.W. Griffith, Determination of glutathione and gluthathione disulfide using glutathione reductase and 2-vinylpyridine, *Anal Biochem*. 106:207 (1980).
21. L.V. Beck , V.D. Rieck, and B. Duncan, Diurnal variation in mouse and rat liver sulfhydryl, *Proc Soc Soc Exp Biol Med*. 97:229 (1958).
22. B.E. Victor, K.L. Schmidt, G.S. Smith, R.L. Reed, D.A. Thompson, and T.A. Miller, prostaglandin-induced gastric mucosal protection against stress injury: Absence of a relationship to tissue glutathione levels, *Ann Surg*. 209:289 (1989).
23. K.L. Schmidt, J.M. Henagan, G.S. Smith, P.J. Hilburn, and T.A. Miller, Prostaglandin cytoprotection against ethanol-induced injury in the rat: A histologic and cytologic study, *Gastroenterology*. 88:649 (1985).

24. P.H. Guth, G. Paulsen, and H. Nagata, Histologic and microcirculatory changes in alcohol-induced gastric lesions in the rat: Effect of prostaglandin cytoprotection, *Gastroenterology.* 87:1083 (1984).

25. E.R. Lacy, and S. Ito, Microscopic analysis of ethanol damage to rat gastric mucosa after treatment with a prostaglandin, *Gastroenterology.* 83:619 (1982).

26. S. Evangelista, C.A. Maggi, and A. Meli, Lack of influence of capsaicin-sensitive sensory fibers on adaptive cytoprotection in rat stomach, *Dig Dis Sci.* 33:1050 (1988).

27. K.A. Sharkey, R.G. Williams, and G.J. Dockray, Sensory substance P innervation of the stomach and pancreas: Demonstration of capsaicin-sensitive sensory neurons in the rat by combined immunohistochemistry and retrograde tracing, *Gastroenterology.* 87:914 (1984).

28. C. Sternini, J.R. Reeve, and N. Brecha, Distribution and characterization of calcitonin gene-related peptide immunoreactivity in the digestive system of normal and capsaicin-treated rats, *Gastroenterology.* 93:852 (1987).

29. H.C. Su, A.E. Bishop, R.F. Power, Y. Hamada, and J.M. Polak, Dual intrinsic and extrinsic origins of CGRP- and NPY-immunoreactive nerves of rat gut and pancreas, *J. Neurosci.* 7:2674 (1987).

30. T. Green, and G.J. Dockray, Characterization of the peptidergic afferent innervation of the stomach in the rat, mouse and guinea-pig, *Neuroscience.* 25:181 (1988).

A BIOBEHAVIORAL PROFILE OF AN ULCER
SUSCEPTIBLE RAT STRAIN

William P. Paré, and Eva Redei

Selye[1,2] not only formalized the stress concept, but made stress the legitimate subject of scientific inquiry. Since his seminal publications, we recognize that stress is a ubiquitous phenomenon that is related to the development of many diseases,[3,4] including mood disorders[5]. Everyone is exposed to stress in one form or another, but not all individuals succumb to the effects of stress. It is also intriguing that, in some cases, a seemingly innocuous stressor can cause profound disturbances in some subjects, but have little effect on others. This problem of subject variability is countered by examining the effects of stress in an individual that is either hyperreactive to stress, or is either predisposed to the illness under study. Therefore stress hyperreactive animal strains are valuable tools in studying the connections between disease states and stress. We have observed that the Wistar Kyoto (WKY) rat strain is vulnerable to stress effects.

The WKY rat strain was developed as the normotensive control strain to the spontaneously hypertensive rat (SHR)[6]. While the WKY rat may be an appropriate control strain for hypertension studies, the strain manifests certain behavioral idiosyncrasies and cannot be considered a normal control strain for behavioral studies[7]. Our investigations have indicated that WKY rats manifest distinct behavioral and physiological responses to stress. These responses suggest the presence of a biobehavioral profile in WKY rats with stress ulcer susceptibility and behavioral depression as major manifestations of this profile.

BEHAVIORAL STUDIES

When WKY rats are observed in the open field test of emotionality, they exhibit little exploratory behavior. WKY rats rapidly adopt an immobile posture in the Porsolt forced swim test. In addition, WKY rats readily acquire a learned helplessness response[8]. The forced swim test and the learned helplessness procedure are putative models of depression[9]. WKY rats score highly in both procedures. Therefore, we have suggested that WKY rats are prone to depressive behavior[10].

While WKY rats typically obtain high score on behavioral tests of depressive behavior, the question remains whether other tests of emotional behavior will also discriminate between WKY rats and rats from other strains. To address this question, WKY, Wistar and Fischer-344 (F-344) rats were exposed to one of three behavioral test of anxiety[11]. Thus, rats from all three strains were exposed to either the defensive-withdrawal test[12], the elevated plus maze[13], or the defensive burying test[14]. Anxiolytic agents typically influence behaviors in these three tests. Following behavioral testing, all rats were then exposed to the ulcerogenic water-restraint procedure[15]. None of these tests were consistently effective in discriminating WKY rats from the other two strains, except the defensive burying test. The response, however, of WKY rats in the defensive burying test was unorthodox. The "anxiety" response in defensive burying involves burying, with bedding material, a shocking prod that protrudes from one of the test enclosure walls. Rats from the other strains invariably showed this burying response after being shocked by the prod, but WKY rats consistently withdrew to the opposite corner of the cage and remained immobile throughout the testing period. We believe that this immobility response is just another manifestation of depressive behavior in WKY rats. In addition, WKY rats in this study had more ulcer when subsequently exposed to restraint stress[11].

WKY rats also perform consistently on different tests of depressive behavior. WKY and Wistar rats were exposed, in a semi-random fashion, to the open field test, the defensive burying test, the Porsolt forced swim test, and the learned helplessness paradigm. The forced swim test was positively correlated with the learned helplessness procedure and, to a lesser degree with defensive burying, but these relationships were observed only in WKY rats, not Wistar rats. We concluded that WKY rats represented a more sensitive strain for detecting possible relationships between putative animal models of depression. Following the conclusion of the behavioral testing, all rats were exposed to water-restraint stress. WKY rats also had more ulcers than Wistar rats[16].

If the predominant coping strategy for WKY rats is freezing and immobility, then WKY rats should be superior in adopting a passive avoidance response, wherein the learned response involves the inhibition of a prepotent active response. We observed this to be the case when WKY, Wistar and F-344 rats were tested in two passive avoidance procedures - a platform step-down procedure and a light chamber - dark chamber one-way avoidance procedure. Rats were assigned to either one of the two procedures and then exposed to water-restraint stress. WKY rats not only adopted these passive avoidance responses faster that Wistar and F-344 rats, but demonstrated a fascinating behavior in the one-way avoidance task. While rats from the other strains would re-enter, during extinction trials, the dark compartment in which shock had been previously experienced, WKY rats would straddle the threshold between the two chambers for the duration of the testing period. We have labelled this behavior, "ambivalence behavior," and propose that it reflects the decision-making paralysis that characterizes depressed patients. WKY rats also had more ulcer following water-restraint stress[17].

The propensity to recall unpleasant memories contributes to the symptomatology of depression[18]. This phenomenon was observed in WKY rats. WKY and Wistar rats were trained on a one-way avoidance task (i.e., the unpleasant event) but were also exposed to grid shock, either prior to (i.e., the proactive treatment) or after (i.e., the retroactive treatment) passive avoidance training. Subsequent test trials revealed that the retention of the unpleasant event was more pronounced for the proactive treatment, and this effect was dramatically greater in WKY rats as compared to Wistar rats.

There are many other symptoms that characterize depression. Reduced appetite and loss of body weight are symptoms of depression[19]. When hungry WKY, Wistar and F-344 rats are provided with the opportunity to acquire a food pellet in the middle of a novel open field, the number of feeding contacts is significantly less in WKY rats as compared to other rats. This fear of feeding in a novel environment, or "hyponeophagia," is significantly greater in WKY rats[20]. Loss of appetite, or decreased feeding following exposure to stress, is frequently been reported as a measure of response to stress[21,22], but the capacity of a novel environment to produce a strain-specific anhedonic response has never been reported.

The prevalence of the female gender in mood disorders[23,24] is also reflected in WKY rats. Previous research on the question of differential sex susceptibility to stress effects had produced conflicting results. But our studies revealed that proestrus-estrus female WKY rats were judged as more depressed, according to the Porsolt forced swim test, as compared to WKY female diestrus rats and WKY male rats. In addition, proestrus-estrus WKY female rats developed more ulcers when exposed to water-restraint stress[25]. These data imply that reactivity to stress may be influenced, not only by the animal strain, but also by sex and estrus factors. The greater vulnerability to stress-ulcer and higher depression scores in WKY proestrus-estrus females provided data congruent with human clinical documentation[26].

Table 1. Summary (Means ± SEM) of Porsolt Forced-Swim Behaviors, and Stomach Ulcer Scores Following Desipramine Administration and Water-Restraint Stress

Treatments		Porsolt forced-swim behavior		Ulcer measures following water-restraint stress	
		Struggling(min)	Floating(min)	No. of Ulcers	Ulcer length(mm)
WKY	Placebo	1.12 ± 0.29	13.80 ± 0.34	16.60 ± 4.32	14.44 ± 2.55
	5 mg *	3.98 ± 1.43	10.10 ± 0.94	5.40 ± 0.24	4.60 ± 0.40
	10 mg	5.03 ± 2.17	8.92 ± 2.15	2.00 ± 0.80	1.14 ± 0.80
	20 mg	9.93 ± 1.48	3.99 ± 1.48	1.80 ± 2.08	0.74 ± 0.48
Wistar	Placebo	5.38 ± 1.39	8.56 ± 1.40	7.60 ± 2.16	4.76 ± 2.13
	5 mg	7.46 ± 0.40	6.48 ± 0.50	0.60 ± 0.40	0.20 ± 0.14
	10 mg	7.63 ± 1.19	6.27 ± 1.22	0	0
	20 mg	11.16 ± 0.50	3.28 ± 0.44	0	0
F-344	Placebo	5.58 ± 0.91	8.44 ± 0.91	8.20 ± 1.75	7.74 ± 1.44
	5 mg	5.86 ± 0.35	9.10 ± 0.36	0.20 ± 0.20	1.10 ± 0.20
	10 mg	6.84 ± 1.09	7.73 ± 1.09	0.60 ± 0.20	0.26 ± 0.26
	20 mg	10.13 ± 0.44	3.93 ± 0.45	.80 ± 0.49	0.32 ± 0.10

* Dose of desipramine in mg/kg body weight administered daily for nine days. Porsolt test was carried out on day seven.

These data strongly suggested that the stress response in WKY rats is characterized by a susceptibility to stress ulcer and the prevalence of depressive behavior. If this is the case, it is reasonable to assume that anti-depressant drugs would not only reduce depressive behavior, but also reduce the vulnerability to stress-ulcer. There are earlier reports which suggest this to be the case[27]. To test this hypothesis we exposed WKY, Wistar and F-344 rats to the anti-depressant, desipramine (DMI), with either 5, 10, or 20 mg/kg/day for nine days. On day seven all rats were tested in the Porsolt forced swim test and on day nine all rats were exposed to water-restraint stress. DMI not only reduced depression in the forced swim test in a dose-dependent fashion, but also significantly reduced ulcer severity[28]. These data are summarized in Table 1.

If we apply the behavioral data for WKY rats to the diagnostic criteria for depression[19], a behavioral profile emerges suggesting the prevalence of depressive behavior in WKY rats (see Table 2). The behavioral data imply that WKY rats are emotionally very reactive to stress and this hyperreactivity is revealed by behavioral signs of depressive behavior.

Table 2. Application of Wistar Kyoto Rat Behaviors and Physiological Responses to DSM-III Criteria for Behavioral Depression

DSM-III Criteria	WYK Behaviors
. Diminished interest or pleasure	Hyponeophagia[20]
. Psychomotor retardation	Immobility in OFT and FST[8,16]
. Indecisiveness	Ambivalence in passive avoidance[17]
also:	
-brooding, obsessive rumination	Memory bias
-prevalence in female	Increase ulcer-depression in female WKY rats[25]
-reversal by anti-depressants	Desipramine reduces immobility[28]

PHYSIOLOGICAL RESPONSES

WKY rats reveal several behavioral responses of stress hyperreactivity, but a distinct physiological response pattern is also observable in these animals. In our earlier studies we reported that ulcerogenic stressors typically produced more ulcers in WKY rats as compared to other rat strains. This effect occurred consistently with either the ulcerogenic procedures of activity-stress[29], prone immobilization[15], or water-restraint stress[8]. In the majority of the behavioral studies reviewed in the previous section, animals were also exposed to the water-restraint ulcerogenic stressor. Table 3 summarizes the ulcer data from all these studies. In all instances WKY rats emerged as the most ulcer-vulnerable strain. Thus stress-ulcer is one of the characteristic physiological responses to stress in WKY rats.

The physiological system most likely involved in our WKY ulcer-depression syndrome is the hypothalamic-pituitary-adrenal (HPA) axis[30,31,32], and the HPA axis in WKY rats is hyperreactive to stress. This statement is based partly on the observation that the adrenocorticotropin (ACTH) response to stress in WKY rats is greater than the similar response in other rat strains. WKY and Wistar rats were surgically prepared with chronic jugular catheters and subsequently exposed to restraint stress. There were no differences in pre-stress basal plasma ACTH or corticosterone (CORT) values between strains. But, the ACTH stress response was significantly greater for WKY rats throughout the stress period. Surprisingly there were no strain differences in CORT. Thus, the exaggerated ACTH response of WKY

Table 3. Summary of Ulcer Data (Mean ±SEM) From Behavioral Studies Wherein Multiple Rat Strains Were Tested on Various Behavioral Test and Subsequently Exposed to the Water-Restraint Ulcerogenic Procedure

Behavioral Studies	Mean (± SEM) cumulative ulcer length-(mm).			
	WKY rats	Wistar rats	F-344 rats	SHR rats
Open field test (OFT) [33]*	85.0 ± 0.6	35.2 ± 5.4	---	60.6 ± 6.5
OFT [8]	13.5 ± 5.2	3.6 ± 1.8	6.8 ± 2.1	3.2 ± 0.9
OFT - gender differences[25]				
Male rats	27.3 ± 7.8	9.0 ± 2.4	11.2 ± 3.8	---
Female rats**	37.9 ± 4.9	13.5 ± 1.9	12.4 ± 2.3	---
Learned helplessness (LH)[8]	24.3 ± 3.8	3.2 ± 1.1	7.3 ± 1.6	27. ± 1.0
Forced swim test (FST)[8]	26.7 ± 6.1	2.9 ± 0.8	4.8 ± 1.7	4.9 ± 1.6
Elevated-plus maze[11]	20.8 ± 3.9	10.8 ± 2.2	16.3 ± 2.2	---
Defensive burying (DB)[11]	19.8 ± 2.3	10.1 ± 1.3	13.2 ± 1.0	---
Defensive withdrawal[11]	17.5 ± 1.8	13.6 ± 0.9	14.7 ± 1.6	---
Step-down passive avoidance[17]	20.8 ± 3.4	13.4 ± 1.7	15.6 ± 2.1	---
One-way passive avoidance[17]	25.4 ± 3.0	16.8 ± 2.8	15.9 ± 4.0	---
FST, OFT, LH, & DB[16]	13.7 ± 1.7	2.7 ± 0.3	---	---

* Ulcers induced by activity-stress procedure
** Tested during proestrus-estrus

rats occurred despite the presence of appropriate CORT levels. In addition, none of the rats from the comparison strain had ulcers, whereas all the WKY rats had ulcers[10].

The exaggerated ACTH response to stress in WKY rats could be attributed to either a pituitary hyperresponsiveness to corticotropin releasing factor (CRF), or a lack of sensitivity to the negative glucocorticoid feedback. One way to address this question involved simply removing the glucocorticoid negative feedback in the HPA-axis loop by adrenalectomy (ADX). Accordingly, WKY, Wistar or F-344 rats received either ADX, sham ADX, or ADX plus CORT replacement (in the form of CORT subdermal pellets), and these rats were exposed to the ulcerogenic water-restraint procedure. Intact WKY rats had dramatically more ulcers and higher anterior ACTH and proopiomelanocortin (POMC) mRNA levels than the other strains. ADX, in the WKY rats increased ulcer incidence, but had little effect on thymus weight, ACTH content or hypothalamic CRF mRNA levels in contrast to the profound, and opposite, effects of ADX on these parameters in the other strains. Furthermore, CORT replacement was without effect in WKY rats, while it reversed the effects of ADX in the other strains[34]. These data are summarized in Table 4. Thus, WKY rats that responded to stress with increased ulcers, already had enhanced synthesis of POMC mRNA. The mechanism underlying this dysregulation of the HPA-axis may be related to decreased efficacy of glucocorticoid negative feedback as manifested by diminished changes in response to ADX and CORT replacement. This glucocorticoid resistance in WKY rats resembles the relative insensitivity to glucocorticoid observed in individuals with endogenous depression[35]. Also, this proposed hyposensitivity of the corticotroph to the inhibitory action of glucocorticoid does not exclude the possibility that the dysregulation of the WKY HPA-axis may be attributable to an increased CRF secretion or an increased sensitivity of the corticotroph to CRF.

Table 4. Mean (± SEM) Ulcer, Thymus and HPA Measures in Adrenalectomized (ADX), Sham ADX, and Corticosterone Replacement (ADX+CORT) Groups Following Restraint Stress in WKY, Wistar and F-344 Rats

Treatment		Ulcer[1]	Thymus[2]	ACTH Content[3]	CRF mRNA[4]	POMC[5]
	Sham	22.0 ± 2.7	104 ± 19	25.2 ± 1.5	1.12 ± .08	7.3 ± 3.2
WKY	ADX	34.6 ± 3.4	110 ± 28	25.8 ± 1.3	1.18 ± .04	20.9 ± 6.3
	ADX+CORT	29.3 ± 4.2	96 ± 19	---	1.13 ± .03	11.2 ± 2.3
	Sham	11.7 ± 3.2	140 ± 15	18.6 ± 1.0	1.23 ± .10	2.5 ± 1.1
Wistar	ADX	1.7 ± 0.3	170 ± 47	26.2 ± 1.1	1.46 ± .27	11.5 ± 3.4
	ADX+CORT	10.6 ± 1.8	87 ± 16	---	1.00 ± .01	2.1 ± 0.4
	Sham	12.3 ± 1.5	104 ± 19	20.9 ± 1.5	1.22 ± .01	2.9 ± 1.1
F-344	ADX	2.2 ± 0.8	110 ± 28	24.4 ± 2.1	2.74 ± .16	10.3 ± 3.0
	ADX+CORT	13.7 ± 3.9	96 ± 19	---	1.33 ± .16	2.0 ± 0.3

[1] Mean cumulative ulcer length - mm.
[2] Thymus weight/body weight ratio - g/g x 1000
[3] ACTH - mg/anterior pituitary
[4] Relative CRF mRNA - CRF mRNA/β-actin mRNA ratio
[5] Relative POMC mRNA - POMC mRNA/β-actin mRNA ratio

DISCUSSION

There are significant differences in the biobehavioral stress response of WKY rats. To summarize, we have observed in the WKY rat: (a) high susceptibility to stress ulcer; (b) depressive behavior, (c) a high level of activity of the HPA axis in response to stress, and (d) a dysregulation of the HPA axis. These observations can form the basis of our hypothesis that WKY rats are endogenously hyperresponsive to stress. This hyperresponsiveness is manifested in a high susceptibility to stress ulcer as well as the hyperreactivity of the HPA axis. Since clinical studies have proposed a possible relationship between ulcer disease and depression[36,37,38], as well as between stressful events and depression[30,31,32], our original hypothesis can be further expanded. Thus, we believe that the WKY's hyperresponsiveness to stress leads to signs of behavioral depression, and this apparently strain-specific proclivity makes this rat strain a potential model of endogenous depression.

There are different kinds of animal models. Each model type emphasizes one aspect of the pathological condition that it represents. Thus, we have behavioral models that mimic the symptoms and signs of the human disorder, mechanistic models that emphasize the neurobiological mechanisms of the disorder, and empirical validity models that attempt to predict pharmacological outcome (for a critical review, see McKinney[39]). Rarely do we find all of these characteristics in the same animal model. Yet when we review the data from the WKY studies, we note that the WKY strain contains elements of the different model types. Thus the strain exhibits behaviors that mimic the behavioral disorder of depression. It reveals physiological processes (i.e., mechanisms) that have been reported in endogenous depression, namely the insensitivity of the glucocorticoid system. Finally, anti-depressants reverse the alleged depressive behavior in WKY rats. The data reviewed in this paper, plus these theoretical considerations, would suggest that the WKY rat strain may indeed serve as a useful model of endogenous depression.

REFERENCES

1. H. Selye, Thymus and adrenals in the response of the organism to injuries and intoxications, *Brit J Exp Pathol.* 17:234 (1936).
2. H. Selye, The general-adaptation-syndrome and the diseases of adaptation, *in:* "Textbook of Endocrinology," Acta Endocrinologica, Inc., Montreal (1949).
3. K.A. Holyroyd, M.A. Appel, and F. Andrasik, *in:* "Stress reduction and prevention," D. Meichenbaum, M.E. Jaremko, eds., Plenum Press, New York (1983).
4. B.H. Natelson, Stress, predisposition and the onset of serious disease: Implications about psychosomatic etiology, *Neurosci Biobehav Rev.* 7:511 (1983).
5. M.J. Horowitz, Stress-response syndrome: Post-traumatic and adjustment disorders, *in:* "Psychiatry Vol 1", J.O. Cavenar, ed., Basic Books, New York (1986).
6. K. Okamoto, and K. Aoki, Development of a strain of spontaneously hypertensive rats, *Jpn Circ J.* 27:282 (1963).
7. W.P. Paré, Stress ulcer and open field behavior of spontaneously hypertensive, normotensive and Wistar rats, *Pavlovian J Biol Sci.* 24:54 (1989).
8. W.P. Paré, Stress ulcer susceptibility and depression in Wistar Kyoto (WKY) rats, *Physiol Behav.* 46:993 (1989).
9. P. Willner, Animal models of depression: An overview, *Pharmacol Ther.* 45:425 (1990).
10. W.P. Paré and E. Redei, Depressive behavior and stress ulcer in Wistar Kyoto rats, *J. Physiol (Paris).* 87:229 (1993).
11. W.P. Paré, The performance of WKY rats on three tests of emotional behavior, *Physiol Behav.* 51:1051 (1992).
12. L.K. Takahashi, N.H. Kalin, J.A. Vanden Burgt, and J.E. Sherman, Corticotropin-releasing factor modulates defensive-withdrawal and exploratory behavior in rats, *Behav Neurosci.* 103:648 (1989).
13. S. Pellows, P. Chopin, S.E. File, and M. Briley, Validation of open:closed arm entries in an elevated plus-maze as a measure of anxiety in the rat, *J Neurosci Meth.* 14:149 (1985).
14. J.P.J. Pinel, L.A. Symons, B.K. Christensen, and R.C. Tess, Development of defensive burying in Rattus norvegicus: Experience and defensive responses, *J Comp Psychol.* 103:359 (1989).
15. W.P. Paré, A comparison of two ulcerogenic techniques, *Physiol Behav.* 44:417 (1988).
16. W.P. Paré, Open field, learned helplessness, defensive burying and forced-swim tests in WKY rats, *Physiol Behav.* 55:433 (1994).
17. W.P. Paré, Passive-avoidance behavior in Wistar-Kyoto (WKY), Wistar, and Fischer-344 rats, *Physiol Behav.* 54:845 (1993).
18. T. Dalgleish, and F.N. Watts, Biases of attention and memory in disorders of anxiety and depression, *Clin Psychol Rev.* 10:589 (1990).
19. American Psychiatric Association, Diagnostic and Statistical Manual of Mental Disorders, 3rd ed.-revised, Washington (1987).
20. W.P. Paré, Hyponeophagia in Wistar Kyoto (WKY) rats, *Physiol Behav.* in press (1994).
21. G.A. Kennett, S.L. Dickinson, and G. Curzon, Enhancement of some 5-HT-dependent behavioral responses following repeated immobilization in rats, *Brain Res.* 330:253 (1985).
22. W.P. Paré, Stress and consummatory behavior in the albino rat, *Psychol Rep.* 16:135 (1965).
23. B.P. Dohrenwend, and B.S. Dohrenwend, Sex differences and psychiatric disorders, *Am J Sociol.* 81:1447 (1976).
24. M.M. Weissman, and G.L. Klerman, Sex differences and the epidemiology of depression, *Arch Gen Psychiat.* 34:98 (1977).
25. W.P. Paré and E. Redei, Sex differences and stress response of WKY rats, *Physiol Behav.* 54:1179 (1993).
26. C. Ernst, and J. Angst, The Zurich study. XII, Sex and depression. Evidence from longitudinal epidemiological data, *Arch Psychiat Clin Neurol.* 241:222 (1992).

27. D.E. Hernandez, and B.G. Xue, Imipramine prevents stress gastric glandular lesions in rats, *Neurosci Lett.* 103:209 (1989).
28. W.P. Paré, Learning behavior, escape behavior, and depression in an ulcer susceptible rat strain, *Integrative Physiol Behav Sci.* 27:130 (1992).
29. W.P. Paré, The influence of food consumption and running activity on the activity-stress ulcer in the rat, *Am J Dig Dis.* 20:262 (1975).
30. H. Anisman, Vulnerability to depression: Contribution of stress, *in:* "Neurobiology of Mood Disorders," R.M. Post, J.C. Ballenger, eds., Williams and Wilkins, Baltimore (1984).
31. G.B. Glavin, Stress and brain noradrenaline: A review, *Neurosci & Biobehav Rev.* 9:233 (1985).
32. J.M. Weiss, and P.G. Simson, Neurochemical mechanisms underlying stress-induced depression, *in:* "Stress and Coping," T.M. Fields, P.M. McCabe, N. Schneiderman, eds., Lawrence Erlbaum Assoc., Hillsdale, NJ (1985).
33. W.P. Paré and G.T. Schimmel, Stress ulcer in normotensive and spontaneously hypertensive rats, *Physiol Behav.* 36:699 (1986).
34. E. Redei, W.P. Paré, F. Aird, and J. Kluczynski, Strain differences in hypothalamic-pituitary-adrenal activity and stress ulcer, *Am J Physiol.* 266:R353 (1994).
35. F. Holsboer, Implications of altered limbic-hypothalamic-pituitary-adrenocortical (LHPA) function for neurobiology of depression, *Acta Psychiat Scan Suppl.* 341:72 (1988).
36. J.H. McIntosh, R.W. Nasiry, D. McNeil, C. Coates, H. Mitchell, and D.W. Piper, Perception of life event stress in patients with chronic duodenal ulcer, *Scand J Gastroenterol.* 20:563 (1985).
37. J.H. Medalie, K.C. Stange, and S.J. Zyzanski, The importance of biopsychosocial factors in the development of duodental ulcer in a cohort of middle-aged men, *Am J Epidemiol.* 136:1280 (1992).
38. D.W. Piper, M. Grieg, J. Thomas, and J. Skinners, Personality patterns of patients with chronic gastric ulcer. Study of neuroticism and extroversion in a gastric ulcer and a control population, *Gastroenterology.* 73:444 (1977).
39. W.T. McKinney, Models of mental disorders, *in:* "A New Comparative Psychiatry", Plenum Press, New York (1988).

THE GASTRIC MUCOSAL BARRIER: A DYNAMIC, MULTIFACTORIAL SYSTEM

Andrew Garner, Paul H. Rowe, and Jeffrey B. Matthews

The mucosal lining of the upper gastrointestinal tract possesses two distinct properties; the ability to withstand autolysis by endogenous secretions, and the ability to rapidly repair superficial damage induced by exogenous irritants. Resistance of the gastric mucosa to autodigestion has been recognized for more than 200 years[1]. However, major advances in understanding the mechanisms involved in acid disposal and recovery from acute injury in the stomach have taken place over the past 10 years[2,3].

Current concepts of gastric mucosal defense developed largely from the studies of Hollander[4] and Davenport[5]. Hollander regarded the gastric mucosal barrier as a two-component system comprised of the layer of mucus plus the subjacent epithelium. He suggested that anatomic integrity of the surface cell layer was the most important factor in resistance to acid-peptic autodigestion. Davenport conceived the mucosal barrier in terms of its ability to impede passive diffusion of hydrogen ions from the lumen of the stomach into the mucosa and diffusion of sodium ions in the reverse direction. The observation that aspirin and other injurious agents increase acid loss and sodium gain led to use of these ion fluxes as indicators of mucosal integrity.

The view of the mucosal barrier as a static entity has now given way to a multifactorial and dynamic system of protection. The various constituents of this defence system have been reviewed in detail elsewhere[6]. Barrier function reflects the interplay between many anatomical and physiological components and these can be divided into two broad categories; extrinsic factors and intrinsic factors. Extrinsic factors pertain to both pre- and post-epithelial sites. The former consists of a surface mucus layer containing bicarbonate and phospholipid, and provides a first-line of protection. Post-epithelial components include interstitial acid-base balance, the microvasculature and various mediators which regulate mucosal function such as eicosanoids, neuropeptides and growth factors. Intrinsic protection relates to inherent properties of the epithelial cells themselves and includes membrane permeability, cytoplasmic pH homeostasis, junctional integrity, restitution (re-epithelialization) and proliferation (mitosis).

In this article, we discuss relative contributions of the surface mucus-bicarbonate barrier and interstitial bicarbonate generation in maintaining mucosal acid-base balance. The relative roles of cellular migration and proliferation in repair of mucosal defects are also highlighted.

Neuroendocrinology of Gastrointestinal Ulceration
Edited by S. Szabó and Y. Taché, Plenum Press, New York, 1995

PRE-EPITHELIAL FACTORS

Mucoid Layer

Mucus glycoprotein is synthesized and secreted by surface cells and mucus neck cells. It is usually present as a thin layer which adheres to the epithelium forming a stable, unstirred boundary zone. The mucus layer acts as a barrier to high molecular weight substances such as pepsin by a gel exclusion effect[7]. Pepsinogen presumably enters the lumen from within the gastric glands under the hydrostatic force accompanying acid secretion. However, continual synthesis and secretion of mucus is necessary since activation to pepsin in the lumen leads to proteolytic degradation of mucus glycoprotein and thus solubilization of the gel. Under normal circumstances, the mucus gel layer also contains phospholipids and small amounts of bicarbonate[8,9]. These components tend to reduce acidity at the epithelial surface by increasing hydrophobicity and by neutralizing hydrogen ions which enter the gel.

The contributions of the so-called 'mucus-bicarbonate' and 'hydrophobic' barriers in protection of the normal stomach have both been disputed[3,10]. Following acute damage, however, a thick and continuous gelatinous layer consisting of mucus, fibrin and cellular debris accumulates within minutes over the injured region[11]. Thus while resting bicarbonate output from the surface cells may appear unable to provide complete protection when luminal pH is highly acidic, the increase in mucosal permeability following superficial damage allows interstitial bicarbonate to freely diffuse from the mucosa into the mucoid cap. Under these circumstances, a protective zone is created beneath which the process of rapid repair can proceed in an environment of neutral pH[12].

Bicarbonate Production

At the end of the last century Pavlov recognized that mucus lining in the stomach was alkaline and suggested it had a role in local neutralization of luminal acid. It has since been demonstrated that the gastric mucosa is capable of secreting small amounts of bicarbonate derived from the metabolic activity of the surface cells[13]. Base is also produced by the parietal cells, the so-called "alkaline tide", in direct proportion to the amount of acid produced. After delivery to the superficial mucosa via the bloodstream, this bicarbonate can enter the lumen of the stomach as a result of transcellular transport via the surface cells or by diffusion via the paracellular route. Extracellular mucus gel maintains an unstirred layer or mixing barrier adjacent to the apical cell membrane. Thus, rather than being immediately dissipated within the lumen, it is thought that secreted bicarbonate remains within the interstices of the gel where it is available for neutralization of hydrogen ions diffusing from the lumen of the stomach into the mucus layer.

Neutralization of acid within the mucus layer is postulated to provide the first-line of defense against autodigestion by maintaining a near neutral microenvironment at the mucosal surface[9,14]. Indeed a standing pH gradient can be demonstrated to exist at the epithelial surface of the stomach by advancing mini-electrodes from the lumen towards the mucosa[15]. However, rates of gastric alkaline secretion are very small compared with total acid output and the pH gradient is readily collapsed when luminal acidity falls below about 20 mM (pH 1.8). As discussed above, superficial damage also induces passive permeability pathways in the cell barrier allowing diffusion of interstitial bicarbonate into the lumen. This most probably occurs as the result of a transient, reversible increase in permeability of the tight junctions as evidenced by the increase in transmucosal urea flux which occurs in the intestine in response to the presence of HCl in the lumen in concentrations as low as 10 mM[16].

POST-EPITHELIAL FACTORS

Acid-Base Balance

Ambient bicarbonate plays a vital role in maintaining the acid-base status of gastric mucosa. The ability of nutrient (blood-side) bicarbonate to protect the epithelium from injury has been demonstrated in a variety of experimental circumstances[17,18]. Thus a deficiency of bicarbonate enhances the susceptibility of gastric mucosa to ulceration by decreasing its ability to withstand a fall in the intramucosal pH. Conversely, parenteral administration of bicarbonate in animals subject to hemorrhagic shock-induced ulceration attenuates or completely prevents subsequent injury. These observations are consistent with the finding that an actively secreting stomach, with its attendant alkaline tide arising from the parietal cells, is capable of tolerating luminal acid loads far better than a resting or acid-inhibited tissue[19].

Intracellular pH is also maintained within fine limits, mainly through operation of the sodium-hydrogen exchanger[20]. This process, which regulates cytoplasmic pH in both surface and glandular cells of the stomach, is augmented in the case of parietal cells by the bicarbonate-chloride exchanger, also present on the basolateral membrane, which prevents excessive alkalinization of the cell during acid secretion and gives rise to the alkaline tide.

Mucosal Blood Flow

Electron microscopy and microvascular casting of the gastric mucosa have revealed the close proximity of fenestrated capillaries to both parietal and surface epithelial cells[21]. As a consequence of the direction of blood flow, the alkaline tide generated by actively secreting parietal cells is transported by the capillaries towards the surface epithelium. Thus, during periods of active acid secretion, the amount of bicarbonate in the blood supplying the surface cells is increased. As a result, the capacity of the interstitium to neutralize acid, either within the mucosa as a result of back-diffusion or within the lumen following an increase in mucosal permeability, is considerably enhanced. The amount of bicarbonate available for uptake by the surface cells themselves via the sodium-dependent anion cotransporter is also increased[22].

A number of studies on experimentally-induced gastric ulceration strongly suggest that impairment of blood flow is a major factor in precipitating mucosal injury[23]. Maintenance of blood flow is critical not only in providing adequate oxygenation but also in counteracting acidosis. The microvasculature is also viewed as the main site of action of exogenous damaging agents, particularly lipophilic substances such as ethanol[24]. By analogy with the influence of blood flow in vivo, mucosal injury or preservation from damage can be reproduced in an isolated stomach preparation by either removing or increasing the bicarbonate content of the nutrient bathing solution[25].

PROLIFERATION AND MIGRATION

Renewal and Repair

The gastric epithelium, in common with the rest of the intestinal tract, undergoes continuous turnover[26]. In the stomach, new cells are formed in a proliferative zone located in the isthmus of gastric glands and the adjacent portion at the base of the pits. The majority of cells migrate towards the surface to replenish cells lost by exfoliation into the lumen of the stomach although a fraction must move down the gland to replace parietal and chief cells. Human gastric epithelial cells replicate at a rate of about one cell per 100 cells per hour and up to two days may be required for these cells to reach the surface; specialized acid and pepsin secreting cells, in contrast, have a diurnal half-life measured in terms of months.

The processes of cellular replenishment and tissue rebuilding which are necessary for peptic ulcer healing are akin to those described for wound healing in other tissues such as the skin[27]. Repair of chronic lesions, like normal mucosal turnover, requires cellular proliferation and is therefore quite distinct from the process of cellular migration which underlies mucosal restitution. Repair of acute and chronic lesions are further distinguished by the fact that while ulcer healing is promoted by antisecretory drugs, restitution occurs at low luminal pH and does not appear to be influenced by inhibition of acid output or by growth factors. Indeed, resistance of the healthy mucosa to low luminal pH is augmented when the parietal cell is actively secreting.

Restitution

The process of re-epithelialization, also termed "restitution", occurs in response to acute, superficial injury and enables rapid restoration of the cellular barrier after widespread exfoliation of surface cells. Restitution of barrier integrity as a result of cell migration is a fundamental property of epithelia and can be readily demonstrated in tissue culture. It has been intensively studied in the stomach since this is a common site of superficial damage in response to oral ingestion of spicy foods, alcohol and various drugs.

Gastric mucosal restitution occurs by migration of undamaged epithelial cells from within the gastric crypts across the basal lamina to cover areas denuded as a result of detachment and exfoliation caused by cellular injury[28]. Complete restoration of epithelial continuity and function, reflected by a return of potential difference and acid secretion, can occur in well under 60 min in vivo but may take up to 4 hr in vitro[29]. Various factors have been demonstrated to inhibit restitution but no agent has been convincingly shown to accelerate the process. Complete restitution is dependent on formation of new junctional complexes and is therefore inhibited in the absence of calcium[30]. It is also inhibited by cytochalasin, reflecting the role of microfilaments in cellular motility, and by anti-laminin antibodies.

THE BARRIER IN PERSPECTIVE

Contrary to the classical view, breaching the gastric mucosal barrier is probably an extremely common occurrence which is caused by the mildest of challenges including ingestion of food. As such, the response may be considered as physiological rather than a pathological event which will inevitably lead to the development of hemorrhagic erosions. However, failure of rapid repair mechanisms may result in progression of superficial injury to more severe damage and eventually to development of a chronic ulcer. While maintenance of a cellular barrier remains at the heart of gastric mucosal homeostasis and protection, it is clear that resistance of the stomach to autodigestion depends on a multitude of factors interacting in a dynamic system. The mucus barrier is able to make-and-break in order to allow entry of acid juice into the lumen of the stomach. Intercellular junctions appear to behave similarly in allowing a flux of alkaline interstitial fluid into the lumen during periods of extreme acidity or acute injury.

Breaches in the mucosal barrier may vary from a transient increase in junctional permeability to wholescale loss of surface cells. The former is associated with increased passive diffusion of ions including a flux of bicarbonate into the lumen while the latter requires extensive restitution in order to re-establish a cellular barrier. Under conditions of maximal acid production, the alkaline tide arising from parietal cells augments bicarbonate levels in the superficial mucosa thereby maintaining acid-base balance within the interstitium and neutrality at the mucosal surface. Under basal acid conditions, it is likely that bicarbonate derived from the metabolic activity of the surface cells and secreted into the mucus gel is sufficient to maintain the pre-epithelial microenvironment close to neutrality.

SUMMARY

The present status of mucosal protection and repair of gastric mucosa is reviewed. The concept of a static, single-locus barrier pioneered by Davenport has been integrated into a multi-component, dynamic model. Bicarbonate in conjunction with blood flow contributes to acid disposal at both pre-epithelial and interstitial sites. In common with other epithelia, cells lining the surface, pits and glands of the stomach display homeostatic mechanisms for maintaining cytoplasmic pH within physiological limits. Continuity of this superficial epithelial cell layer can be rapidly re-established following acute injury as a result of migration of viable cells from within the gastric pits and glands. Although acute repair should be clearly distinguished from the process of wound healing required to repair deep ulcers, failure of the mechanism of reepithelialization (restitution) could predispose to the development of a chronic lesion.

ACKNOWLEDGEMENT

We thank Mrs. Hoda Mansour for secretarial assistance.

REFERENCES

1. J. Hunter, On the digestion of the stomach after death, *Phil Trans Roy Soc.* 62:447 (1772).
2. A. Garner, and P.E. O'Brien, "Mechanisms of Injury, Protection and Repair of the Upper Gastrointestinal Tract", John Wiley & Sons, Chichester, U.K. (1991).
3. A. Allen, G. Flemstrom, A. Garner, and E. Kivilaakso, Gastroduodenal mucosal protection, *Physiol Rev.* 73:823 (1993).

4. F. Hollander, The two-component mucus barrier, *Arch Intern Med.* 93:107 (1954).

5. H.W. Davenport, The gastric mucosal barrier, *Digestion.* 5:162 (1972).

6. A. Garner, A. Allen, J. Gutknecht, A. Yanaka, P.J. Goddard, W. Silen, E.R. Lacy, P. Bauerfiend, M. Starlinger, and J.L. Wallace, Mechanisms of gastric mucosal defence, *Eur J Gastroenterol Hepatol.* 2:165 (1990).

7. A. Allen, Gastrointestinal mucus, *in:* "Handbook of Physiology, The Gastrointestinal System: Salivary, Gastric and Hepatobiliary Secretions," J.G. Forte, ed., *Am Physiol Soc.* Bethesda (1989).

8. B.A. Hills, B.D. Butler, and L.M. Lichtenberger, Gastric mucosal barrier: hydrophobic lining to the lumen of the stomach, *Am J Physiol.* 244:G561 (1983).

9. G. Flemstrom, and A. Garner, Gastroduodenal bicarbonate transport: characteristics and proposed role in acidity regulation and mucosal protection, *Am J Physiol.* 242:G183 (1982).

10. J.L. Wallace, Gastric resistance to acid: is the "mucus-bicarbonate barrier" functionally redundant?, *Am J Physiol.* 256:G31 (1989).

11. L.A. Sellers, A. Allen, and M.K. Bennett, Formation of a fibrin-based gelatinous coat over repairing rat gastric epithelium after acute ethanol damage: interaction with adherent mucus, *Gut.* 28:835 (1987).

12. J.L. Wallace, G.W. McKnight, The mucoid cap over superficial gastric damage in the rat, *Gastroenterology.* 99:295 (1990).

13. G. Flemstrom, Active alkalinization by amphibian gastric fundic mucosa in vitro, *Am J Physiol.* 233:E1 (1977).

14. A. Allen, and A. Garner, Gastric mucus and bicarbonate secretion and their possible role in mucosa protection, *Gut.* 21:249 (1980).

15. E.M. Quigley, and L.A. Turnberg, pH of the microclimate lining human gastric and duodenal mucosa in vivo: studies in control subjects and in duodenal ulcer patients, *Gastroenterology.* 92:1876 (1987).

16. J.M. Wilkes, A. Garner, and T.J. Peters, Mechanisms of acid disposal and acid-stimulated alkaline secretion by gastroduodenal mucosa, *Dig Dis Sci.* 33:361 (1988).

17. R. Schiessel, A. Merhav, J.B. Matthews, L.A. Fleischer, A. Barzilai, and W. Silen, The role of nutrient bicarbonate in protection of amphibian gastric mucosa, *Am J Physiol.* 239:G536 (1980).

18. E. Kivilaakso, High plasma bicarbonate protects gastric mucosa against acute ulceration in the rat, *Gastroenterology.* 81:921 (1981).

19. E. Kivilaakso, and W. Silen, Pathogenesis of experimental gastric mucosal injury, *New Engl J Med.* 301:364 (1979).

20. T.E. Machen, A.M. Paradiso, Regulation of intracellular pH in the stomach, *Ann Rev Physiol.* 49:19 (1987).

21. B.J. Gannon, J. Browning, and P.E. O'Brien, The microvascular architecture of the glandular mucosa of the rat, *J Anat.* 133:667 (1982).

22. K. Takeuchi, A. Merhav, W. Silen, Mechanism of luminal alkalinization by bullfrog fundic mucosa, *Am J Physiol.* 243:G377 (1982).

23. M. Starlinger, and R. Schiessel, Bicarbonate delivery to the gastroduodenal mucosa by the blood: its importance for mucosal integrity, *Gut.* 29:647 (1988).

24. S. Szabo, J.S. Trier, A. Brown, and J. Schnoor, Early vascular injury and increased vascular permeability in gastric mucosal injury caused by ethanol in the rat, *Gastroenterology.* 88:228 (1985).

25. E. Kivilaakso, A. Barzilai, R. Schiessel, R. Cross, and W. Silen, Ulceration of amphibian gastric mucosa, *Gastroenterology.* 77:31 (1979).

26. M. Lipkin, P. Sherlock, and B. Bell, Cell proliferation kinetics in the gastrointestinal tract of man: II cell renewal in the stomach, ileum, colon and rectum, *Gastroenterology.* 45:721 (1963).

27. T.K. Hunt, The principles of wound healing, *in:* "Mechanisms of Peptic Ulcer Health," F. Halter, A. Garner, and G.N.J. Tytgat, eds., Kluwer, Dortrecht (1991).

28. W. Silen, and S. Ito, Mechanisms for rapid reepithelialization of the gastric mucosal surface, *Ann Rev Physiol.* 47:217 (1985).

29. E.R. Lacy, Gastric mucosal defence after superficial injury, *Clin Invest Med.* 10:189 (1987).

30. J. Critchlow, D. Magee, S. Ito, K. Takeuchi, and W. Silen, Requirements for restitution of the surface epithelium of frog stomach after mucosal injury, *Gastroenterology.* 88:237 (1985).

NEW APPROACHES TO GASTROPROTECTION:
CALCIUM MODULATORS

Gary B. Glavin and Arleen M. Hall

Despite impressive developments in gastrointestinal pharmacology over the last several years, the treatment of gastroduodenal disease remains palliative and the search for new therapeutic agents continues. One focus of our investigation has been directed towards calcium modulators and their effects in several animal models of gastric function and pathology.

Early in vitro observations suggested that calcium was important for gastric acid secretion, since pentagastrin- and vagal-stimulated acid secretion were reduced in the absence of calcium[1]. Subsequently, Sewing and Hannemann[2] observed that verapamil blocked basal, histamine-stimulated and dibutyryl cyclic AMP-stimulated acid secretion from isolated parietal cells, although the mechanism whereby verapamil exerted these effects was unclear.

Several small human studies were attempted following these initial in vitro findings, however, no consistent effects were observed. For example, two groups found no effect of verapamil on either basal, pentagastrin- or histamine-stimulated gastric secretion[3,4] while a second group found that another calcium channel blocker, nifedipine, significantly attenuated both basal and pentagastrin-stimulated acid output in man[5].

With this work in mind, several groups began testing calcium channel blockers in a variety of animal models of gastric function and disease. Initial findings indicated that verapamil, as well as nifedipine, reduced restraint stress-induced gastric erosions[6,7,8], and indomethacin-induced gastric mucosal injury[9], while at the same time, exacerbating 100% ethanol-induced gastric mucosal injury[10,11]. Subsequently, Glavin[12,13] tested a representative of each of three classes of calcium channel blockers, dihydropyridines, phenylalkylamines, and benzothiazepines, as well as a calcium "agonist", using nitrendipine, verapamil, diltiazem, and CGP28392, respectively. All compounds, including the calcium agonist reduced restraint stress-induced gastric erosions. In the ethanol-involved ulcer model, all calcium antagonists exacerbated gastric damage, however, verapamil and nitrendipine did so in a biphasic manner. A low dose of nitrendipine significantly reduced ethanol-induced gastric injury, while all higher doses worsened these lesions. Verapamil, at the lowest and highest doses worsened, while the middle dose reduced, ethanol-induced gastric mucosal injury.

Neuroendocrinology of Gastrointestinal Ulceration
Edited by S. Szabo and Y. Taché, Plenum Press, New York, 1995

Subsequently, we showed that adrenalectomy, but not vagotomy, reverses the lesion-worsening effect of nitrendipine and verapamil such that under this condition, the latter compounds are protective against ethanol-induced gastric mucosal injury[14]. Dexamethasone replacement once again restored the exacerbating effect of nitrendipine and verapamil on ethanol-induced gastric lesions[14]. Adrenal medullectomy was without effect[15], suggesting that adrenal corticoids modulate the worsening effect of calcium channel blockers on ethanol-induced gastric mucosal damage.

When tested in animal models of gastric acid secretion, calcium channel blockers exhibit consistent anti-secretory effects. For example, in conscious rats with gastric cannulae, verapamil and nifedipine were shown to reduce basal acid output[6,12,16]. Glavin[12] extended these observations to include anti-secretory effects for diltiazem, nitrendipine, and, again surprisingly, the calcium channel agonist, CGP28392. Especially noteworthy was the antisecretory effect of nitrendipine, which, at all doses tested, virtually abolished gastric acid output (both volume and concentration).

Since nitrendipine exerted the most striking effects, in terms of reducing stress-induced gastric erosions and in blocking gastric secretion, we decided to explore possible mechanisms of action for this compound[12]. Rats were pretreated with a non-ulcerogenic, but cyclooxygenase-inhibiting dose of the non-steroidal anti-inflammatory agents, indomethacin or sodium meclofenamate prior to receiving nitrendipine and restraint-cold stress. Neither agent abolished the protective effect of nitrendipine, whereas that of verapamil, diltiazem, and CGP28392, could be completely reversed with such pretreatment. These data suggested that the gastroprotective action of nitrendipine is not mediated through prostaglandin activity in the gut, whereas that of the other calcium blockers and CGP28392, may involve prostaglandins, a suggestion recently confirmed by others[17].

Since tissue non-protein sulfydryls may mediate gastric protection induced by a variety of agents[18], we tested the ability of a sulfhydryl alkylator, N-ethylmaleimide, to reverse nitrendipine-induced gastroprotection. When given either prior to or following nitrendipine treatment, N-ethylmaleimide failed to reverse nitrendipine's protective effect in terms of significantly reducing stress-induced gastric erosions, suggesting that tissue sulfhydryls do not mediate the action of nitrendipine[12].

We then tested whether the effects of nitrendipine were specific to experimental gastric ulcers, by examining the action of nitrendipine in a model of experimental cysteamine-induced <u>duodenal</u> ulcer[19]. Again, we observed a significant protective effect of nitrendipine against duodenal ulcers, and, at higher doses, this compound virtually abolished duodenal lesions[13]. Since cysteamine produces a massive depletion of somatostatin in both brain and duodenal mucosa, and since somatostatin is a significant modulator (inhibitor) of many "aggressive" factors in the gut[20], including histamine-, gastrin-, and cholinergically-stimulated acid secretion, we decided to examine whether nitrendipine exerted some of its protective action through somatostatin. While not significantly affecting levels of somatostatin-like immunoreactivity (SLI) when given to non-treated rats, nitrendipine significantly protected duodenal mucosal levels of SLI in the presence of cysteamine. When given prior to cysteamine, nitrendipine resulted in levels of SLI not significantly different from those seen in control (non-cysteamine treated) rats. Thus, while not increasing SLI levels, nitrendipine nonetheless protected against cysteamine-induced depletion, and for this reason, we have used the term "preserved" to describe nitrendipine's effect on SLI in the face of cysteamine challenge and to, at least in part, account for the action of nitrendipine to reduce experimental duodenal ulcerogenesis[13].

Since basal gastric acid output was markedly inhibited by nitrendipine, we chose to examine its effects on acid secretion more closely, by testing its ability to affect acid secretion stimulated by selective agonists at receptors known to exist on gastric parietal cells. We stimulated acid secretion by the muscarinic cholinergic agonist, bethanechol; the gastrin agonist, pentagastrin; and the histamine H_2 agonist, dimaprit. Nitrendipine blocked <u>only</u> bethanechol-stimulated acid output, and did not substantially affect gastric secretion induced by the other secretagogues[12]. This finding is suggestive of vagal involvement in the actions of nitrendipine and we are pursuing this result. However, recent suggestions that calcium channel antagonists inhibit the parietal cell H^+ and K^+-ATPase must also be considered when explaining the mechanisms underlying the antisecretory actions of these compounds[21].

The actions of the calcium agonist, CGP28932, remain puzzling. This compound reduced <u>both</u> stress- and ethanol-induced gastric damage, although its stress ulcer-reducing effects were always less dramatic than those seen with all three calcium antagonists. CGP28392 also reduced basal gastric output, but to a lesser degree than seen with verapamil and nitrendipine, which almost completely blocked acid secretion. Finally, the ability of CGP28392 to block stress-induced gastric lesions was reversed by pretreatment with indomethacin, similar to the effect seen with verapamil and diltiazem, but unlike that of nitrendipine, the protective effects of which were not blocked by prostaglandin synthesis inhibition.

Both CGP28392 and nitrendipine act at "L"-type calcium channels, which <u>may</u> exist in several isoforms[22]. Interestingly, co-administration of the agonist, CGP28392 together with the antagonist, nitrendipine, results in complete retention of the gastroprotective effects of nitrendipine, suggesting that, in a competitive situation, antagonist binding/action is preferred, at least in the gut (Glavin, in preparation).

Recent evidence suggests that receptors for calcium channel antagonists are subject to both acute and chronic regulation, since chronic treatment with these agents produces both up and down-regulation of calcium antagonist receptors[22]. Moreover, disease states also modify calcium antagonist receptor number and/or affinity, partially explaining why administration of calcium channel blockers has a greater effect in disease states than in control conditions. We therefore examined the effects of chronic pre-treatment with nitrendipine, on subsequent experimental ulcerogenesis. Rats were given daily administration of nitrendipine (8.0, 16.0 or 32.0 mg/kg i.p.) or vehicle for 7 days prior to exposure to either restraint-cold stress or 100% ethanol as described previously[12]. Such chronic treatment with nitrendipine resulted in a complete reversal of its acute effects; that is, chronic nitrendipine resulted in very little protection against stress-induced gastric erosions and very little worsening of ethanol-induced gastric mucosal damage. These findings suggest that chronic exposure to nitrendipine resulted in a "down-regulation" of its receptor/binding site, thereby blunting the effects seen in the acutely-treated animal (Glavin, in preparation).

Very recently[23] we showed that the calcium chelators, EGTA and EDTA are protective against both ethanol-induced and stress-induced gastric mucosal injury and are powerfully anti-secretory in both conscious basal acid output as well as in pylorus-ligated rats. Adrenalectomy and vagotomy abolish the gastroprotective effects of EGTA, while those of EDTA remained intact. We are currently exploring the hypothesis that calcium chelators exert gastroprotection through a novel mechanism - inhibition of calcium-dependent protease (calpain) activity in the gut[24], in light of our initial finding that EGTA and EDTA significantly block ethanol-induced release of gastric calpain (Glavin, in preparation).

Table 1 presents a compilation of the gastrointestinal effects of various calcium-modulating compounds. In summary, it appears that there is a role for calcium-modulating compounds in gastrointestinal disease. Especially significant is the marked gastroprotective profile of nitrendipine. The use of calcium channel antagonists in human gastric disease has been attempted[3,5], and, in light of the profound effects of nitrendipine in experimental ulcerogenesis, may continue on a larger scale (Defeudis and Christen, 1989). Whether these effects are specific to acute gastric mucosal injury, as indicated in the previously mentioned studies, or whether they can be generalized to chronic gastroduodenal ulcer disease, remains to be determined and will be examined in this laboratory in the future.

Table 1. Effects of Calcium Modulators on Gastric Acid Secretion and Gastric Mucosal Injury

Compound	Gastric Acid Secretion	Ethanol-Induced Gastric Mucosal Injury	Stress-Induced Gastric Mucosal Injury
Antagonists			
diltiazem	↓	↑	↓
nifedipine	↓	↑	↓
nitrendipine	↓↓	↑↑	↓↓
verapamil	↓		
Agonists			
CGP28392	↓	↓	↓
Chelators			
EDTA	↓	↓↓	↓↓
EGTA	↓	↓	↓

Abbreviations: ↓↓ = large decrease; ↑↑ = large increase; ↓ = small decrease; ↑ = small increase.

SUMMARY

Calcium is important in regulating many physiologic functions, including stimulus-secretion coupling in the gut. The observation that hypercalcemia is associated with excess gastric secretion and, possibly, frank ulcer disease, prompted an examination of several calcium channel blockers as well as calcium chelators in models of gastric mucosal injury and gastric secretion. Nitrendipine, and other calcium channel blockers, are powerful antisecretory agents, but exert opposite effects on ethanol-and stress-induced gastric lesions, exacerbating and reducing damage in these models, respectively. Calcium chelators, such as EGTA and EDTA reduce gastric secretions and diminish gastric mucosal injury produced by both stress and ethanol. It is suggested that manipulation of gastric calcium may represent a novel pharmacotherapeutic approach in gastroduodenal ulcer disease.

ACKNOWLEDGEMENTS

This work was supported by the Medical Research Council of Canada.

REFERENCES

1. I. Szelenyi, Calcium, histamine and pentagastrin: Speculations about the regulation of gastric acid secretion, *Agents Actions.* 10:187 (1980).
2. K. Sewing, and H. Hannemann, Calcium channel antagonists verapamil and gallopamil are powerful inhibitors of acid secretion in isolated and enriched guinea pig parietal cells, *Pharmacology.* 27:9 (1983).
3. E. Aadland, and A. Berstad, Effect of verapamil on basal and pentagastrin-stimulated gastric acid secretion, *Clin Pharmacol Ther.* 34:399 (1983).
4. R. Levine, S. Petokas, A. Starr, and R. Eich, Effect of verapamil on basal and pentagastrin-stimulated gastric acid secretion, *Clin Pharmacol Ther.* 34:399 (1983).
5. R. Caldara, E. Masci, C. Barbieri, M. Sorghi, V. Piepoli, and A. Tittobello, A., Effect of nifedipine on gastric acid secretion and gastrin release in healthy man, *Eur J Clin Pharmacol.* 28:677 (1985).
6. G. Glavin, Verapamil and nifedipine effects on gastric acid secretion and ulcer formation in rats, *J Pharm Pharmacol.* 40:514 (1988).
7. C. Ogle, C. Cho, C., Tong, M., and M. Koo, The influence of verapamil on the gastric effects of stress in rats, *Eur. J. Pharmacol.* 112:399 (1985).
8. R. Wait, A. Leahy, J. Vee, and T. Pollock, Verapamil attenuates stress-induced gastric ulceration, *J Surg Res.* 38:424 (1985).
9. B. Ghanayem, H. Matthews, and R. Maronpot, Calcium channel blockers protect against ethanol- and indomethacin-induced gastric lesions in rats, *Gastroenterology.* 92:106 (1987).
10. J. Esplugues, R. Brage, J. Cortijo, M. Marti-Bonmati, G. Hernandez, B. Sarria, and J. Esplugues, Differential effects of verapamil on various gastric lesions in rats, *Pharmacology.* 36:69 (1988).
11. M. Koo, C. Cho, and C. Ogle, Verapamil worsens ethanol-induced gastric ulcers in rats, *Eur. J. Pharmacol.* 120:355 (1986).
12. G. Glavin, Calcium channel modulators: Effects on gastric function, *Eur J Pharmacol.* 160:323 (1989).
13. G. Glavin, Nitrendipine reduces experimental gastric and duodenal ulceration: Possible role of somatostatin, *Dig Dis Sci.* 34:1313 (1989).
14. G. Glavin, and C. Ogle, Adrenalectomy, but not vagotomy, reverses the worsening effect of Ca^{2+} blockers on ethanol-induced gastric lesions in rats, *Eur J Pharmacol.* 205:171 (1991).
15. G. Glavin, and C. Ogle, Adrenal demedullectomy does not affect the ability of nitrendipine to worsen ethanol-induced gastric lesions, *Med Sci Res.* 19:405 (1991).
16. M. Koo, C. Cho, and C. Ogle, Luminal acid in stress ulceration and the antiulcer action of verapamil in rat stomachs, *J Pharm Pharmacol.* 38:845 (1986).
17. C. Gutierrez-Cabano, Prostaglandins and sulfhydryls mediate gastric protection induced by verapamil in rats, *Dig Dis Sci.* 38:2043 (1993).
18. S. Szabo, J. Trier, and P. Frankel, Sulfhydryl compounds may mediate gastric cytoprotection, *Science.* 214:200 (1981).
19. S. Szabo, and C. Cho, From cysteamine to MPTP: Structure-activity studies with duodenal ulcerogens, *Toxicol Pathol.* 16:205 (1988).
20. G. Glavin, and A. Hall, Brain-gut relationships: Gastric mucosal defense is also important, *Acta Physiol Hung.* 80:107 (1992).
21. J. Nanoi, R. King, D. Kaplan, and R. Levine, Mechanisms of gastric proton pump inhibition by calcium channel antagonists, *J Pharmacol Exp Ther.* 252:1102 (1990).
22. T. Godfraind, and S. Govoni, Increasing complexity revealed in regulation of Ca^{2+} antagonist receptor, *Trends Pharmacol Sci.* 10:297 (1989).
23. G. Glavin, and S. Szabo, Effects of the calcium chelators EGTA and EDTA on ethanol- or stress-induced gastric mucosal lesions and gastric secretions, *Eur J Pharmacol.* 233:269 (1993).
24. R. Mellgren, Calcium-dependent proteases: An enzyme system active at cellular membranes. *FASEB J.* 1:110 (1987).
25. F. Defeudis, and H. Christen, State- and use-dependent calcium antagonists offer hope in control of GI disorders, *Trends Pharmacol Sci.* 10:170 (1989).

EFFICACY OF DOPAMINERGIC AGENTS IN PEPTIC ULCER HEALING AND RELAPSE PREVENTION - FURTHER INDICATION OF THE IMPORTANCE OF STOMACH DOPAMINE IN THE STRESS ORGANOPROTECTION CONCEPT

P. Sikiric, I. Rotkvic, S. Mise, S. Seiwerth,
Z. Grabarevic, M. Petek, R. Rucman, V. Zjacic-Rotkvic,
M. Duvnjak, V. Jagic, E. Suchanek, A. Marovic, M. Banic,
T. Brkic, M. Hanzevacki, J. Separovic, S. Djacic and
V. Simicevic

Ulcer disease is often referred to as a multifactorial overall disturbance of the body[1]. The intriguing complexity of ulcer etiology is currently in the focus of many different scientific approaches. However, in spite of these efforts, the problem of how to incorporate such a theoretical recognition into a meaningful integrative therapeutic management concept, remains to be solved. One possible solution may lie in dopamine research.

The Importance of Dopamine: A "Stomach" View

Since the early demonstrations that dopamine could be protective against stress-induced gastric lesions[2], and that its agonists are strongly effective in reducing cysteamine-induced duodenal ulcer in rats[3], an increasing number of reports has confirmed the beneficial activity of dopamine[4-12]. Dopamine's protective effects could be reduced by antagonists that are also known to augment various gastroduodenal lesions[4-12], due to their inherent ulcerogenic capacity in both rats and mice[5,6]. The same data were obtained after either peripheral[1-9] or central[10-12] application of dopaminergic compounds. Subsequent work confirmed the hypothesis that dopamine's beneficial effect involves many different organs[1,7,13-16], an activity shared also by prostaglandins and somatostatin[17-20]. This research emphasized the concept of organoprotection wherein dopamine is one of the important endogenous protective mediators[21,22] in addition to its role in disease[23]. In all the beneficial effects of dopamine, gastroduodenal protection appears to be the most prominent[1,7,13-16]. This may be due to the widespread presence of dopamine in the body as well as its physiological significance[1-16,21,22] and the hypothesis is fully supported by dopamine system changes in the stomach during the beginning of life threatening conditions, such as during stress. Notably, the stomach is specifically vulnerable to stress[24]. The

Neuroendocrinology of Gastrointestinal Ulceration
Edited by S. Szabo and Y. Taché, Plenum Press, New York, 1995

increased number of dopamine receptors found in the stomach mucosa (as opposed to no changes in several brain regions[10] considered to be essential for the initiation of the stress coping response), is a strong indicator for the stomach to be seen as an initiating site for dopamine involvement in the stress-coping response[21,22] (Figure 1). Supporting this concept is the finding of increased dopamine receptor binding in patients with duodenal ulcer disease[25].

With regard to possible clinical implications of dopamine system management as an integrative therapeutic approach, additional support has been forthcoming. Along with the established use of dopamine-like agents in many other conditions[26,27], clear evidence for their superior efficacy in peptic ulcer disease has been provided[23,28-30]. Dopamine agents such as bromocriptine and amantadine (unlike H_2 blockers) result in increased initial healing in duodenal ulcer patients[23,28-30] and also effectively decrease relapse rates in many patients[23]. These findings will be reviewed briefly in the present report.

Efficacy of Dopaminergic Agents in Ulcer Disease Therapy

The clinical trial was performed in the General Hospital "Sv. Duh", Zagreb, and in the Clinical Hospital "Firule", Split, Croatia. A total of 124 patients (84 males, 40 females) aged 20-75 years, entered into the treatment trial (four weeks), and 120 finished it. All patients with initially healed ulcer (100) were thereafter included in the relapse prevention study for an additional six month period. Each patient had endoscopically proven one or more duodenal ulcers that were more than 5 mm in diameter.

All patients signed an informed consent prior to inclusion in the study, but they were not informed which treatment they had received during the trial. They were randomly assigned to each one of the following groups (31 patients in each group): i. bromocriptine (Bromergone "Lek", Ljubljana, Slovenia) 2.5 mg two times daily; ii. amantadine (Symetrel, "Pliva", Zagreb, Croatia) 100 mg nocte; iii. cimetidine (Belomet, "Belomet", Koprivnica, Croatia) 800 mg nocte; iv. famotidine (Ulfamid, "Krka", Novo Mestro, Slovenia) 40 mg nocte. All drugs were administered for a four week period. No maintenance therapy was recommended after completion of the four-week period. According to the schedule of the trial, the endoscopist was "blind", and medication with steroids, non-steroidal antiinflammatory drugs as well as calcium channel-blocking drugs was not allowed. The occasional consumption of antacid (Gastal, tablets, "Pliva", Zagreb, Croatia) was suggested.

A high healing rate was seen for all drugs at four weeks, but slightly better results were noted in the groups treated with dopamine-like drugs (90% for amantadine and 87% for bromocriptine vs. 77% for cimetidine and 80% for famotidine). In terms of relapse prevention for six months, a relatively low incidence was noted for both H_2 blockers (34.7% for cimetidine and 25% for famotidine). Thus, the lower incidence observed in both groups treated with dopamine-like agents - either amantadine (11.7%) or bromocriptine (7.7%) - appears promising. However, subsequent statistical analysis showed that this difference in the initial ulcer healing rate was not significant. It is noteworthy that in terms of relapse prevention rate, these differences did reach significance, particularly relative to cimetidine (Tables 1 and 2). The therapeutic effectiveness, as subjectively evaluated by the patients, is shown in Table 3. The effectiveness of all drugs is presented as relative values based on the incidence of side effects. No statistically significant difference was observed among the treatment groups. It is important to indicate that "high effectiveness" was the most frequent evaluation.

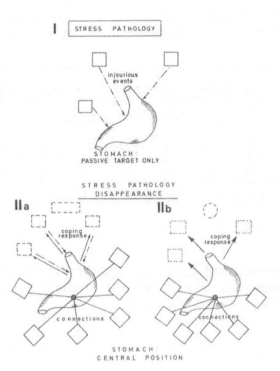

Figure 1. Hypothesis. The stomach aspects in conventional stress terms (I) and in terms of the stomach/stress/organoprotection hypothesis (IIa,IIb). I. Stomach taken as only a passive target. Stomach stress response is considered only because of the local (negative) significance. *Mediators: CRF, ACTH, corticosteroids, catecholamines, hypothalamus-hypophysis-adrenal gland axis, brain-gut axis*[10-12,24]. The stomach, specifically affected, is initiating, organizing and mediating a counteracting organoprotective response. IIb. This physiological organoprotective response could be amplified by administration of organoprotective agents and full organoprotection homeostasis is restored. *Mediators: organoprotectors (PGs, dopamine, somatostatin, BPC), the axes between stomach (G-I tract) and the brain and other organs, the potential as the largest neuroendocrine organ in the body*[21,22]. This hypothesis is in keeping with the physiological significance of the hypothesized organoprotectors and their beneficial effects[1,7,13-22,67-72], gastrointestinal tract connections[54-61], and especially in light of dopamine and ulcer disease as reflecting an overall disturbance of the body (stress disturbance) and in line with the efficacy of dopaminergic compounds in peptic ulcer healing and relapse prevention[23] - a new integrative therapeutic management.
Modified from ref. 21,22.

No significant differences between treated groups were observed regarding the ulcer symptoms such as epigastric pain, vomiting, and heartburn. There were no bleeding episodes during either the acute treatment or posthealing periods. The consumption of antacids was low in both initial treatment and posthealing periods, and no significant differences between the investigated groups were observed.

All patients with relapse who were offered an "open" treatment with the different drugs (amantadine, bromocriptine, cimetidine, famotidine) responded well to the corresponding therapy and healed within four weeks.

The physical examination before and after medication or during the posthealing period did not reveal any abnormalities in any treated groups. Laboratory tests revealed a small increase in serum creatinine in seven patients treated with cimetidine during the acute treatment phase. In the initial treatment period, all noted side effects (with the exception of side effects causing withdrawal of four patients) were

Table 1. Effectiveness After Four Weeks of Treatment with Amantadine, Bromocriptine, Famotidine, and Cimetidine in 120 Patients Subjected to Second Endoscopy[*]

Local status	Amantadine		Bromocriptine		Famotidine		Cimetidine	
	No.	%	No.	%	No.	%	No.	%
Duodenal ulcer								
Healed	27	90	26	87	24	80	23	77
Not healed	3	10	4	13	6	20	7	23
Total	30	100	30	100	30	100	30	100

[*]Chi-square test: NS. Four patients (one patient in each group) were excluded from the trial because of side effects.
Modified from ref. 23, 28-30.

Table 2. Effectiveness in Relapse Prevention Six Months After Healing Period Following Discontinuation of Four Weeks of Treatment with Amantadine, Bromocriptine, or Cimetidine and Famotidine

Duodenal ulcer	Dopamine Agonists[*]				H_2 Receptor Blockers			
	Amantadine[†]		Bromocriptine[†]		Famotidine		Cimetidine	
	No.	%	No.	%	No.	%	No.	%
No relapse	24	88.9	24	92.3	18	75	15	65.3
Relapse	3	11.1	2	7.7	6	25	8	34.7
Total	27	100.0	26	100.0	24	100.0	23	100.0

[*]Chi-square test vs. cimetidine: $P<0.05$.
[†]Chi-square test vs. H_2 receptor antagonists: $P<0.01$.
Modified from ref. 23, 28-30.

relatively mild. The overall evidence of side effects was 20% (six patients) in the amantadine and bromocriptine groups, 13.3% (four patients) in the groups treated with famotidine, and 16.7% (six patients) in the cimetidine group. Some unwanted effects (e.g., nausea, fatigue, vomiting) were more frequently observed in the cimetidine- or bromocriptine-treated groups, but the differences were not significant. No side effects persisted after discontinuation of the initial treatments.

Clinical Efficacy of Dopamine Agents and the Central Position of the Stomach: A "Gut-Brain" Concept

With respect to initial healing, although no double blind regimen was used, the effectiveness of H_2 blockers (cimetidine and famotidine) is strongly indicated by the data from this study. The healing rates for both cimetidine and famotidine were comparable to other studies reported in the literature[31].

Table 3. Therapeutic Effectiveness as Subjectively Evaluated by Patients Finishing the Trial[*]

Therapeutic effectiveness	Amantadine		Bromocriptine		Famotidine		Cimetidine	
	No.	%	No.	%	No.	%	No.	%
High	22	77.3	17	56.7	21	70	18	60
Moderate	7	23.4	9	30	8		9	30
None	1	3.3	4	13.3	1		3	10
Total	30	100.0	30	100.0	30		30	

[*]Chi-square test: NS.
Modified from ref. 23, 28-30.

Regarding the relapse after discontinuation of acute treatment, the results of our study are consistent with findings obtained in other similar controlled trials[32]. In addition, the total number of patients followed over six months was similar to those enrolled in other studies [33-41]. The relapse rate in our patients treated with the H_2 blockers was similar[40] or even lower[33,38,41] than the relapse rate previously reported by other groups[32]. Therefore, the significantly lower number of relapsing patients in the groups treated with the dopamine-like compounds (particularly relative to cimetidine) assumes importance. It should be emphasized that such a low number is also important considering the high relapse rate observed in patients treated with H_2 blockers and with other drugs including tripotassium dicitrate bismuthate[32,34,36].

Additional benefits concerning the efficacy of the dopamine-like drugs tested in our study include a simple daily regimen and the fact that both the incidence of side effects and the dropout rate were not different from those obtained in patients treated with H_2 blockers. As indicated above, these results are consistent with our previous clinical studies with bromocriptine in peptic ulcer patients[28] but are in contrast to an early report on the use of bromocriptine in ulcer patients[42]. However, it should be emphasized that both bromocriptine and amantadine were used in our peptic ulcer patients at much lower doses than those used in conditions such as Parkinson's disease[43,44]. The antiulcer dose of bromocriptine was similar to that used in the treatment of less severe disturbances such as ablation, polycystic ovarian disease, and premenstrual syndrome[26,27]. However, in these disorders a relatively high dropout rate in patients treated with bromocriptine has been noted due to severe side effects[26,27]. With regard to amantadine, the fact that only minor side effects were observed during initial treatment is probably related to its administration once daily at bedtime.

Our results are also supported by the disappointing findings of ulcer treatment with dopamine antagonists[45,46]. Moreover, haloperidol (a D_2 dopamine receptor antagonist), in addition to ulcer induction in animals[5,6] as with other dopamine receptor blockers [5,6], has also been found to provoke ulcers in humans[1].

Various effects thought to be beneficial have been proposed as an explanation for the described dopamine protective activity. Along with the possible local significance of dopamine[21,22], and dopamine storage in the gastrointestinal tract[47], a non-acid-related mechanism has been proposed to account for these findings. Although a stimulation of mucosal alkaline secretion was found[48], the influence of dopamine and its agonists on gastric acid secretion has produced conflicting results[4,6-8,49-51]. In analogy with Robert's prostaglandin cytoprotection[19], our hypothesis was termed "dopamine cytoprotection"[7,11,50,51]. Using both indomethacin and dopamine agents (both *in vivo* and *in vitro*)[7,50], a prostaglandin-independent beneficial mechanism for bromocriptine and amantadine has been suggested. This hypothesis considers the stomach mucosa as completely unable to sustain anaerobic metabolism even for a short period of time and further, that the mucosa requires a continuous supply of glucose and oxygen[52]. Thus, a strong vasodilatory effect of dopamine in the gastrointestinal mucosa[53] appeared to be essential. This hypothesis considers the stomach as a "central" organ in stress and coping responses[21,22], as well as being part of a much more complex overall integrative response to stress. The stomach mucosa is profoundly affected during the earliest stages of life-threatening conditions such as stress[21,22]. For example, metabolic changes and blood redistribution away from the gastric mucosa ("centralization") could be a trigger for the initiation of the stress coping response (for example, an increase in dopamine receptors in the stomach mucosa[21,22]) (Figure 1).

In addition to the described local significance of dopamine, several additional points have to be underscored[21,22]. Of note, the stomach appears to be fully connected with many different organs with well-defined axes[54-61]. The stomach has a specific innervation and particular connections with the brain[61], as well as a major representation in the brain cortex[61]. Notable is the efficacy of dopamine in the treatment of shock[62], the presence of a disturbed dopamine/prolactin axis in peptic ulcer patients[63], the increased ulcer incidence in Parkinson's disease patients[64] as well as animal models of Parkinson's disease and the general dopamine efficacy in managing either condition[63-65]. These data, taken together with the aforementioned efficacy of dopaminergic agonists after either peripheral or central application[1-12] as well as the significance of dopamine in organoprotection[1,7,13-16], provide a compelling network of evidence for dopamine as an important link between the gut and the brain[21,22], and favors use of dopamine agents in further therapy. Emphasizing the gastrointestinal tract as the largest neuroendocrine organ in the body[21,22] will improve our understanding of the concept originally described as the "brain-gut axis"[10-12].

Future Perspectives

A complex interaction with dopamine has been demonstrated for a variety of peptides, including somatostatin[1,66] and neurotensin[10]. Recently, we isolated a new peptide from gastric juice, coded BPC, which has a wide range of mucosal and organoprotective effects[21,22,67-70]. A 15 amino acid peptide, named BPC 157, exhibits no sequence homology with known gut peptides[21,22,67-70], and is thought to be essential for the activity of the entire, larger peptide, and has been synthesized. The protective properties of this peptide have been independently investigated and confirmed by others[71,72]. Its action is complex and is probably mediated by the hormones of the adrenal, parathyroid, thyroid and ovarian glands[67] as well as by the dopamine system[22]. A significant pool of BPC[22] has been immunohistochemically demonstrated in the stomach as well as in the brain. This suggested to us that some of the organoprotective effects of dopamine are due to its interaction with BPC 157 and is an intriguing area for further research.

SUMMARY

Recently, we demonstrated that dopamine-like compounds (e.g., bromocriptine and amantadine) are equally effective as H_2 blockers in peptic ulcer healing and may offer an advantage over H_2 blockers concerning their efficacy in preventing duodenal ulcer relapse. This evidence is considered in the context of the importance of gastric dopamine, the concept of ulcer disease as an overall disturbance of the body, and with the hypothesized central role of the stomach in the stress coping response. The effectiveness of dopamine in shock, as well as its efficacy after either peripheral or central application in gastroduodenal lesions has been particularly emphasized. Along with the dopaminergic compounds used in many other conditions, and their demonstrated efficacy in different organ lesion prevention (organoprotection), dopamine agonist application is proposed as a new and integrative therapeutic management for ulcer disease that arises from an overall disturbance of the body. Dopamine interaction with other protective systems (e.g., peptides, somatostatin, neurotensin, and a new organoprotective peptide, BPC) is also discussed.

REFERENCES

1. S. Szabo, and J.L. Neumeyer, Dopamine agonists and antagonists in duodenal ulcer disease, *in:* "ACS Symposium Series," C. Kaiser, W. Kebabian, eds., American Chemical Society, Washington (1983).
2. H.H. Lauterbach, and P. Mattes, Effects of dopamine on stress ulcer of rats, *Eur Surg Res.* 9:258 (1977).
3. S. Szabo, Dopamine disorder in duodenal ulceration, *Lancet.* 2:880 (1979).
4. P. Sikiric, I. Rotkvic, S. Mise, S. Krizanac, E. Suchanek, M. Petek, V. Gjuris, M. Tucan-Foretic, I. Udovicic, T. Anic, and I. Balen, The influence of dopamine agonists and antagonists on gastric lesions in mice, *Eur J Pharmacol.* 144:237 (1987).
5. P. Sikiric, J. Geber, E. Suchanek, D. Ivanovic, V. Gjuris, J. Aleksic, P. Reic, and A. Marovic, The role of dopamine in the formation of gastric ulcers in rats, *Eur J Pharmacol.* 112:127 (1985).
6. P. Sikiric, J. Geber, D. Ivbanovic, E. Suchanek, V. Gjuris, M. Tucan-Foretic, S. Mise, B. Cvitanovic, and I. Rotkvic, Dopamine antagonists induce gastric lesions in rats. *Eur J Pharmacol.* 131:105 (1986).
7. P. Sikiric, I. Rotkvic, S. Mise, S. Krizanac, V. Gjuris, J. Jukic, E. Suchanek, M. Petek, I. Udovicic, L. Kalogjera, J. Geber, M. Tucan-Foretic, M. Duvnjak, M. Philipp, I. Balen, and T. Anic, The influence of dopamine agonists and antagonists on indomethacin lesions in stomach and small intestine in rats. *Eur J Pharmacol.* 158:61 (1988).
8. D.E. Hernandez, R.C. Orlando, J.W. Adcock, K.S. Patrick, C.B. Nemeroff, and A.J. Prange, Prevention of stress induced gastric ulcers by dopamine agonists in the rat, *Life Sci.* 35:2453 (1984).
9. N.S. Parmar, M. Tariq, and A.M. Ageel, Effect of bromocriptine a dopamine receptor agonist on experimentally induced gastric ulcers in albino rats, *Life Sci.* 35:2035 (1984).
10. D.E. Hernandez, Neuroendocrine mechanisms of stress ulceration: focus on thyrotropin releasing hormone (TRH), *Life Sci.* 39:279 (1986).
11. G.B. Glavin, and A.M. Hall, Brain-gut relationships: gastric mucosal defense is also important, *in:* "Cell Injury and Protection in the Gastrointestinal Tract: From Basic Science to Clinical Perspectives," G.Y. Mózsik, A. Pár, G. Csomós, M. Kitajima, M. Kondo, C.J. Pfeiffer, K.D. Rainsford, P. Sikiric, S. Szabo, eds., Akadémiai Kiadó, Budapest (1992).
12. P.G. Henke, and A. Ray, Stress ulcer modulation by limbic system structures, *in:* "Cell Injury and Protection in the Gastrointestinal Tract: From Basic Science to Clinical Perspectives," G.Y. Mózsik, A. Pár, G. Csomós, M. Kitajima, M. Kondo, C.J. Pfeiffer, K.D. Rainsford, P. Sikiric, S. Szabo, eds., Akadémiai Kiadó, Budapest (1992).

13. P. Sikiric, I. Rotkvic, S. Mise, S. Krizanac, V. Gjuris, V. Jagic, E. Suchanek, M. Petek, I. Udovicic, J. Geber, M. Tucan-Foretic, B. Cvitanovic, D. Ivanovic, and A. Marovic, The influence of dopamine receptor agonists and an antagonist on acute pancreatitis in rats, *Eur J Pharmacol.* 147:321 (1988).

14. P. Sikiric, I. Rotkvic, S. Seiwerth, Z. Grabarevic, S. Mise, M. Duvnjak, B. Artukovic, K. Dernikovic, M. Dodig, Z. Djermanovic, I. Senecic, S. Gilljanovic, L. Kalogjera, Z. Rogina, P. Djaja, and M. Uglesic, The differences between dopamine gastric and spleen protection in restraint stress, *Dig Dis Sci.* 35:1040 (1990).

15. P. Sikiric, I. Rotkvic, S. Seiwerth, Z. Grabarevic, M. Duvnjak, V. Jagic, B. Artukovic, Z. Djermanovic, M. Dodig, I. Senecic, S. Giljanovic, D. Erceg, L.J. Komericki, M. Banic, T. Brkic, and T. Anic, The differences between dopamine gastric and testis protection in restraint stress, *Dig Dis Sci.* 35:1040 (1990).

16. P. Sikiric, S. Seiwerth, Z. Grabarevic, I. Rotkvic, V. Jagic, S. Mise, M. Duvnjak, B. Artukovic, H. Brkic, S. Vukovic, M. Banic, T. Brkic, Z. Djermanovic, M. Dodig, M. Bacic, A. Marovic, and M. Uglesic, Comparison between the effects of dopamine drugs on gastric and liver as well as duodenal and liver lesions induced in two suitable rat models, *Dig Dis Sci.* 35:1567 (1990).

17. S. Szabo, and K.H. Usadel, Cytoprotection-organoprotection by somatostatin: gastric and hepatic lesions, *Experientia.* 38:254 (1982).

18. K.H. Usadel, Hormonal and nonhormonal cytoprotective effect by somatostatins, *Hormone Res.* 29:83 (1988).

19. A. Robert, Cytoprotection by prostaglandins, *Gastroenterology.* 77:761 (1979).

20. A. Robert, C. Lancaster, A.S. Olafsson, S. Gilbertson-Beadling, and W. Zhang, Gastric adaptation to the ulcerogenic effect of aspirin, *Exp Clin Gastroenterol.* 1:73 (1991).

21. P. Sikiric, M. Petek, I. Rotkvic, R. Rucman, H. Krnjevic, S. Seiwerth, S. Grabarevic, E. Suchanek, V. Jagic, S. Mise, M. Duvnjak, B. Turkovic, and I. Udovicic, Hypothesis: stomach stress response, diagnostic and therapeutic value - a new approach in organoprotection, *Exp Clin Gastroenterol.* 1:15 (1991).

22. P. Sikiric, M. Petek, R. Rucman, S. Seiwerth, Z. Grabarevic, I. Rotkvic, V. Jagic, B. Mildner, M. Duvnjak, and N. Lang, A new gastric juice peptide, BPC-an overview of stomach/stress/organoprotection hypothesis and BPC beneficial effects, *J Physiol. (Paris)* 87:313 (1993).

23. P. Sikiric, I. Rotkvic, S. Mise, M. Petek, R. Rucman, S. Seiwerth, V. Zjacic-Rotkvic, M. Duvnjak, V. Jagic, E. Suchanek, Z. Grabarevic, T. Anic, T. Brkic, Z. Djermanovic, M. Dodig, A. Marovic, and D.E. Hernandez, Dopamine agonists prevent duodenal ulcer relapse: A comparative study with famotidine and cimetidine, *Dig Dis Sci.* 36:905 (1991).

24. H. Selye, A syndrome produced by diverse noxious agents, *Nature.* 138:32 (1936).

25. D.E. Hernandez, C.H. Walker, J.E. Valenzuela, and G.A. Mason, Increased dopamine receptor binding in duodenal mucosa of duodenal ulcer patients, *Dig Dis Sci.* 34:543 (1989).

26. K.Y. Ho, and M.O. Thorner, Therapeutic application of bromocriptine in endocrine and neurological diseases, *Drugs.* 36:67 (1988).

27. H. Roehrich, C.A. Dackis, and M.S. Gold, Bromocriptine, *Med Res Rev.* 7:243 (1987).

28. P. Sikiric, S. Mise, I. Rotkvic, E. Suchanek, J. Geber, V. Gjuris, M. Tucan-Foretic, D. Ivanovic, B. Cvitanovic, and A. Marovic, Bromocriptine in duodenal ulcer patients - a preliminary report, *in:* "Xth International Congress of Pharmacology," Sydney, August 23-27, Abstracts, p. 1527 (1987).

29. P. Sikiric,, I. Rotkvic, S. Mise, M. Petek, R. Rucman, and S. Seiwerth, Dopamine agonists and H_2 receptors antagonists in duodenal ulcer treatment - a comparative blind study, *Dig Dis Sci.* 34:1320 (1989).

30. I. Rotkvic, P. Sikiric, S. Mise, M. Petek, R. Rucman, S. Seiwerth, V. Zjacic-Rotkvic, N. Lang, Z. Grabarevic, V. Jagic, T. Brkic, and M. Banic, Dopamine agonists and H_2 receptor antagonists in duodenal ulcer treatment - follow-up, *Dig Dis Sci.* 35:1567 (1990).

31. K.W. Sommerville, and M.S.J. Langman, Newer anti-secretory agents for peptic ulcer, *Drugs.* 25:315 (1983).

32. G. Dobrilla, P. Vallaperta, and S. Amplatz, Influence of ulcer healing agents on ulcer relapse after discontinuation of acute treatment: a pooled estimate of controlled clinical trials, *Gut.* 29:181 (1988).

33. D.F. Martin, S.J. May, D.E.F. Tweedle, D. Hollandrer, M.M. Ravenstroft, and J.P. Miller, Difference in relapse rates of duodenal ulcer healing with cimetidine or tripotassium dicitrato bismuthate, *Lancet.* 1:7 (1981).

34. J.K. Kang, and D.W. Piper, Cimetidine and colloidal bismuth in treatment of chronic duodenal ulcer. Comparison of initial healing and recurrences after healing, *Digestion.* 23:73 (1982).

35. D.R. Shreeve, H.J. Klass, and P.E. Jones, Comparison of cimetidine and tripotassium dicitrato bismuthate in healing and relapse of duodenal ulcer, *Digestion.* 28:96 (1983).

36. F.I. Lee, I.M. Samlof, and M. Hardman, Comparison of tripotassium dicitrate bismuthate tablets and ranitidine in healing and relapse of duodenal ulcers, *Lancet.* 1:1299 (1985).

37. I. Hamilton, H.J. O'Connor, N.C. Wood, I. Bradbury, and A.T.R. Axon, Healing and recurrence of duodenal ulcer after treatment with tripotassium dicitrato bismuthate (TDB) tablets or cimetidine, *Gut.* 27:106 (1986).

38. J. Hansky, M.G. Korman, G.T. Schmidt, A.I. Stern, and R.G. Shaw, Relapse rate of duodenal ulcer after healing with cimetidine or Mylanta II, *Gastroenterology.* 78:1179 (1980).

39. P. Vezzadini, C. Sternini, G. Bonora, C.P. Botti, C. Lugli, and G. Labo, Relapse rates of duodenal ulcer after healing with prenzipine or cimetidine: A double-blind study, *Scand J Gastroenterol.* 17(S78):100 (1982).

40. J. Dzieniszewski, J. Pokora, and Z. Knapik, Duodenal ulcer relapse rate in treatment with pirenzipine and ranitidine, *Dig Dis Sci.* 31:210 (1986).

41. E.K. Hentschel, K. Schulze, and W. Dufek, Relapse rate of duodenal ulcer after treatment with sucralfate or cimetidine, *Wien Klin Wochenschr.* 96:153 (1984).

42. G.O. Cowan, Bromocriptine and gastric acid output, *Lancet.* 1:425 (1977).

43. D. Parkes, Amantadine, *Adv Drug Res.* 8:1 (1974).

44. D. Parkes, Bromocriptine, *Adv Drug Res.* 12:247 (1977).

45. R.M. Pinder, R.N. Brodgen, P.R. Sawyer, T.M. Speight, R. Spencer, and G.S. Avery, Metoclopramide: A review of its pharmacological properties and clinical use and therapeutic efficacy in peptic ulcer disease, *Drugs.* 12:81 (1976).

46. R.A. Harrington, C.W. Hamilton, R.N. Brodgen, J.A. Linkewich, J.A. Romankiewicz, and R.C. Heel, Metoclopramide: An updated review of its pharmacological properties and clinical use, *Drugs.* 25:451 (1983).

47. L.A. Orloff, M.S. Orloff, N.W. Bunner, and J.H. Walsh, Dopamine and norepinephrine in the alimentary tract changes after chemical sympathectomy and surgical vagotomy, *Life Sci.* 36:1625 (1985).

48. G. Flemström, B. Säfsten, and G. Jedstedt, Stimulation of mucosal alkaline secretion in rat duodenum by dopamine and dopaminergic compounds, *Gastroenterology.* 104:825 (1993).

49. J.L. Willems, W.A. Buylaert, R.A. Lefebvre, and M.G. Bogaert, Neuronal dopamine receptors on autonomic ganglia and sympathetic nerves and dopamine receptors in the gastrointestinal system, *Pharmacol Rev.* 37:165 (1985).

50. P. Sikiric, I. Rotkvic, S. Seiwerth, Z. Grabarevic, V. Jagic, M. Petek, E. Suchanek, R. Rucman, Z. Djermanovic, S. Mikulandra, M. Duvnjak, T. Brkie, M. Dodig, M. Banic, and D.E. Hernandez, A preliminary report on dopamine cytoprotection: an in vivo and in vitro study, *Exp Clin Gastroenterol.* 1:133 (1991).

51. J. Wallace, Mucosal defense: New avenues for treatment of ulcer disease, *Gastroenterol Clin N Amer.* 19:87 (1990).

52. R. Menguy, L. Desbaillets, and Y.F. Masters, Mechanism of stress ulcer: influence of hypovolemic shock on energy metabolism in the gastric mucosa, *Gastroenterology.* 66:46 (1974).

53. R. Kullman, W.R. Breull, J. Reinsberg, K. Wassermann, and A. Konopatzki, Dopamine produces vasodilation in specific regions and layers of the rabbit gastrointestinal tract, *Life Sci.* 32:2115 (1983).

54. M.R. Enochs, and L.R. Johnson, Effect of hypophysectomy and growth hormone on serum and antral gastrin levels in rat, *Gastroenterology.* 70:727 (1976).

55. G.H. Greely, Entero-insular axis: incretin candidates, *in:* "Gastrointestinal Endocrinology," J.C. Thompson, G.H. Greeley, P.L. Rayford, C.M. Townsend, eds., McGraw-Hill Book Company, New York (1987).

56. C.V. Grijalva, and D. Novin, The role of the hypothalamus and dorsal vagal complex in gastrointestinal function and pathophysiology, *Ann NY Acad Sci.* 597:207 (1990).

57. M. Marx, C.W. Cooper, and J.C. Thompson, Gastrin-calcium-calcitonin axis. *in:* "Gastrointestinal Endocrinology," J.C. Thompson, G.H. Greeley, P.L. Rayford, C.M. Townsend, eds., McGraw-Hill Book Company, New York (1987).

58. G.A. Mason, and D.E. Hernandez, The role of the hypothalamic-pituitary-thyroid axis in stress gastric ulcers, *Ann NY Acad Sci.* 597:239 (1990).

59. C.J. Pfeiffer, J.C. Keith, C.H. Cho, S. DeRolf, D.C. Pfeiffer, and H.P. Misra, Gastric and cardiac organoprotection by lidocaine, *Acta Physiol Hung.* 73:129 (1989).
60. L. Buéno, and M. Gue, Role of hypothalamic factors in the genesis of upper gut motility disorders associated with stress, *Exp Clin Gastroenterol.* 2:79 (1992).
61. P.G. Henke, and A. Ray, Review: The limbic brain, emotions and stress ulcers, *Exp Clin Gastroenterol.* 1:287 (1991).
62. AMA Department of Drugs, Agents used in hypotension and shock, *in:* "AMA Drug Evaluation," AMA Department of Drugs, eds., Publishing Sciences Group Inc., Littleton (1977).
63. I Rotkvic, D. Hrabar, H. Krpan, M. Banic, T. Brkic, M. Duvnjak, V. Zjacic, and P. Sikiric, Inhibition and stimulation of prolactin release. Delayed response in duodenal ulcer patients, *Dig Dis Sci.* 37:1815 (1992).
64. R. Strang, The association of gastro-duodenal ulceration and Parkinson's disease, *Med J Austral.* 1:842 (1965).
65. S. Szabo, A. Brown, G. Pihan, H. Dali, and J.L. Neumeyer, Duodenal ulcer induced by MPTP (1-methyl-4-phenyl-1,2,3,6-tetrahydropyridine), *Proc Soc Exp Biol Med.* 180:567 (1985).
66. P. Sikiric, I. Rotkvic, S. Mise, S. Seiweth, Z. Grabarevic, M. Petek, and R. Rucman, Dopamine, PGs and somatostatin in gastric protection in rats, *Eur J Pharmacol.* 183:2026 (1990).
67. P. Sikiric, M. Petek, R. Rucman, I. Rotkvic, S. Seiwerth, Z. Grabarevic, V. Jagic, B. Turkovic, B. Mildner, M. Duvnjak, A. Cviko, M. Kolega, M. Dodig, A. Sallmani, S. Djacic, D. Lucinger, and D. Erceg, The significance of the gastroprotective effect of body protection compound (BPC): modulation by different procedures. *in:* "Cell Injury and Protection in the Gastrointestinal Tract: From Basic Science to Clinical Perspectives," G.Y. Mózsik, A. Pár, G. Csomós, M. Kitajima, M. Kondo, C.J. Pfeiffer, K.D. Rainsford, P. Sikiric, S. Szabo, eds., Akadémiai Kiadó, Budapest (1992).
68. P. Sikiric, S. Seiwerth, Z. Grabarevic, R. Rucman, P. Petek, I. Rotkvic, B. Turkovic, V. Jagic, B. Mildner, M. Duvnjak, and Z. Danilovic, Hepatoprotective effect of BPC 157, a 15-aminoacid peptide, on liver lesions induced by either restraint stress or bile duct and hepatic artery ligation or CCl_4 administration. A comparative study with dopamine agonists and somatostatin. *Life Sci.* 53:PL291 (1993).
69. P. Sikiric, K. Gyires, S. Seiwerth, Z. Grabarevic, R. Rucman, M. Petek, I. Rotkvic, B. Turkovic, I. Udovici, V. Jagic, B. Mildner, M. Duvnjak, and Z. Danilovic, The effect of pentadecapeptide BPC 157 on inflammatory, non-inflammatory, direct and indirect pain and capsaicin neurotoxicity, *Inflammopharmacology.* 2:121 (1993).
70. P. Sikiric, S. Seiwerth, Z. Grabarevic, M. Petek, R. Rucman, B. Turkovic, I. Rotkvic, V. Jagic, M. Duvnjak, S. Mise, S. Djacic, J. Separovic, M. Veljaca, A. Sallmani, M. Banic, and T. Brkic, The beneficial effect of BPC 157, a 15 amino acid peptide BPC fragment, on gastric and duodenal lesions induced by restraint stress, cysteamine and 96% ethanol in rats. A comparative study with H_2 receptor antagonists, dopamine promoters and gut peptides, *Life Sci.* 54:PL63 (1994).
71. G. Mózsik, P. Sikiric, and M. Petek, Gastric mucosal preventing body protective compound (BPC) on the development of ethanol and HCl-induced gastric mucosal injury, *Exp Clin Gastroenterol.* 1:87 (1991).
72. W. Paré, The effect of a new gastric juice peptide, BPC, on the classic stress triad in stress procedures, *Exp Clin Gastroenterol.* 4: (1994) in press.

DIETARY FACTORS INFLUENCING GASTROINTESTINAL ULCERATION: THE LUMINAL REGULATORY SYSTEM

Francisco Guarner, Jaime Vilaseca, Antonio Salas,
Rosa Rodriguez and Juan-R. Malagelada

Eicosanoids are mediators of defensive and inflammatory processes in the gut mucosa, and their role in the pathogenesis of gastrointestinal ulceration has been documented in a large number of studies. The local eicosanoid system is modulated by intraluminal, neural and hormonal factors. Among the intraluminal factors, the diet might have a significant relevance in the regulation of mucosal eicosanoid biosynthesis, since the dietary intake of precursor fatty acids could directly influence the rate and pattern of eicosanoid generation.

Marine lipids have a high proportion of long-chain polyunsaturated fatty acids called omega-3 because of the double bond located between the third and fourth carbon atom from the methyl end of the fatty acid chain. Omega-3 fatty acids are scarce in the normal Western diet, which mainly includes polyunsaturated fatty acids from the omega-6 series. Recent evidence that chronic intake of fish-derived omega-3 fatty acids inhibits neutrophil and monocyte functions[1,2] suggests that omega-3 fatty acids have antiinflammatory properties. A hypothetical mechanism that has been proposed to explain these observations is the competition between eicosapentaenoic acid (EPA, 20:5 omega-3) and arachidonic acid (AA, 20:4 omega-6) as precursors of eicosanoid synthesis. Eicosapentaenoic acid competitively inhibits AA metabolism and is transformed into eicosanoid analogs with diverse biological activities when compared to AA-derivatives[3,4].

Thus, the aim of our experimental study was to investigate the influence of a fish oil diet on eicosanoid biosynthesis by the gastrointestinal mucosa. Secondly, we tested the effect of the fish oil diet on the development of mucosal ulcerations induced either by acid hypersecretion (cysteamine-induced duodenal ulcers) or by inflammatory injury (hapten model of granulomatous colitis).

MODIFICATION OF MUCOSAL EICOSANOID SYNTHESIS BY A FISH OIL DIET

Two experimental diets were prepared using a rodent low-fat diet supplemented at 8% by weight with either sunflower or cod liver oil, as sources of omega-6 or omega-3 polyunsaturated fatty acids, respectively. Oils were added to the standard mixture during the manufacturing process (Letica, Barcelona, Spain). The final fatty acid content of the pellets was determined by gas chromatography (Table 1).

Two groups of male Sprague-Dawley rats were fed with the experimental diets[5]. Both diets were well tolerated by the laboratory animals. During the first week on the diet, daily food intake averaged 18.8 ± 2.1 g in the sunflower group and 19.8 ± 1.8 g in the cod liver group. Animals fed the sunflower diet had a body weight gain of 22.9 ± 5.1 g/week on the first week, which was similar to that achieved by cod liver fed rats (21.3 ± 3.6). There was no significant difference in body weight between the two experimental groups after four weeks on diet. Fecal fat was investigated in serial samples from both groups of rats, and was found to be undetectable in every instance.

Plasma fatty acid levels are shown in Figure 1. Sunflower fed rats had significantly higher levels of linoleic acid (18:2 omega-6) and AA (20:4 omega-6) than cod liver fed animals. As expected, the cod liver group showed high levels of omega-3 fatty acids (EPA, 20:5 omega-3, and docosahexaenoic, 22:6 omega-3) which were almost absent in sunflower rats.

After four weeks on either diet, generation of PGE_2, 6-keto-$PGF_{1\alpha}$, TXB_2, LTB_4 and LTC_4 by gastric and colonic mucosa specimens was significantly reduced in the cod liver diet fed rats when compared to the sunflower group. On the other hand, EPA derived eicosanoids were undetectable in sunflower fed rats, while the cod liver group showed significant generation of PGE_3 and LTC_5 by both gastric and colonic mucosa. Interestingly, generation of LTC_5 by the colonic mucosa of rats from the cod liver group was quantitatively higher than the generation of LTC_4. Thus, these data indicated that changes in the dietary intake of precursor fatty acids would modify the eicosanoid system in the gastrointestinal tract.

Table 1. Fatty Acid Content of the Chow Pellets

Fatty Acid	Sunflower Diet	Cod Liver Diet
14:0	0.33	2.88
16:0	4.76	10.68
16:1 omega-7	0.22	5.49
18:0	4.22	3.51
18:1 omega-9	24.34	13.69
18:1 omega-7	1.04	3.52
18:2 omega-6	51.51	12.21
18:3 omega-3	0.78	1.22
18:4 omega-3	-	1.10
20:1 omega-9	0.24	7.57
20:4 omega-6	0.03	0.26
20:5 omega-3	-	5.95
22:1 omega-11	-	5.04
22:6 omega-3	-	6.91
Total	88.77	84.90

Values are expressed as mg per g of diet.

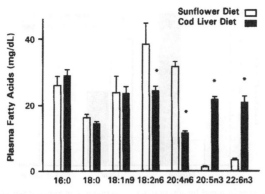

Figure 1. Plasma fatty acid levels after four weeks on sunflower or cod liver oil supplemented diet. Five rats per point (mean ± sem; * p< 0.05 between sunflower and cod liver fed rats). For simplicity only major fatty acids are represented. Data reproduced from reference 12.

CYSTEAMINE-INDUCED DUODENAL ULCERS

The cysteamine model of duodenal ulcers was first described by Selye and Szabo[6]. A single injection of cysteamine to laboratory rats induces acute mucosal damage in the duodenum mediated by an increase in gastric acid output and a reduction of duodenal bicarbonate release, among other factors. We used this model to test the effect of the fish oil diet on the development of acute mucosal ulcerations[7]. After four weeks on cod liver or sunflower diet, matched groups of rats were subjected to the following protocols: (a) duodenal damage induced by 400 mg/kg s.c. cysteamine in nine rats killed 24 h after cysteamine and in 16 rats followed for up to 14 days; (b) gastric acid and duodenal bicarbonate secretion after 100 mg/kg i.v. cysteamine in anesthetized rats.

Duodenal lesions were assessed microscopically and scored as erosions, ulcers or perforated/penetrated ulcers. As shown in figure 2, duodenal lesions assessed 24 h after cysteamine were deeper in sunflower fed rats when compared to the cod liver group. In addition, the outcome of the rats followed for up to 14 days showed a significantly higher rate of epithelialization in rats fed with the cod liver diet than in rats fed with the sunflower diet (figure 3). Gastric acid output was significantly stimulated by cysteamine in anesthetized rats from both groups (p<0.01), but the increase was more marked in the sunflower group (6.1 ± 2.8 to 26.6 ± 1.1 μmol/h) than in cod liver rats (4.2 ± 0.8 to 16.1 ± 2.6, p<0.05 versus sunflower group). There were no differences between both groups in basal and 10 mM HCl stimulated duodenal bicarbonate secretion, and after cysteamine both groups of animals failed to release duodenal bicarbonate in response to 10 mM Hcl.

In summary, the fish oil diet reduced the increase in gastric acid output stimulated by cysteamine and decreased the severity of the duodenal lesions. Experimental data by other investigators suggested that prostaglandins derived from omega-3 fatty acids have more potent antisecretory action than those derived from arachidonic acid[8]. In addition, recent experimental work suggests that dietary polyunsaturated fatty acids enhance mucosal eicosanoid generation and mucosal resistance to acid[9].

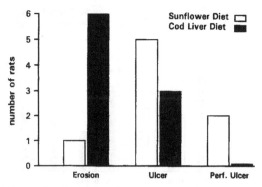

Figure 2. Duodenal lesions induced by 400 mg/kg s.c. cysteamine in rats fed with either the sunflower or the cod liver oil supplemented diet. Lesions were assessed microscopically and scored as erosions, ulcers or perforated ulcers. The difference between both distributions is significant (p<0.05, Chi-square test). Data reproduced from reference 7.

CHRONIC ULCERATIVE COLITIS INDUCED BY TNB

In another series of experiments[10], we used the trinitrobenzene sulfonic acid (TNB) model of granulomatous colitis which features chronic ulcerations of the colonic mucosa that persist for up to eight weeks[11]. Colonic inflammatory lesions were induced by a single intracolonic administration of the hapten TNB (50 mg diluted in 10% ethanol), in rats fed with the experimental diets for at least four weeks. On days 0, 3, 20, 30, 40 and 50 after TNB, the luminal release of eicosanoid mediators was measured by intracolonic dialysis, as described elsewhere[12]. On days 20, 30 and 50, six to ten rats per group were killed and the distal colon was removed to assess mucosal damage macroscopically and histologically according to the criteria stated in Table 2.

As shown by Figure 4, luminal release of PGE_2 and TXB_2 changed as time elapsed from TNB instillation. For PGE_2, peak release occurred earlier (around day 3 after TNB) than for TXB_2 (around day 20 after TNB). For PGE_2, there were no significant differences between both experimental groups throughout the study,

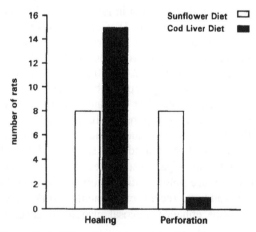

Figure 3. Outcome after a single 400 mg/kg s.c. cysteamine injection to rats fed either sunflower or cod liver diet. Rats were followed for up to 14 days after cysteamine and killed for microscopical assessment of duodenal damage. Animals with perforation died during the first week of the follow up. The difference between groups is significant (p<0.01, Fisher's exact test). Data reproduced from reference 7.

although mean levels in the dialysis fluid from sunflower fed rats were always above those detected in cod liver fed rats. In contrast, luminal release of TXB_2 markedly increased from day 3 to days 20 and 30 in sunflower fed rats but this change did not occur in cod liver fed animals. Hence, significant differences between sunflower and cod liver group occurred on days 20 and 30. Changes in the intracolonic release of LTB_4 were also different in both groups of rats. In the sunflower group, levels increased from day 0 (0.2 ± 0.03 ng/ml) to day 3 (6.0 ± 1.3) and persisted high on day 20 (5.8 ± 1.8). Cod liver fed rats showed a similar pattern (day 0: 0.1 ± 0.02; day 3: 6.4 ± 1.0), but mean levels on day 20 (3.6 ± 0.6) were lower than on day 3, suggesting that the inflammatory activity had decreased.

Figure 5 depicts the mean macroscopic and histological scores obtained from rats sacrificed on days 20, 30 and 50 after TNB. On day 20, macroscopic damage scores were significantly lower in cod liver animals than in sunflower group, but there was no difference in the histological scores. With the progression of the inflammatory colitis, differences between both experimental groups became more evident and cod liver fed rats showed lower macroscopic and histological lesion scores than sunflower fed rats. By day 50, inflammation and ulcerations were almost absent in cod liver fed animals, while the sunflower group still showed flaring mucosal lesions. Thus, the initial injury was similar in both groups of rats, but the development of chronic inflammatory lesions in the colon was mitigated by the cod liver diet.

In summary, these experiments suggest that dietary supplementation with a fish oil rich in omega-3 fatty acids palliates the progression of chronic ulcerative lesions in the TNB-model of inflammatory colitis, and shortens the course of the disease. The beneficial effects are associated with a reduction of the luminal release of thromboxane and leukotriene B_4 during the chronic stage of the inflammatory lesions.

CONCLUSIONS

Changes in the pattern of eicosanoid synthesis by the gastrointestinal mucosa can be induced by altering the dietary intake of their precursor fatty acids. These changes may be relevant for the expression of diseases involving gastrointestinal ulceration.

SUMMARY

Eicosanoids are major mediators of defensive and inflammatory processes in the gut mucosa. We have investigated whether precursor fatty acids present in the diet can modulate mucosal eicosanoid synthesis. Rats fed with a cod liver oil supplemented diet showed high omega-3 and low omega-6 plasma fatty acid levels when compared to rats fed with a sunflower oil diet. Synthesis of omega-6 fatty acid derived eicosanoids by gastric and intestinal mucosa was found to be lower in the cod liver group. Generation of omega-3 fatty acid derived eicosanoids (PGE_2 and LTC_5) was only observed in cod liver rats.

Duodenal lesions induced by cysteamine were deeper in sunflower fed rats. Cysteamine-stimulated acid secretion was also higher in these rats, whereas changes in duodenal bicarbonate were similar in both groups. In another series of experiments, we used the TNB model of inflammatory colitis which features chronic ulcers of the colonic mucosa that persist for up to 8 weeks. Luminal release of eicosanoid mediators, as measured by intracolonic dialysis, was lower in cod liver than in sunflower group particularly during the chronic stage of the disease. Macro-and

Table 2. Criteria for Assessment of Colonic Lesions

Macroscopic Scores		
Adhesions	None	0
	Minimal	1
	Involving several bowel loops	2
Strictures	None	0
	Mild	2
	Severe, proximal dilatation	3
Ulcers	None	0
	Linear ulceration < 1 cm	1
	Two linear ulcers < 1 cm	2
	More sites of ulceration or on large ulcer > 1 cm	3
Wall Thickness	Less than 1 mm	0
	1 - 3 mm	1
	More than 3 mm	2
	Maximum Score	10
Histologic Score		
Ulceration	No ulcer, epithelization	0
	Small ulcers < 3 mm	1
	Large ulcers > 3 mm	0
Inflammation	None	0
	Mild	1
	Moderate	2
	Severe	3
Depth of the Lesion	None	0
	Submucosa	1
	Muscularis propria	2
	Serosa	3
Fibrosis	None	0
	Mild	1
	Severe	2
	Maximum Score	10

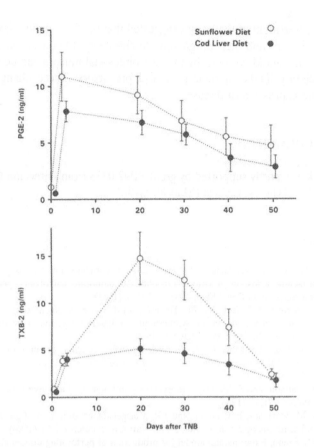

Figure 4. Changes in luminal release of PGE_2 (upper panel) and TXB_2 (lower panel) after the induction of colitis by intracolonic administration of TNB on day 0. Values represent ng per ml of fluid obtained after 1 h of *in vivo* dialysis of the colon. Ten rats per point (mean ± SEM) but on days 40 and 50, in which seven rats from sunflower group and six rats from cod liver group were subjected to dialysis. Data reproduced from reference 12.

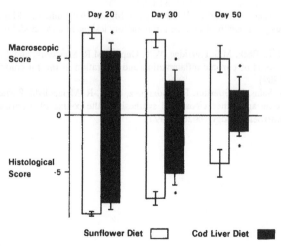

Figure 5. Morphological lesion scores of colonic damage induced by TNB, assessed according to the criteria shown in Table 2. Means ± SEM are shown (* $p < 0.05$ between sunflower and cod liver fed rats). Data reproduced from reference 12.

microscopical assessment of the lesions suggested that the fish oil diet diminishes the severity of the lesions and their progression to chronicity. In conclusion, changes in the pattern of eicosanoid synthesis by the gastrointestinal mucosa can be induced by altering the dietary intake of their fatty acid precursors. These changes may be relevant for the expression of disease.

ACKNOWLEDGEMENTS

This work was partly supported by grant PB92-0733 from Dirección General de Investigación Científica y Técnica (Madrid, Spain).

REFERENCES

1. T.H. Lee, R.L. Hoover, J.D. Williams, *et al.*, Effect of dietary enrichment with eicosapentaenoic and docosahexaenoic acids on in vitro neutrophil and monocyte leukotriene generation and neutrophil function, N Eng J Med. 1985; 312:1217 (1985).
2. S. Endres, R. Ghorbani, V.E. Kelley, *et al.*, The effect of dietary supplementation with n-3 polyunsaturated fatty acids on the synthesis of interleukin-1 and tumor necrosis factor by mononuclear cells, N Eng J Med. 320:265 (1989).
3. P. Needleman, A. Raz, M.S. Minkes, J.A. Ferrendeli, and H. Sprecher, Triene prostaglandins: prostacyclin and thromboxane biosynthesis and unique biological properties, Proc Natl Acad Sci USA. 76:944 (1979).
4. T. Terano, J.A. Salmon, and S. Moncada, Biosynthesis and biological activity of leukotriene B_5, Prostaglandins. 27:217 (1984).
5. R. Rodríguez, M. Martínez, F. Guarner, and J.R. Malagelada, Modification of gut mucosal eicosanoid synthesis by a fish oil diet, J Clin Nutr Gastroenterol. 5:11 (1990).
6. H. Selye, and S. Szabo, Experimental model for production of perforating duodenal ulcers by cysteamine in the rat, Nature. 244:458 (1973).
7. F. Guarner, J. Vilaseca, A. Salas, R. Rodríguez, and J.R. Malagelada, Reduction of cysteamine-induced duodenal ulcers by dietary fish oil, Eur J Gastroenterol Hepatol. 3:239 (1991).
8. B.J.R. Whittle, S. Moncada, and J.R. Vane, Biological activities of some metabolites and analogues of prostacyclin, in: "Medicinal Chemistry Advances," F.G. De Las Heras, S. Vega, eds., Pergamonr Press, Oxford (1981).
9. A. Lugea, A. Salas, F. Guarner, and J.R. Malagelada, Influence of the dietary fat intake on the duodenal resistance to acid, Gut. 34:1303 (1993).
10. J. Vilaseca, A. Salas, F. Guarner, R. Rodríguez, M. Martínez, and J.R. Malagelada, Dietary fish oil reduces progression of chronic inflammatory lesions in a rat model of colitis, Gut. 31:539 (1990).
11. G.P. Morris, P.L. Beck, M.S. Herridge, W.T. Depew, M.R. Szewezuk, and J.L. Wallace, Hapten induced model of chronic inflammation and ulceration in the rat colon, Gastroenterology. 96:795 (1989).
12. J. Vilaseca, A. Salas, F. Guarner, R. Rodríguez, and J.-R. Malagelada, Participation of thromboxane and other eicosanoid synthesis in the course of experimental inflammatory colitis, Gastroenterology. 98:269 (1990).

LIST OF CONTRIBUTORS

Amann, R.
Department of Experimental and Clinical Pharmacology, University of Graz, Graz, Austria

Bakke, Hans-Kristian
Department of Biological and Medical Psychology, Section for Physiological Psychology, University of Bergen, Arstadveien 21, N-5009, Bergen, Norway

Banic, M.
Department of Pharmacology, University of Zagreb, Salata 11, POB 916, 41000 HR, Zagreb, Republic of Croatia

Beck, Paul
Department of Biology, Queen's University, Kingston, Ontario, Canada, K7L 3N6

Bell, R.A.
Department of Chemistry, McMaster University, Hamilton, Ontario, Canada, L8N 3Z5

Bodis, B.
First Department of Medicine, Medical University of Pecs, H-7643, Pecs, Hungary

Bonaz, Bruno
CURE/Gastroenteric Biology Center, VA Wadsworth Medical Center, Department of Medicine and Brain Research Institute, UCLA, Los Angeles, California, USA, 90073

Brkic, T.
Department of Pharmacology, University of Zagreb, Salata 11, POB 916, 41000 HR, Zagreb, Republic of Croatia

Bryan, Robert
Division of General Surgery, Department of Surgery, The Milton S. Hershey Medical Center, The Pennsylvania State University, Hershey, Pennsylvania, USA, 17033

Burks, Thomas F.
Department of Pharmacology, Medical School, University of Texas Health Science Center, Houston, Texas, USA, 77025

Cho, C.H.
Department of Pharmacology, University of Hong Kong, 5 Sassoon Road, Hong Kong

Djacic, S.
Department of Pharmacology, University of Zagreb, Salata 11, POB 916, 41000 HR, Zagreb, Republic of Croatia

Duvnjak, M.
Department of Pharmacology, University of Zagreb, Salata 11, POB 916, 41000 HR, Zagreb, Republic of Croatia

Evangelista, Stefano
Department of Pharmacology, Malesci Pharmaceuticals, via N. Porpora 22/44. 50144, Firenze, Italy

Fallone, Carlo A.
Department of Biology, Queen's University, Kingston, Ontario, Canada, K7L 3N6

Fandriks, Lars
Department of Physiology and Centre for Gastroenterologic Research, University of Goteborg, Medicinaregatan 11, S-413 90 Goteborg, Sweden

Folkman, J.
Department of Surgery, Children's Hospital and Harvard Medical School, Boston, MA, USA

Garamszegi, M.
First Department of Medicine, Medical University of Pecs, H-7643, Pecs, Hungary

Garner, Andrew
Department of Pharmacology and Therapeutics, Faculty of Medicine and Health Sciences, United Arab Emirates University, P.O. Box 17666, Al Ain, United Arab Emirates

Glavin, Gary B.
Departments of Pharmacology and Therapeutics and Surgery, Faculty of Medicine, University of Manitoba, Winnipeg, Manitoba, Canada, R3E 0W3

Grabarevic, Z.
Department of Pharmacology, University of Zagreb, Salata 11, POB 916, 41000 HR, Zagreb, Republic of Croatia

Guarner, Francisco
Digestive System Research Unit, Hospital General Vall d'Hebron, Autonomous University of Barcelona, 08035 Barcelona, Spain

Guth, P.
CURE/UCLA Gastroenteric Biology Center, University of California at Los Angeles and VA Wadsworth Medical Center, Los Angeles, CA, USA

Hall, Arleen M.
Department of Pharmacology and Therapeutics, Faculty of Medicine, University of Manitoba, Winnipeg, Manitoba, Canada, R3E 0W3

Hanzevacki, M.
Department of Pharmacology, University of Zagreb, Salata 11, POB 916, 41000 HR, Zagreb, Republic of Croatia

Heinemann, A.
Department of Medicine, University of Graz, Graz, Austria

Henagan, Julia M.
Department of Surgery, The University of Texas Medical School, Houston, Texas, USA, 77030

Henke, Peter G.
Neuroscience Laboratory, St. Francis Xavier University, P.O. Box 5000, Antigonish, Nova Scotia, Canada, B2G 2W5

Herridge, Margaret S.
Department of Biology, Queen's University, Kingston, Ontario, Canada, K7L 3N6

Holzer, Peter
Dept. of Experimental and Clinical Pharmacology, University of Graz, Universitatsplatz 4, A-8010, Graz, Austria

Ishikawa, Toshio
National Center of Neurology and Psychiatry, Ichikawa, Chiba 272, Japan

Jagic, V.
Department of Pharmacology, University of Zagreb, Salata 11, POB 916, 41000 HR, Zagreb, Republic of Croatia

Jocic, M.
Department of Experimental and Clinical Pharmacology, University of Graz, Graz, Austria

Jonson, Claes
Department of Physiology and Centre for Gastroenterologic Research, University of Goteborg, Medicinaregatan 11, S-413 90 Goteborg, Sweden

Karadi, O.
First Department of Medicine, Medical University of Pecs, H-7643, Pecs, Hungary

Kauffman, Gordon L.
Division of General Surgery, Department of Surgery, The Milton S. Hershey Medical Center, The Pennsylvania State University, Hershey, Pennsylvania, USA, 17033

Keenan, Catherine M.
Gastrointestinal Research Group, University of Calgary, Calgary, Alberta, Canada, T2N 4N1

Kiraly, A.
First Department of Medicine, Medical University of Pecs, H-7643, Pecs, Hungary

Kitajima, Masaki
Department of Surgery, Keio University, 35 Shinanomachi, Shinjuku, Tokyo 160, Japan

Koo, M.W.L.
Department of Pharmacology, University of Hong Kong, 5 Sassoon Road, Hong Kong

Kusstatscher, S.
Departments of Pathology, Brigham and Women's Hospital and Harvard Medical School, Boston, MA, USA

Lippe, I. Th.
Department of Experimental and Clinical Pharmacology, University of Graz, Graz, Austria

Livingston, E.
CURE/UCLA Gastroenteric Biology Center, University of California at Los Angeles and VA Wadsworth Medical Center, Los Angeles, CA, USA

Lopez, Rafael A.
Department of Surgery, The University of Texas Medical School, Houston, Texas, USA, 77030

Malagelada, Juan-R.
Digestive System Research Unit, Hospital General Vall d'Hebron, Autonomous University of Barcelona, 08035 Barcelona, Spain

Mantellini, Paola
Gastroenterology Unit, Department of Clinical Pathophysiology, University of Florence, Firenze, Italy

Marovic, A.
Department of Pharmacology, University of Zagreb, Salata 11, POB 916, 41000 HR, Zagreb, Republic of Croatia

Matthews, Jeffrey
Department of Surgery, Beth Israel Hospital, Boston, MA, USA

Miller, Thomas
Department of Surgery, The University of Texas Medical School, Houston, Texas, USA, 77030

Mise, S.
Department of Pharmacology, University of Zagreb, Salata 11, POB 916, 41000 HR, Zagreb, Republic of Croatia

Morris, Gerald P.
Department of Biology, Queen's University, Kingston, Ontario, Canada, K7L 3N6

Mozsik, Gyula
First Department of Medicine, Medical University of Pecs, H-7643, Pecs, Hungary

Murison, Robert
Department of Biological and Medical Psychology, Section for Physiological Psychology, University of Bergen, Arstadveien 21, N-5009, Bergen, Norway

Nagata, Mitsuhiro
Department of Pharmacology, Kochi Medical School, Nankoku, Kochi 783, Japan

Nagy, L.
First Department of Medicine, Medical University of Pecs, H-7643, Pecs, Hungary

Ogle, C.W.
Department of Pharmacology, University of Hong Kong, 5 Sassoon Road, Hong Kong

Okuma, Yasunobu
Department of Pharmacology, Kochi Medical School, Nankoku, Kochi 783, Japan

Osumi, Yoshitsugu
Department of Pharmacology, Kochi Medical School, Nankoku, Kochi 783, Japan

Pabst, M.
Department of Histology and Embryology, University of Graz, Graz, Austria

Paré, William P.
Eastern Research and Development Office, VA Medical Center, Perry Point, Maryland, USA, 21902

Peskar, B.A.
Department of Experimental and Clinical Pharmacology, University of Bochum, Bochum, Germany

Peskar, B.M.
Department of Experimental and Clinical Pharmacology, University of Bochum, Bochum, Germany

Petek, M.
Department of Pharmacology, University of Zagreb, Salata 11, POB 916, 41000 HR, Zagreb, Republic of Croatia

Prior, T.
Intestinal Disease Research Programme, McMaster University, Hamilton, Ontario, Canada, L8N 3Z5

Rangachari, P.K.
Intestinal Disease Research Programme, McMaster University, Hamilton, Ontario, Canada, L8N 3Z5

Raybould, H.
CURE/UCLA Gastroenteric Biology Center, University of California at Los Angeles and VA Wadsworth Medical Center, Los Angeles, CA, USA

Redei, Eva
Department of Pharmacology, University of Pennsylvania, Philadelphia, Pennsylvania, USA, 19104

Renzi, Daniela
Gastroenterology Unit, Department of Clinical Pathophysiology, University of Florence, Firenze, Italy

Rodriguez, Rosa
Digestive System Research Unit, Hospital General Vall d'Hebron, Autonomous University of Barcelona, 08035 Barcelona, Spain

Rotkvic, I.
Department of Pharmacology, University of Zagreb, Salata 11, POB 916, 41000 HR, Zagreb, Republic of Croatia

Rowe, Paul
Department of Surgery, General Hospital, Eastbourne, U.K.

Rucman, R.
Department of Pharmacology, University of Zagreb, Salata 11, POB 916, 41000 HR, Zagreb, Republic of Croatia

Salas, Antonio
Digestive System Research Unit, Hospital General Vall d'Hebron, Autonomous University of Barcelona, 08035 Barcelona, Spain

Sandor, Zs.
Departments of Pathology, Brigham and Women's Hospital and Harvard Medical School, Boston, MA, USA

Schmidt, Karmen L.
Department of Pathology and Laboratory Medicine, The University of Texas Medical School, Houston, Texas, USA, 77030

Seiwerth, S.
Department of Pharmacology, University of Zagreb, Salata 11, POB 916, 41000 HR, Zagreb, Republic of Croatia

Separovic, J.
Department of Pharmacology, University of Zagreb, Salata 11, POB 916, 41000 HR, Zagreb, Republic of Croatia

Shanahan, Fergus
Department of Medicine, University College Cork, Cork Regional Hospital, Cork, Ireland

Sikiric, P.
Department of Pharmacology, University of Zagreb, Salata 11, POB 916, 41000 HR, Zagreb, Republic of Croatia

Simicevic, V.
Department of Pharmacology, University of Zagreb, Salata 11, POB 916, 41000 HR, Zagreb, Republic of Croatia

Smith, Gregory
Department of Surgery, The University of Texas Medical School, Houston, Texas, USA, 77030

Suchanek, E.
Department of Pharmacology, University of Zagreb, Salata 11, POB 916, 41000 HR, Zagreb, Republic of Croatia

Suto, G.
First Department of Medicine, Medical University of Pecs, H-7643, Pecs, Hungary

Szabo, Sandor
Departments of Pathology, Brigham and Women's Hospital and Harvard Medical School, 75 Francis Street, Boston, Massachusetts, USA, 02115

Taché, Yvette
CURE/Gastroenteric Biology Center, VA Wadsworth Medical Center, Department of Medicine and Brain Research Institute, UCLA, Los Angeles, California, USA, 90073

Tornwall, Michael S.
Department of Surgery, The University of Texas Medical School, Houston, Texas, USA, 77030

Toth, Gy.
First Department of Medicine, Medical University of Pecs, H-7643, Pecs, Hungary

Vilaseca, Jaime
Digestive System Research Unit, Hospital General Vall d'Hebron, Autonomous University of Barcelona, 08035 Barcelona, Spain

Vincze, A.
First Department of Medicine, Medical University of Pecs, H-7643, Pecs, Hungary

Wachter, Ch.
Department of Experimental and Clinical Pharmacology, University of Graz, Graz, Austria

Wallace, John L.
Gastrointestinal Research Group, University of Calgary, Calgary, Alberta, Canada, T2N 4N1

Walsh, J.
CURE/UCLA Gastroenteric Biology Center, University of California at Los Angeles and VA Wadsworth Medical Center, Los Angeles, CA, USA

Weiner, Herbert
Department of Psychiatry and Biobehavioral Science, University of California, Los Angeles, California, USA, 90024-1759

Wolfe, M.
Department of Medicine, Harvard Medical School, Boston, MA, USA

Wong, H.
CURE/UCLA Gastroenteric Biology Center, University of California at Los Angeles and VA Wadsworth Medical Center, Los Angeles, CA, USA

Xing, Lianping
Division of General Surgery, Department of Surgery, The Milton S. Hershey Medical Center, The Pennsylvania State University, Hershey, Pennsylvania, USA, 17033

Yang, Hong
CURE/Gastroenteric Biology Center, VA Wadsworth Medical Center, Department of Medicine and Brain Research Institute, UCLA, Los Angeles, California, USA, 90073

Yokotani, Kunihiko
Department of Pharmacology, Kochi Medical School, Nankoku, Kochi 783, Japan

Yoneda, Masashi
CURE/Gastroenteric Biology Center, VA Wadsworth Medical Center, Department of Medicine and Brain Research Institute, UCLA, Los Angeles, California, USA, 90073

Zjacic-Rotkvic, V.
Department of Pharmacology, University of Zagreb, Salata 11, POB 916, 41000 HR, Zagreb, Republic of Croatia

INDEX

Lightning Source UK Ltd.
Milton Keynes UK
UKOW06f2209190215

246578UK00004B/194/P